Rec'd. DEC 1 1987

GLASS: SCIENCE AND TECHNOLOGY

VOLUME 2

Processing I

Contributors

Michael Cable
H. Dislich
Charles Hendricks
W. C. Hynd
George W. Keller
Richard M. Klein
R. K. Shukla
N. N. SinghDeo
R. W. Wilson
Jerzy Zarzycki

GLASS: SCIENCE AND TECHNOLOGY

Edited by **D. R. UHLMANN**

DEPARTMENT OF MATERIALS SCIENCE
AND ENGINEERING
MASSACHUSETTS INSTITUTE OF TECHNOLOGY
CAMBRIDGE, MASSACHUSETTS

N. J. KREIDL

SANTA FE, NEW MEXICO

VOLUME 2
Processing I

1984

ACADEMIC PRESS, INC.
(Harcourt Brace Jovanovich, Publishers)

Orlando San Diego New York London
Toronto Montreal Sydney Tokyo

COPYRIGHT © 1984, BY ACADEMIC PRESS, INC.
ALL RIGHTS RESERVED.
NO PART OF THIS PUBLICATION MAY BE REPRODUCED OR
TRANSMITTED IN ANY FORM OR BY ANY MEANS, ELECTRONIC
OR MECHANICAL, INCLUDING PHOTOCOPY, RECORDING, OR ANY
INFORMATION STORAGE AND RETRIEVAL SYSTEM, WITHOUT
PERMISSION IN WRITING FROM THE PUBLISHER.

ACADEMIC PRESS, INC.
Orlando, Florida 32887

United Kingdom Edition published by
ACADEMIC PRESS, INC. (LONDON) LTD.
24/28 Oval Road, London NW1 7DX

Library of Congress Cataloging in Publication Data
(Revised for vol. 2)

Glass--science and technology (New York, N.Y.)
 Glass--science and technology.

 Includes bibliographies and indexes.
 Contents: v. 1. Glass-forming systems--v. 2.
Processing I-- --v. 5. Elasticity and strength in
glasses--[etc.]
 1. Glass--Collected works. I. Uhlmann, D. R. (Donald
Robert) II. Kreidl, N. J.
TP848.G56   666'.1   80-51
ISBN 0-12-706702-7 (v. 2: alk. paper)

PRINTED IN THE UNITED STATES OF AMERICA

84 85 86 87   9 8 7 6 5 4 3 2 1

Contents

LIST OF CONTRIBUTORS ix
PREFACE xi

Chapter 1 Principles of Glass Melting
Michael Cable

I.	Introduction	1
II.	Glass Melting	5
III.	Bubbles in Glass Melts	16
IV.	The Homogenizing of Glass Melts	28
V.	Volatilization from Glass Melts	36
VI.	Conclusion	40
	References	40

Chapter 2 Flat Glass Manufacturing Processes
W. C. Hynd

I.	The Development of Flat Glass	46
II.	Basic Science of Flat Glass Processes	50
III.	Rolled Glass Processes	58
IV.	Flat Drawn Sheet Processes	70
V.	The Float Process	83
VI.	Glass for Radiation Control	100
	References	106

Chapter 3 Container Manufacture
George W. Keller

I.	History	107
II.	Furnaces	114
III.	Forming Machines	125
IV.	Raw Material and Glass Analytical Procedures	127
V.	Coatings	129
VI.	Pollution Control and Its Effect on Industry	133
	References	136

Chapter 4 Tubing and Rod Manufacture
R. W. Wilson

I.	Danner Process	138
II.	Updraw Process	140
III.	Vello and Downdraw Processes	141
IV.	Tubing/Rod Drawing Operations	144

Chapter 5 Glass Spheres
Charles Hendricks

I.	Introduction	149
II.	Rayleigh Technique	150
III.	Commercial Processes for Producing Glass Spheres	152
IV.	Production of High-Precision Spheres by the Drop-Generator Process	154
V.	Conclusion	167
	References	167

Chapter 6 Solder Glass Processing
N. N. SinghDeo and R. K. Shukla

I.	Introduction	170
II.	Solder Glass Requirements	171
III.	Solder Glass Evolution	172
IV.	Cerdip Process	180
V.	Glass Properties and Processing Effects	186
VI.	Moisture Outgassing	194
VII.	Soft Error	200
VIII.	Future Perspectives	204
	References	205

Chapter 7 Processing of Gel Glasses
Jerzy Zarzycki

I.	Introduction	209
II.	Historical Outline	210
III.	Gel Preparation	214
IV.	Gel Drying	221
V.	Densification Process	231
VI.	Special Features of Gel-Produced Glasses	244
VII.	Conclusion	245
	References	245

Chapter 8   Coatings on Glass
H. Dislich

I.	Introduction and Overview	252
II.	Vacuum Processes and Products	256
III.	Spray Processes and Products	259
IV.	CVD Processes and Products	262
V.	Wet Reduction Processes and Products	265
VI.	Dip Coating Processes and Products	267
VII.	Leaching Processes and Products	277
VIII.	Specialties	279
IX.	Concluding Remarks	281
	References	282

Chapter 9   Optical Fiber Waveguides
Richard M. Klein

I.	Introduction	285
II.	Fiber Characteristics	286
III.	Fiber Processing	298
IV.	Summary	334
	References	335

MATERIALS INDEX   341
SUBJECT INDEX   347

List of Contributors

Numbers in parentheses indicate the pages on which the authors' contributions begin.

MICHAEL CABLE (1), *Department of Ceramics, Glasses, and Polymers, University of Sheffield, Sheffield S10 2TZ, England*

H. DISLICH (251), *Schott Glaswerke, 6500 Mainz, Federal Republic of Germany*

CHARLES HENDRICKS (149), *Lawrence Livermore Laboratory, Livermore, California 94550*

W. C. HYND (45), *Pilkington Brothers P.L.C., St. Helens WA10 3TT, England*

GEORGE W. KELLER* (107), *Technical Development Department, Glass Containers Corporation, Fullerton, California 92634*

RICHARD M. KLEIN (285), *Optical Fiber and Components Department, GTE Laboratories, Inc., Waltham, Massachusetts 02254*

R. K. SHUKLA (169), *Intel Corporation, Santa Clara, California 95051*

N. N. SINGHDEO[†] (169), *Exel Microelectronics, Inc., San Jose, California 95131*

R. W. WILSON (137), *Corning Glass Works, Corning, New York 14831*

JERZY ZARZYCKI (209), *Laboratory of Materials Science and CNRS Glass Laboratory, University of Montpellier, 34060 Montpellier Cedex, France*

---

* Present address: Container General Corporation, Chattanooga, Tennessee 37410.
† Present address: Indy Electronics, Inc., Manteca, California 95336.

Preface

Processing has received widespread recognition as a vital link in the materials chain, and the processing of glasses is an area that has seen important, even drastic, changes occur during this generation. Because of the central role of processing in glass technology and our desire to provide coverage of both fundamental principles and salient applications across a broad spectrum, it has been decided to devote two volumes to the coverage of this topic. Each volume will represent a mixture of principles and applications, and taken *in toto,* it is hoped that the two volumes will provide a perspective on glass processing that is unavailable from conventional sources.

The applications covered in the present volume range from wide ribbons of flat glass produced in a continuous process from a glass tank to small hollow glass spheres produced in a batch process from solutions. The volume of materials represented by respective processes covers a similar range, from 100 to 200 tons to a few pounds per day, with technical requirements that are tailored to specific applications.

The chapters in the two volumes on glass processing will be concerned exclusively with oxide glasses. At a later date we plan to organize companion volumes that will treat the processing of metallic and polymer glasses.

CHAPTER 1

Principles of Glass Melting

Michael Cable

DEPARTMENT OF CERAMICS, GLASSES, AND POLYMERS
UNIVERSITY OF SHEFFIELD
SHEFFIELD, ENGLAND

I. Introduction	1
A. Choice of Glass Composition	2
B. Choice of Raw Materials	4
II. Glass Melting	5
A. Energy Requirements	5
B. Batch Mixing	7
C. Batch Heating	8
D. The Chemistry of Melting	9
E. Control of Oxidation	12
F. Decolorizing	15
III. Bubbles in Glass Melts	16
A. Fining	16
B. Reboil and Foaming	24
IV. The Homogenizing of Glass Melts	28
A. Sources of Inhomogeneity	28
B. Diffusion	28
C. Flow in Glass Melts	30
V. Volatilization from Glass Melts	36
A. Introduction	36
B. Theory	36
C. Experimental Data	38
VI. Conclusion	40
References	40

I. Introduction

Glass making is one of mankind's oldest skills requiring fire. Glasses are generally considered aesthetically pleasing materials, and glass objects have been made for their artistic appeal from very early times. Containers have been produced for about 3500 years and now represent one of the most important sectors of the industry, having attained such a position soon after the invention of blowing with a hollow blow pipe around 1 A.D. The forming of molten glass into various objects by manual and oral techniques also fascinates many people by its apparent ease and deceptive simplicity. Firms which produce their wares by hand and are willing to admit visitors nearly always attract more than they can accommodate and glass workers at exhibitions always find that crowds gather to watch them.

The basic operations involved in forming glass into useful or artistic objects appear trivially simple and have been practised for about 2000 years, but many subtleties and complications arise when one begins to study the limits of various processes and the ways in which the efficiency of making particular products can be improved. The two apparently simple basic operations are taking viscous glass and making it flow into the desired shape under the influence of gravity or deliberately applied stresses, then cooling it until it cannot deform under its own weight. A third stage of controlled cooling to achieve the desired residual stresses in the finished object is usually added. Slow cooling to produce the smallest stresses that can be achieved is called annealing. Traditionally glasses have been melted and made ready for working by people other than those who worked them. The glass blowers knew nothing of the exertions of the melters or teasers but were free to decline to work the glass if they thought it unfit. Similar distinctions tend to persist today even though glass quality can have very important effects on forming processes.

In principle glasses are homogeneous isotropic solids, normally made by cooling a melt. It is thus natural for engineers and scientists who know no better to assume that glass melts are, in practice, these ideal materials. Unfortunately, as glass technologists will readily admit, these ideal glass melts are almost unknown. Inhomogeneities in the glass are well understood to be closely connected with certain defects in the products but the possible relation between inhomogeneity and other types of defect, or overall efficiency of production, has long been, and remains, a matter of controversy. To begin any study of a forming process or problem by assuming that the glass can always be considered homogeneous and isotropic will sometimes prove a very dubious assumption. It is easy for engineers to forget this and to fail to recognize that the continuous production of good quality glass for them to work may demand as much skill and experience as the efficient operation of the forming processes with which they may be chiefly concerned.

A. CHOICE OF GLASS COMPOSITION

One of the most attractive properties of glasses is the ability to obtain a specific value of one property by suitable choice of composition. Unfortunately the range of compositions suitable in practice may be restricted because several properties, not just one, need to be specified. Most properties of glasses can be regarded as being uniquely defined by the glass composition (although some, such as density and refractive index, can be affected by thermal history, especially during annealing). Composition–property relations are often complex even for simple glasses but very different for different properties, and so particular values of several properties can be obtained by using multicomponent glasses; this is the reason why container and flat glass compositions nearly all contain Na_2O, K_2O, CaO, MgO, Al_2O_3, and SiO_2.

From a mathematical point of view, if n properties are to be specified there must be n degrees of freedom, that is to say $m = n + 1$ constituents. If $n = m - 1$ there is one solution or one particular composition; if $n < m - 1$ a range of compositions is possible and additional factors, such as ease of melting or batch cost, may be considered; if $n > m$ no solution is formally possible but the discrepancies between actual and desired properties can be minimized by optimization techniques familiar to chemical and control engineers. Some important properties, such as chemical durability and liquidus temperature, need to be within particular limits rather than having specific values and this relaxes the constraints. An exact value of thermal expansion is required for a glass to be sealed to a metal but not for a glass to be used for containers: precise optical properties are needed for optical glasses but not for decorative lead crystal products. Although pure silica is stronger than most other glasses, effective strength generally depends more on the properties of the surface than on the bulk composition. Strength, the most important of all properties, thus rarely needs to be considered when choosing glass composition, although compositions may sometimes be chosen to give a high Young's modulus. The next most important property is chemical durability, which depends strongly on composition; durability must always be considered when choosing glass compositions for practical purposes. Karlsson and his colleagues (1983) have recently devised some very elegant programs for optimizing composition–property relations for glasses in terms of oxide composition and also batch cost. Minimum batch cost does not necessarily lead to the lowest total cost but is a constraint worthy of consideration.

Much useful information about composition–property relations may be found in the literature, for example, in Morey's classic work (1954). The most comprehensive and admirable compendium is by Mazurin and his colleagues (1973–1981): there are, so far, six volumes but they exclude systems with more than three components and thus nearly all glasses of commercial importance, except for silica. As the data are presented in tables or figures, their being in Russian should not prove a serious obstacle.* Volf (1961) discussed the composition used for several important types of commercial products and many individual fragments of data are to be found in papers and patents. Composition–property relations are an obvious field for computer analysis and prediction but lack of sound theoretical models can make it difficult to attain the accuracy desired, especially for extrapolation rather than interpolation. Despite this, some notably successful exercises have been published; for example, the viscosity–temperature relations for $Na_2O-K_2O-CaO-MgO-Al_2O_3-SiO_2$ glasses (with some other minor constituents) of Lakatos, Johansson, and Simmingsköld (1972–1981). Success is made easier by keeping to a restricted range of compositions. Although Scholze's (1965) excellent book is mainly concerned with glasses having no more than three components, it does bring together models that have been proposed for the calculation of several properties of more complex compositions; these include density, refractive index, dispersion, thermal expansion, specific heat, thermal conductivity, surface tension, and elastic moduli. Turkdogan's (1983) recent book is largely concerned with slags but includes data on other binary and ternary melt systems of interest to metallurgists as well as glass technologists; it will be a widely valued additional source of information.

The final user of a glass may easily fail to recognize that the manufacturing process can also limit high-temperature properties, especially the viscosity–temperature curve and liquidus temperature. Some processes require a specific viscosity at the beginning of operations. If this corresponds to a temperature somewhat below the liquidus devitrification may be a serious problem. Such was the situation in the early days of sheet glass drawing; the difficulties were resolved, as demonstrated by Swift (1947), by putting about 3.5% MgO into the glass.

B. Choice of Raw Materials

Glass composition determines viscosity and other properties but this does not tell us the ease of melting the glass. The choice of raw materials can have very important effects on the ease of melting and glass quality; thus the total cost of melting the glass is not always minimized by using the cheapest raw materials, even though raw materials usually represent the biggest item in the list of running costs. Both physical and chemical properties of the batch affect the ease of melting, fining, and homogenizing a glass melt, and the judicious use of minor constituents not needed to produce the required room-temperature properties can be very beneficial.

---

* An English edition is now being planned.

Fining agents which aid bubble removal are the most obvious example but minor constituents may be needed to control oxidation (hence color) or dissolved gases. The choice of raw materials can also influence volatilization and corrosion in the furnace and the seriousness of atmospheric pollution problems. The best choice of raw materials must depend on all of these considerations and involves many complex chemical and physical processes. No simple rules allow the best choice to be made without trial and error experimentation. This is one important reason why glassmakers rightly tend to be very conservative about making changes in their operations but offers an important challenge to improve understanding and make possible more accurate prediction. West-Oram (1979) published a good review of raw materials for the glass industry, and many examples of how chemical and physical properties of raw materials affect melting were mentioned by Cable (1969). Addition of refractory oxides such as Al_2O_3 or ZrO_2 in that form is rarely desirable; use of more reactive compounds is usually wiser.

II. Glass Melting

A. Energy Requirements

Glasses can rarely be melted at temperatures below the liquidus (often 950–1150°C) and are usually melted at considerably higher temperatures (1400–1600°C), at which the efficient use of thermal energy is very important. Although it has little practical relevance it is easy to estimate the minimum heat required to make any particular glass. The glass could not be made unless sufficient heat were supplied for three purposes:

(1) to supply the necessary heat of reaction,
(2) to raise the glass or raw materials (and intermediate reaction products) to the maximum temperature used, and
(3) to raise the gases evolved by reactions to the temperatures at which those reactions become rapid.

The same temperature may be assumed for (2) and (3), but this is not inevitably the case. The data needed for accurate calculations are frequently not available, and estimates involve approximations such as ignoring heats of mixing of different stoichiometric silicate melts. The most extensive discussion of such calculations is by Kröger (1953) and Kröger *et al.* (1958). Table I shows some fairly typical results of such calculations.

Two points are obvious: heat of reaction is rarely more than 25% of the total heat required, and neither heat of reaction nor total heat is any

TABLE I

MINIMUM ENERGY REQUIREMENTS (kJ/kg) FOR GLASS MELTING

Type of glass	Heat of reaction	Total energy to raise to 1200°C	1500°C
Lead crystal	400	1830	2250
Borosilicate	415	1820	2255
Container	475	2120	2620
Flat	700	2410	2925

guide to ease of melting. Of the glasses listed lead crystal is the easiest to melt but heat-resistant borosilicate glass is, by far, the most difficult.

The thermal efficiency of glass melting operations is largely determined by rates of reaction, and not by simple equilibrium considerations. This is why the obvious logical implication of Table I, that using the lowest feasible temperature will save energy, is quite false. Making good quality glass requires completion of the initial reactions, dissolution of residual quartz grains, removal of gas bubbles (fining), and mixing of the inhomogeneous liquid to achieve sufficient homogeneity. Most of these are assisted by high temperatures. The upper limit to the desirable range of temperatures is likely to be set by the adverse effects of higher temperatures on evaporation of volatile constituents from the melt or increased corrosion of furnace refractories (including those in regenerators or recuperators); occasionally the difficulties of attaining very high temperatures also become important.

Good current practice in the glass container industry is equivalent to a fuel consumption of about 5250 kJ/kg, and the glass industry in Europe has a long history, more than 60 years, of striving to use fuel as economically as possible: Garstang (1971) has written an excellent review of this subject, showing developments from about 1920 to 1970. A paper by Barton (1982) may be regarded as a useful addendum to this.

The sol–gel method of making glasses at lower temperatures will rarely offer energy savings in practice. More energy will usually have been used (and paid for) in making the special raw materials than would be used in conventional melting when this is feasible. The sol–gel method nevertheless may have many interesting applications for special purposes and with particularly refractory glasses, as may chemical vapor deposition, the review of which by Schultz (1979) may be recommended. Scherer and

TABLE II

TYPICAL AVERAGE MASS FRACTIONS $\bar{M}$ AND MAXIMUM VARIANCES S_0^2 FOR COMPONENTS OF TYPICAL GLASS BATCH

Material	$\bar{M}$	S_0^2
Sand	0.61	0.238
Soda	0.21	0.167
Lime	0.17	0.141
Refining agent		
Sulfate	0.04	0.038
Arsenic	0.01	0.010

Schultz (1983) have recently published another excellent review of all the known unconventional ways of producing glasses.

B. Batch Mixing

A glass often requires a batch of widely differing proportions of five or more separate materials having different mixing characteristics (size, shape, density, coefficient of friction, etc.). Such mixtures are inherently difficult to mix well, but a large body of experience shows that care taken over batch mixing has considerable benefits in both laboratory and commercial operations. Lynn (1932) long ago made a simple but well-designed set of experiments to show how easily batch may segregate, even if once well mixed; the only defect in his work is that he failed to define the size of sample used. There is no point in using a sample so small that addition or removal of one particle of any material is detectable, a point emphasized by Manring and Bauer (1962). The minimum useful sample size will often be 10–20 g. Little of practical use has been written about the theoretical aspects of mixing multicomponent masses like glass batch, almost certainly because the subject is complex. Some points of importance may be gleaned from a detailed study by Fletcher (1963) of 50-g samples taken from five factories. It is usual to treat multicomponent mixtures as if they were binary ones, i.e., as "A + the rest." Calculations of the maximum variances (S_0^2) from Fletcher's data on this basis shows how important it is to consider which components to determine and what limits to set, see Table II.

Poole (1963) wrote a characteristically forthright paper about some of the problems that can easily arise in a glass factory and how simple measures will often lead to considerable improvements. A common mistake is to mix too long, allowing gravitational forces to initiate segregation. One or two minutes is frequently sufficient, but optimum mixing time must depend on the particular conditions. Care must be taken to minimize segregation between mixer and charging into the furnace.

C. Batch Heating

Glass-making reactions involving silica and sodium carbonate can be detected in the solid state down to about 450°C but only become sufficiently rapid to be important above about 575°C. Decomposition of Na_2CO_3 by reaction with silica becomes almost instantaneous above about 1000°C. The first essential stage of glass melting may thus be considered to involve heating the batch to around 1000°C and supplying the necessary heat of reaction. As glass batch is an excellent thermal insulator at room temperature, supplying the necessary heat is by no means easy on a large scale.

Suppose that batch were being charged at 100 kg/min (120 tonne/day) into a furnace 4.0 m wide as a continuous blanket 0.10 m thick, advancing at about 0.125 m/min. Consider this blanket to be suddenly heated from 25 to 1400°C on both top and bottom as it enters the furnace. If it remains undisturbed and its thermal properties remain constant the time t taken for the center to reach 1000°C are given by $\alpha t/l^2 \simeq 0.6$, where α is the thermal diffusivity of the batch and l is half the thickness of the batch layer. Taking a reasonable value of $\alpha = 2.5 \times 10^{-3}$ cm^2/sec gives the time to reach 1000°C as $t = 100$ min, during which time the batch would travel 12.5 m down the furnace. This ridiculous answer is not correct, because α increases both as temperature rises and as reaction occurs and l does not remain constant, but indicates that batch heating deserves careful attention. Jack and Jacquest (1958) considered in detail the implications of such a simple model and their paper is worth reading; however, note that the abscissa of their Fig. 4 is incorrectly labeled. Some of their conclusions are qualitatively valid.

The apparent thermal conductivities of batch materials and of mixed batch were estimated by Kröger and Eligehausen (1959) by the standard concentric cylinder method for thermal conductivity. At steady state the radial heat flow from the heater inside the inner tube must be

$$J_0 = -2\pi r\lambda \, (dT/dr). \tag{1}$$

If temperatures are measured at r_1 and r_2 (e.g., the surfaces of the annular space filled with the material being studied)

$$2\pi\lambda(T_1 - T_2) = J_0 \ln(r_2/r_1) \tag{2}$$

and the conductivity λ may be evaluated if J_0 is known. Once reaction has begun heats of reaction and convective flow complicate the true meaning of such results for glass batch but the results still give a good insight into

TABLE III

Typical Effective Thermal Conductivities (kJ/°C m h) of Batch and Glass

Temperature (°C)	Conductivity Batch	Conductivity Glass
200	0.99	4.25
400	1.18	5.6
600	1.37	8.1
800	2.90	12.2
1000	13.6	19.7
1200	44.6	44.6

heat transfer in batch between about 25 and 1200°C. Table III gives some fairly typical values. From 200 to about 700°C the apparent conductivity of either batch or glass doubles. At still higher temperatures both increase more rapidly; the value for glass is ten times higher at 1200 than at 200°C and batch increases by a factor of about 40 over the same range. The steep increase in apparent conductivity at higher temperatures is due to radiation and the values become dependent on the size of the sample used to make the measurement. These results help to make clear why the predicted "melting length" of 12.5 m in our original calculation would, in practice, probably only be about 2 m. Large-scale glass manufacturers devote considerable effort to trying to find optimum batch charging techniques, that is to say those giving fastest melting down of the batch and minimum segregation during melting.

Some authors have discussed models of glass melting only in terms of heat transfer. For example, Pugh (1968) assumed a steady state model having a constant layer of freshly melted glass with all the heat supplied being consumed as heat of reaction at the glass–batch interface. Rather similar assumptions were made by Routt *et al.* (1983) to estimate the melting rate in an all-electric melter. A very comprehensive discussion of energy balances was published by Cooper (1980). These models have much of value to teach us but can never be sufficient. Heat is supplied to enable *chemical* reactions to take place and the practical limits of glass melting practice frequently depend on the kinetics of quartz dissolution, fining, or homogenizing, which are largely determined by these chemical processes.

D. The Chemistry of Melting

Sodium carbonate is the most reactive major component of many batches and quartz the least reactive. Good insight into the early stages of melting many glasses would thus be obtained by detailed knowledge of the reactions that can occur between Na_2CO_3 and SiO_2. Most of the techniques available have been used but knowledge is still incomplete. The earliest solid state reactions occurring at temperatures up to about 700°C have been studied by isothermal thermogravimetric methods, but the theoretical models usually used to interpret the results, the Tammann–Jander and the Ginstling and Brounshtein equations, are very crude approximations of the conditions actually involved and do not lead to satisfactory estimates of diffusivities although the reactions clearly are controlled by diffusion. The shortcomings of these theoretical models have been discussed by the author in lecture courses for about fifteen years but never otherwise published; most of the shortcomings are rather obvious on careful reflection. Some of the best experimental work by this technique was done by Harrington et al. (1962) but some of their deductions are suspect because of the shortcomings of the theory used. Extensive studies of the Na_2CO_3–$CaCO_3$–SiO_2 system were made by Kröger and his colleagues (Kröger, 1948, 1949, 1953; Kröger and Marwan, 1955, 1956, 1957; Kröger and Vogel, 1955; Kröger and Ziegler, 1952, 1953, 1954).

Between about 700 and 1000°C the system becomes more complex, involving solid, liquid, and very large volumes of gases which, among other things, encourage vigorous local convection in the reacting mass. Differential thermal analysis provides valuable insight into reactions at temperatures up to about 1000°C but is very difficult to interpret quantitatively. The most valuable contributions by this technique have been made by Wilburn and Thomasson (1958), Dawson and Wilburn (1965), and Wilburn et al. (1965). Few other standard techniques provide useful information in this temperature range; Cable and Martlew (1971, 1984) attempted to overcome some of the problems by studying the dissolution of vitreous silica rods in melts of pure Na_2CO_3, Na_2CO_3 + $CaCO_3$, Na_2CO_3 already reacted with small proportions of silica, etc., at temperatures between about 875 and 1200°C.

The general picture that emerges is an initial reaction to form sodium metasilicate in a wide range of conditions.

$$Na_2CO_3 + SiO_2 \longrightarrow Na_2SiO_3 + CO_2 \uparrow . \tag{3}$$

Reaction occurs readily in the solid state above about 575°C. As most glasses have $Na_2O : SiO_2$ ratios of 1 : 4 or more, this leaves a large excess of silica. At around 700°C there is an exothermic DTA peak chiefly due to

$$Na_2SiO_3 + SiO_2 \longrightarrow Na_2Si_2O_5 . \tag{4}$$

Liquid is first found at about 780°C, at the $Na_2Si_2O_5$–SiO_2 eutectic, but this still leaves excess unreacted silica. It is unusual to be able to identify specific further steps, but these initial reactions produce compositions with 50, 33, and about 24 mole % Na_2O. At higher temperatures the residual silica grains dissolve in a liquid which may be very inhomogeneous but has an average composition close to that of the final glass. Rates of silica reaction may now be several orders of magnitude slower than for the first reaction and silica dissolution now produces liquid of equilibrium composition according to the phase diagram; between 1200 and 1400°C this has between about 19 and 14% Na_2O in Na_2O-SiO_2 melts. It should now be clear why normal glass melting techniques usually produce inhomogeneous melts. As these various liquids often have very different viscosities and somewhat different densities, gravity can rapidly lead to large-scale segregation with the densest liquid at the bottom. The vigorous evolution of gases as carbonates and other compounds decompose can prevent this segregation in the early stages but can also act in the opposite way by carrying sand grains to the surface of the melt by a kind of froth flotation. As calcium carbonate can easily dissolve in molten Na_2CO_3 [tending to form $Na_2Ca(CO_3)_2$], reactions in $Na_2CO_3-CaCO_3-SiO_2$ batches tend to differ only slightly from those in $Na_2CO_3-SiO_2$ batches.

Small proportions of other constituents, chiefly Na_2SO_4 but also $NaNO_3$, NaCl, etc., can significantly affect the degree of segregation occurring during melting. Thomasson and Wilburn (1960) studied some of these reactions by differential thermal analysis. Choosing the right batch to minimize the subsequent problems of fining and homogenizing is one of glass technologists' important skills which still largely rely on experience and intuition.

The last stage of melting, dissolving the small proportion of sand grains not already completely reacted, is a much slower process, undoubtedly controlled by diffusion in the liquid, as demonstrated by Cooper and Kreider (1967) and Hlaváč and Nademlynská (1969). In many cases the batch-free time, or time to dissolve all the batch materials, comprises roughly 15% for heating the batch to reaction temperature, 20% for vigorous reactions, and 60% for dissolution of residual sand grains.

Hot-stage microscopy has been used by several workers to gain a better understanding of some of the complex reactions occurring in glass melting (see Dietzel and Flörke, 1959; Buss, 1962; Conroy *et al.*, 1963).

Empirical studies of how various factors such as batch particle size or temperature affect batch-free time are generally useful to glassmakers even though results cannot be scaled up quantitatively. The earliest extensive investigations of this kind were by Potts and his colleagues (1939, 1944) and by Preston and Turner (1940, 1941). Other very interesting results are in a neglected paper by Frölich (1946) and the work of Boffé and Letocart (1962). When materials are closely graded and of similar sizes, or the others are finer than the sand, decreasing sand particle size improves melting rate, fining, and homogeneity. However, all of these are only true down to about 0.1-mm grain size in soda–lime–silica glasses. With finer sand there may or may not be the expected continued improvement in melting rate but the melt becomes full of tiny seed which are very difficult to remove even when fining agents are used. This effect of very fine sand does not affect all types of glass, and finer sands may be used in some processes. When sands are graded the situation is more complex: the coarsest few percent will determine the batch-free time, as demonstrated by Boffé and Letocart (1962), while the finest fraction may tend to produce fine seed and impair fining. The chemistry and physics of these effects must be largely concerned with the surface area of silica available for wetting, reaction, and bubble nucleation but are little understood. It is fortunate that sand deposits laid down from flowing water have the majority of their grains in the most desirable size range (about 0.10–0.5 mm). Fine sand can become coated with alkali during preparation of pellets and finer sand may sometimes be used in pellets than in normal batch.

The questions already referred to are not the only ways in which chemistry is important in glass melting. Dissolved gases remaining from batch reactions or absorbed from the furnace atmosphere are important in fining, foaming, and reboil; they may also affect some properties. The oxidation of the melt may need also to be controlled, especially when producing colored or decolorized glasses; these are discussed below.

This section is relatively brief because the chemistry of glass melting reactions is complex, largely heterogeneous, and rather poorly understood. Standard and readily available techniques of investigation leave important gaps in our understanding of the whole process. New and better tools could contribute a great deal.

E. Control of Oxidation

On an atomic scale a silicate glass may be thought of as a rather open structure of oxygen ions held together by much smaller silicon and other cations which occupy many of the smaller interstices. Its chemical and many other properties are likely to be determined by the behavior of these oxygen ions. The classic random network hypothesis of silicate glass structure, due to Zachariasen (1932) and reinforced by the early x-ray work of Warren and Biscoe (1938), is that network-forming oxygens occupy strongly bonded sites at the corners of adjacent SiO_4^{4-} tetrahedra and that these bridging oxygens (denoted O^0) are too strongly bound to enter into chemical reactions. Addition of modifiers such as soda then breaks some of these bonds, leaving tetrahedra with a negative charge which needs to be balanced by a neighboring cation. This may be represented by

$$\rightarrow Si-O-Si \leftarrow + Na_2O \longrightarrow 2(\rightarrow Si-O^-) + 2Na^+ \quad (5)$$

In such a structure a very small proportion of the singly bonded oxygens (O^-) may break free to become free oxygen ions (O^{2-}) which should be extremely reactive. It is thus attractive to hope that several kinds of chemical processes could be defined in terms of the effective concentration of free oxygen ions which, it is presumed, may be defined by a suitable equilibrium constant for the equation

$$O^0 + O^{2-} \rightleftharpoons 2O^- \quad (6)$$

This equation is certainly qualitatively correct in indicating that increasing the alkali content of a glass makes the melt more reactive and more oxidizing: a particular activity of free oxygen ions a_O^{2-} ought to represent equilibrium with a specific oxygen partial pressure p_{O_2} with which the melt would be in equilibrium. This could be thought the equivalent of pH in ordinary aqueous chemistry.

Some of the obvious reactions important in glass melting to which these ideas might be applied are the solubilities of CO_2, H_2O, and SO_3 or $SO_2 + O_2$ mixtures as indicated by equations

$$CO_2(g) + O^{2-}(m) \rightleftharpoons CO_3^{2-}(m), \quad (7)$$

$$H_2O(g) + O^{2-}(m) \rightleftharpoons 2(OH^-)(m), \quad (8)$$

$$SO_2(g) + \tfrac{1}{2}O_2(g) + O^{2-}(m) \rightleftharpoons SO_4^{2-}(m). \quad (9)$$

These equations are quite successful in describing the effects of varying gas partial pressures for a particular melt composition but fail when glass composition is varied, the three systems showing quite different behavior. These shortcomings were discussed by Cable (1982). The traditional approach to chemical equilibrium in glass melts and their thermodynamic interpretation has, nevertheless, been based on pursuing these simplistic ideas; perhaps there has been too great an influence carried over directly from high-temperature metallurgy.

Molten metals are very fluid and rapidly tend toward equilibrium; the description of their behavior in terms of the simplest possible species also usually corresponds closely to reality, so such assumptions form a sound basis for both describing equilibria and understanding their behavior. However, thinking of silicate glasses as if they were mixtures of pure oxides such as Na_2O, CaO, Al_2O_3, and SiO_2 gives very little insight into their chemical behavior, although logically permissible. The finding that activity coefficients γ are in the range $\gamma < 10^{-2}$, as is often true for Na_2O, etc., is best thought to show that the formal model has little connection with reality and suggests that a better model should be devised.

Rigid adherence to the traditional ideas at once leads to the paradox already implied, one far more important to glass technology than the Kauzmann paradox recently reexamined by Zelinski et al. (1983). This may be exemplified by considering the behavior of iron dissolved in glasses. As might be expected, iron can exist as either ferrous or ferric ions both of which color glasses appreciably but affect ultraviolet, visible, and infrared absorption in quite different ways. Understanding the behavior of iron is very important in fields such as decolorizing and making heat-absorbing glasses. The simplest equation that could be written to describe ferrous–ferric equilibria is

$$2Fe^{2+}(m) + \tfrac{1}{4}O_2(g) \rightleftharpoons 2Fe^{3+}(m) + O^{2-}(m). \tag{10}$$

This equation successfully describes the dependence of the $Fe^{2+}:Fe^{3+}$ ratio on oxygen partial pressure (although the evidence is much scantier than might be supposed). The paradox is that it completely fails to describe the effect of glass composition. According to Eqs. (5) and (6), adding more alkali to a glass should increase (O^{2-}) and hence, according to Eq. (10), should shift the equilibrium in favor of forming more Fe^{2+}; this is the opposite of what happens.

The paradox is easily resolved by considering appropriate complexes, as is standard practice in aqueous solutions and molten salts. This important point was made by Budd (1966), who also emphasized that to assume activity approximately equal to concentration requires the use of appropriate ionic complexes. Karlsson (1977) showed how the paradox can be resolved for iron by assuming complex formation of the type

$$Fe^{2+} + \tfrac{1}{4}O_2 + \tfrac{1}{2}(2n-1)O^{2-} \rightleftharpoons FeO_n^{(3-2n)-}, \quad 0 < n < 4. \tag{11}$$

The appropriate value of n may vary with composition in any particular system, and Karlsson showed how the data then available could be interpreted by this model for iron in lithium, sodium, and potassium silicates. Rather surprisingly, these attractive ideas have not been widely welcomed. The recent book by Paul (1982) provides an excellent review of the conventional approach to these questions. Although largely concerned with slags and other oxide melts of particular interest to metallurgists, the recent book by Turkdogan (1983) is a valuable addition to the literature on chemical processes in melts.

Disarray among the theoreticians does not absolve the glass technologist from the practical need to control the state of oxidation of his glass.

This is essential when making "colorless" flint glass, in which the tint due to iron must be minimized; or amber, which requires reduction of sulfate to sulfide while preventing excessive reduction of ferric to ferrous; or colored glasses containing variable valence elements such as Mn, Cr, or Cu; it also has an important influence on fining.

Decomposition of raw materials, chiefly carbonates, produces about 1300 liters of gas at 1400°C per liter of melt during melting and more than 99.9% of this may easily be lost during the initial vigorous melting reactions. At this stage the rapid gas evolution and the vigorous stirring that it causes might be expected to make the majority of the melt come close to equilibrium with these gases. Decomposition reactions do not all occur simultaneously, and so there are some effects of time and temperature on the composition of the gases; the melt is also very inhomogeneous and after homogenizing might not be in equilibrium with the same gas mixture. However, the general principle that oxidation can be controlled by appropriate additions to the batch still holds. Relatively oxidizing conditions can be maintained by compounds that give off oxygen during decomposition (e.g., Pb_3O_4, KNO_3) and reducing conditions by carbon, metallic tin, and so on. Once the rapid gas evolution is over, the furnace atmosphere begins to influence the melt, but the high viscosity of most glass melts and the low diffusivity of most species in them mean that small laboratory melts need very long times to achieve equilibrium with the surrounding atmosphere. Larger pot or tank melts, in which there is more vigorous convective flow, do not always need much longer. Considerable control over oxidation can be achieved by careful choice of batch materials and furnace atmosphere, the latter generally being more important in tank furnaces than in pot or laboratory melts. The reliable literature on these topics is disappointingly meager, but a very useful empirical method of estimating the relative oxidizing power of various batch materials has been worked out in recent years (see Simpson and Myers, 1978).

F. DECOLORIZING

Decolorizing is generally considered to be partly chemical and partly physical: the inevitable tint due to iron found in most glasses can be minimized by chemical interaction with other materials in the batch and the melt; the residual tint is then made as near to neutral gray as desired by deliberately adding other coloring agents. The physical part of decolorizing works because the eye is more sensitive to tint than to brightness under ordinary viewing conditions.

Chemical decolorizing involves converting as much of the iron as possible to the ferric state, which absorbs strongly in the ultraviolet (around 230 nm) but very weakly in the visible in soda–lime–silica glasses. Ferrous ions have a strong absorption in the near-infrared, at about 1000 nm; hence their use in heat absorbing glasses, but the tail of this extends into the visible and causes rather more absorption there than ferric ions. Change in oxidation alone can thus alter the tint between, typically, pale yellow–green and rather deeper blue–green. Adding another variable valence element capable of interacting with iron can achieve this result without other steps to oxidize the melt. Manganese (purple when oxidized, almost colorless when reduced) was used for several hundred years but went out of fashion about 60 years ago partly because it can have a disagreeable residual tint but also because ultraviolet radiation can eventually reverse the process, oxidizing the manganese and reducing the iron to make the glass pale pink. Tour guides in some European cities still possessing large expanses of old windows (e.g., Amsterdam) tell wondrous fairy tales about this *solarization* of these windows. Arsenic was a very effective colorless replacement for manganese until it became subject to stringent restrictions on use. Cerium is often now favored although not inherently superior. The literature on decolorizing is deficient; Paul (1982) does not discuss the topic although he did excellent work on the mutual interaction of redox pairs (Paul and Douglas, 1966), but Hlaváč (1983) has a good concise account and his Fig. 65 correctly describes the important results of Johnston (1965), while Paul's (1982) Fig. 5.3 does not.

Physical decolorizing has traditionally used very low concentrations of cobalt (blue) and selenium (pink) to produce the desired tint; not all customers want a neutral gray. The very low concentrations needed make control rather difficult and the chemistry of the selenium as well as its volatility make it rather unstable and thus variable. Other oxides are sometimes used in other glasses; nickel and neodymium have been used in lead and borosilicate glasses. Although written almost 40 years ago, the book by Weyl (1951) on colored glasses remains an extremely valuable source of information on nearly all aspects of the subject.

III. Bubbles in Glass Melts

A. Fining

1. Introduction

Vast quantities of gases are evolved during the early stages of glass melting. Much of the gas is evolved while solid batch particles are still present, and so nucleation and escape of the majority of the gas occur relatively easily. Nucleation of gas bubbles continues, but at a rapidly decreasing rate, as liquid phases form and may not cease, in the body of the melt, until all the batch particles have dissolved. At that stage the glass may be opaque because of innumerable small bubbles. Although very numerous these are usually so small, average diameter no more than about 0.25 mm, that they may represent no more than about 1% by volume of the melt; these residual bubbles (seed) must nevertheless be removed before the glass will be acceptable for most purposes. Quality standards vary according to use but usually involve both size and number; very severe specifications are set for flat glass and optical glasses or optical fibers. Fining is the elimination of these bubbles.

2. Rise to the Surface

The viscosity of most glass melts is so high, even at maximum melting temperatures, that rise to the surface would be very slow to remove seed from static melts. If the melt is flowing, horizontal velocity components affect the path of bubbles but not the time they take to rise to the surface. Vertical velocity components in the melt may assist or hinder rise to the surface, according to direction, and conservation of mass means that every upward flow, which assists rise to the surface, will normally be balanced by an equal downward flow elsewhere. Stirring of the melt is thus most unlikely to accelerate the rise and removal of all seed.

Jebsen-Marwedel (1936) long ago made calculations showing how velocity of rise in a static melt would vary with bubble size and temperature for a typical glass. The obvious implication was that rise to the surface could not account for the observed rates of fining. In an investigation which aroused the author's interest in fining Bastick (1956) confirmed that the typical fining behavior is very close to an exponential decay of number N of seed with time t:

$$N = N_0 \exp(-at). \tag{12}$$

Although Bastick found that the intercept N_0 appeared to be independent of temperature it must be emphasized that this extrapolation into the realms of batch heating and rapid reaction is unreal. Extrapolation to much longer times and values of N less than about $1/cm^3$ is also erroneous; the rate of disappearance (or slope a) usually decreases toward the end of fining. The main point made by Bastick was that a varied much more rapidly with temperature than changes in the viscosity of the melt or gas density (for the same bubble population) would predict, assuming removal by rise to the surface. Neither Jebsen-Marwedel nor Bastick discussed the actual size distributions of seed in melts (although Bastick had unpublished data on these) and so did not make quantitative comparisons between theoretical model and experiment. Dubrul (1956) showed that fining in full-scale pot melts followed the same general law as in small laboratory melts and found that viscosity was an unreliable guide to fining behavior, a point also made by Lyle (1956).

The theoretical model used by Jebsen-Marwedel and Bastick assumed that the velocity of rise of a bubble through the melt was given by Stokes's law, which applies strictly to an isolated solid sphere in an infinite body of liquid, and its application to a swarm of bubbles in a small crucible involves several important and rather dubious assumptions. When most of the bubbles are of diameter $x < 1$ mm and their concentrations is <1% by volume, the most important error is that caused by the assumption that tangential flow v_θ is zero at the surface of the sphere; this is true for a solid sphere but not for a fluid one. When both the density and viscosity of the fluid inside the sphere are negligible compared with those of the external fluid the predicted velocity v_B of rise for a bubble is one and a half times that given by Stokes's law v_S:

$$v_B = \tfrac{3}{2}v_S = \rho g x^2/12\eta, \tag{13}$$

where ρ and η are the density and viscosity of the molten glass. The validity of this equation has recently been confirmed by Jucha *et al.* (1982). Typical glass properties give v_B in the range about $(0.05-0.50)x^2$ mm/sec, values much too small to account for observed rates of removal of bubbles of diameter $x < 0.2$ mm or thereabouts.

The first detailed comparison of experimental results with predictions of rise to the surface was made by Cable (1958a). In that work total numbers of seeds N and size distribution histograms (column width 40 μm) were both measured. These results were used to show how the number of seeds n in any particular size range varied with time. In all cases these relations were much closer to exponential,

$$\log n = \log n_0 - bt, \tag{14}$$

than the simple linear relations suggested by the model.

Although larger bubbles disappeared more quickly than small ones, the dependence of b on diameter x was much less than the $b \propto x^2$ implied by the rise to the surface model. The rate of disappearance b of a particular diameter of seed varied with temperature, size of sand grains used in the batch, or addition of fining agents, and showed no tendency to go to zero as $x \to 0$. The most important result was that observed rates of disappearance could be accounted for entirely by rise to the surface for $x > 0.25$ mm, but as $x \to 0$ the observed rate became increasingly greater than predicted. Some other mechanism was necessary to account for the removal of small seed. The two obvious possibilities were either that bubbles grew, accelerating rise to the surface, or shrank and dissolved, making rise to the surface unnecessary. A significant point not emphasized then or later is that, although fining agents gave faster fining, there seemed to be no essential qualitative difference between the behavior without or with fining agents. This work was extended by Cable (1960, 1961). Similarly detailed data have been reported by only a few other authors; these include Cable and Naqvi (1975) and Okamura (1966).

Quite a number of other workers have found approximately exponential relations between number of seeds and time and investigated various parameters. Many of these were summarized by Cable (1970). Among the most notable results are those of Gehlhoff *et al.* (1930) showing very similar effects of temperature and additions of sulfate on fining of small (100-cm^3) laboratory melts and full size (270-liter) pots, and an investigation of the size of the melt (from 45 to 6 × 10^5 g) by Cable *et al.* (1968b). It is generally accepted that the phenomena are similar in small and large pots and, by a considerable extension of credibility, probably justifiable, in tank furnaces. However, no adequate rules exist for quantitative scaling up; if such could be developed they would represent a very useful advance. The suggestion of Bunting and Bieler (1969) that scaling up could be considered as a matter of heat transfer alone is certainly too bold a simplification.

Although coalescence plays a vital part in the early stages of gas evolution it cannot normally be a major factor in removing bubbles during the later stages of fining, when observation has become reasonably easy. Attention has therefore been concentrated on the possibilities of growth or dissolution of bubbles as the other important fining mechanism. As most of melting and fining takes place in systems far from equilibrium it is neither necessary nor wise, having accepted that mass transfer between melt and bubble is possible, to insist that only growth or only dissolution can be involved. Changes that occur with time or temperature could easily make both important in the same melt at different stages.

3. Gas Composition in Seed

The most obvious evidence for mass transfer between bubble and melt is the way in which the composition of the gas in individual bubbles taken from one melt can change with time of melting (at either constant or varying temperature). The earliest evidence was published by Appen and Polyakova (1938) and extended by the data in Slavyanskii's book (1957). These results were an outstanding achievement with the primitive and inaccurate methods of analysis used; later work has not challenged their main findings. In the past 20 years quite a number of workers have made more detailed investigations by gas chromatography or mass spectrometry, each technique having its characteristic advantages and shortcomings. If a glass is melted from a batch with no fining agents the seeds are found to contain only carbon dioxide. With appropriate additions of fining agent (As_2O_3 or Sb_2O_3) the composition of the gas changes from CO_2 to O_2. In larger melts, such as those studied by Slavyanskii, there is a change from O_2 to N_2 near the end of fining. Most melts presumably tend to follow the cycle $CO_2 \rightarrow O_2 \rightarrow N_2$, as demonstrated by Apak (1975) in 20-kg melts.

The way in which fining and rate of change of gas composition depend on arsenic addition was studied for a simple soda–lime–silica glass by Cable *et al.* (1968a); similar studies were made by Cable and Naqvi (1975) for antimony. The general pattern of behavior is that small additions which do not cause a change in composition of the seed do not improve fining; larger additions which cause the $CO_2 \rightarrow O_2$ change accelerate fining, but even larger additions slow down $CO_2 \rightarrow O_2$ conversion and also begin to impair fining. The best fining appears to accompany the most rapid $CO_2 \rightarrow O_2$ conversion. Excellent and very valuable work on the compositions of seed taken from different parts of tank furnaces was done by Mulfinger (1974, 1976). The techniques now used to analyze seed should make the likelihood of contamination by air or other external gases very remote, but there is some possibility of adsorption or desorption of gases onto or from freshly fractured surfaces if a large surface area is formed when the bubble is released. Some authors have in the past assumed N_2:Ar ratios to be a reliable guide to whether the nitrogen in bubbles came from air. However, the relative solubilities, diffusivities, and saturations of these two gases are unlikely to be so matched as to maintain the initial ratio for very long in bubbles that change size appreciably, so this is a rather dubious guide. In some cases gases present at high temperatures might react together, condense, or react with the surface of the bubble during cooling and thus evade detection at room temperature; this must not be forgotten when seeking to interpret data obtained at room temperature.

It is far from certain that analyses such as those just discussed show the typical change in composition, with time, of one particular bubble. However, the range of compositions found for single bubbles taken at one particular time is much smaller than the changes that occur during the course of fining. It is therefore reasonable to assume that the observed behavior is qualitatively similar to that in any individual bubble. Without other information it is not possible to say whether these changes occur as bubbles grow or dissolve.

4. Observation of Individual Bubbles

Greene and his students (Greene and Gaffney, 1959; Greene and Kitano, 1959; Greene and Lee, 1965; Greene and Platts, 1969) performed very valuable model experiments, studying the dissolution of oxygen and later SO_2 + O_2 bubbles in several glasses at temperatures between about 1000 and 1200°C. The bubbles were held approximately stationary by using a cylindrical sample which was rotated about its horizontal axis. Until these results were published it was widely considered unthinkable that growth or dissolution of bubbles could be important in fining.

Unfortunately it is very difficult to observe individual bubbles directly under conditions typical of the relatively early stages of glass melting which must largely govern their behavior, and it is by no means certain that the experiments that are possible correctly reflect typical melting from batch. Despite careful work by a number of skilled workers it is still not clear what is typical. Solinov and Pankova (1965) observed the growth of bubbles in a soda–lime–silica melt with various proportions of fining agents but these were bubbles >1 mm in diameter, presumably formed after the completion of normal melting and fining. Their results showed *slower* growth when arsenic was used than in melts with no fining agent. They also mentioned that small seed dissolved but did not trouble to study them. Cable and Haroon (1970) injected CO_2 bubbles into a melt at 1200°C, then recovered them after a few minutes and measured both size and composition. The results showed that the $CO_2 \rightarrow O_2$ change was accompanied by *decrease* in size but they were not convinced that this was typical of what happens during fining. Conroy *et al.* (1963) observed rapid resorption of bubbles on cooling somewhat melts fined with sulfate.

A considerable body of data has been assembled and described by Němec (1974,* 1977, 1980). Additions of As_2O_3 (1 and 2 wt. %) greater than those giving optimum fining (~0.5% As_2O_3) made bubbles of diameter greater than about 1 mm grow at 1400°C but shrink considerably during cooling to 1150°C. Without fining agent this shrinkage on cooling did not occur; neither did the growth at a constant 1400°C. The second paper extended these observations to include 0.25 and 0.50% As_2O_3 as well as up to 9% As_2O_3; in these experiments bubble growth was observed for all arsenic additions and showed a maximum at about 3% As_2O_3, but most of the observations were again for bubble diameters between 0.5 and 2 mm. The only experimental data in his last contribution are some additional data for growth of similar bubbles in a melt fined with sulfate at 1470°C and in one fined with chloride at 1530°C.

5. *Theoretical Models for Bubble Growth or Dissolution*

It is the duty of experimenters to seek a sound basis for the interpretation of their results, but soundly based theories need to be based on clearly established qualitative premises. Unfortunately, fining is so complex that several important assumptions which are commonly made are of

---

* The ordinate of Fig. 3 is incorrectly labeled; it presumably should be as in Fig. 4.

dubious validity. Even when these assumptions are made a quantitative model for the growth or dissolution of gas bubbles is surprisingly difficult to develop. The only assumptions that are obviously true are that the seeds are almost perfectly spherical, contain several gases which can be exchanged with the melt, and rise freely within the melt. It is natural to assume almost instantaneous equilibrium at the gas–melt interface and hence control by diffusion in the surrounding melt. This view is very strongly supported by the observation of Greene and Lee (1965) that stopping the rotation of their sample and allowing the bubble to rise freely considerably accelerated its rate of solution. The suggestion of one pair of authors that bubble dissolution may be controlled by reaction at the interface was based on an unspecified but obviously inadequate set of assumptions.

The simplest theoretical model, which might apply in fact to some experiments, is an isolated spherical bubble in conditions of spherical symmetry. An analytical solution is available for this case for a bubble containing only one component which has a very low solubility. This solution for the rate of change of radius a with time t is, by analogy with heat conduction,

$$\frac{da}{dt} = \frac{D(c_i - c_\infty)}{c_s}\left[\frac{1}{a} + \frac{1}{\sqrt{\pi Dt}}\right], \qquad (15)$$

where D is the diffusivity of the gas in the melt, c_s is the concentration (density) of gas inside the sphere, and c_i and c_∞ are the dissolved concentrations in the melt at the interface and in the bulk far from the bubble. This model is rigorously true only for a sphere of constant size, hence its restriction to very low solubility; large errors may be introduced by using it for higher solubilities or for bubbles containing several gases, leading c_i to vary with time. It is possible to adapt this equation to take into account, approximately, the effect of surface tension (see Epstein and Plesset, 1950), and this rather inadequate model was used by Cable (1959, 1961) to gain some insight into the behavior to be expected of oxygen or carbon dioxide bubbles in glass melts. He also then recognized that a better model dealing with the diffusion of at least two gases was needed.

Analytical solutions for one-component growth from zero size which take the radial flow of the liquid into account are possible and were examined in detail by Scriven (1959). The relation between size and time is always of the form

$$a = 2\beta\sqrt{Dt}, \qquad (16)$$

where the growth rate constant β depends on both solubility and the volume change caused by transfer of matter across the bubble–liquid interface. Scriven's analysis has recently been extended by Frade (1983) to one-component spheres with concentration-dependent diffusivity and to multicomponent spheres for which composition is independent of time, which represents the asymptotic behavior of all such systems.

Neither growth from finite size nor dissolution can be solved analytically for cases of real interest; numerical methods or approximations must be used. Readey and Cooper (1966) briefly examined both growth and dissolution by finite-difference computations; their work was expanded by Cable and Evans (1967). Unfortunately the methods of computing then used were rather primitive and the results are not as precise as was hoped. In subsequent years numerous approximations were proposed and compared with one another but not with precise predictions. For example, Rosner and Epstein (1972) proposed a solution giving the asymptotic behavior

$$a = 2\phi\sqrt{Dt}, \tag{17}$$

where ϕ is the solubility parameter defined by Scriven (1959). This is very good for high solubility ($\phi > 10$) but poor for lower solubilities.

It is difficult to find models for dissolving spheres which are acceptable for moderate or high solubilities, although several approximations are very satisfactory for low solubilities ($\phi < 0.01$), but numerical techniques capable of giving accurate results were developed by Duda and Vrentas (1969, 1971).

Published work on multicomponent bubbles has been unsatisfactory, first because the methods used were relatively unsophisticated, and also because only a few specific examples were computed. The only work so far done in which theory and experiment were shown to confirm each other was that of Griffin (1971), who both measured and computed the behavior of $O_2 + CO_2$ bubbles in water. Water was chosen because reasonably accurate solubility and diffusivity data were available and a slight adjustment to initial size a_0 was the only variable used to get a good fit. Further study of the theory of diffusion-controlled growth or dissolution of stationary multicomponent bubbles is unlikely to be necessary following the excellent work of Frade (1983), who has extended Scriven's analytical work and devised accurate and efficient numerical methods for both growing and dissolving bubbles. Frade confirmed that the techniques developed by Weinberg and Subramanian (1980) give good results. Among Frade's interesting findings are the fact that growing bubbles always tend toward an asymptotic composition but dissolving bubbles do not. He also found it difficult to produce the observed $CO_2 \rightarrow O_2 \rightarrow N_2$ change in bubble composition for growing bubbles by using what appear to be appropriate values of solubility and diffusivity data for these gases in glass melts.

There is a very regrettable shortage of reliable solubility and diffusivity data for gases in glass melts. Solubility and actual dissolved concentrations are of vital importance as they determine whether the gas tends to diffuse into or out of the bubble; diffusivity affects only the rate or the time scale. In fining the bubbles rise freely instead of being held stationary. The typical conditions (size of bubble, viscosity of melt) mean that classic forced-convection models, such as that used by Němec (1980), are of very dubious validity in fining, which is unfortunate. Further work such as that of Onorato *et al.* (1981) on freely rising bubbles is needed but awkward problems are involved.

Despite much effort by many talented workers our understanding of fining is very incomplete. It is by no means certain that the assumptions now generally accepted will prove a sufficient base when fuller understanding becomes possible. The traditional ideas about the effect of temperature on $As^{3+} \rightleftharpoons As^{5+}$ or $Sb^{3+} \rightleftharpoons Sb^{5+}$ equilibria may not be sufficient to explain the fining action of these oxides or to show why sulfate and halides can be efficient fining agents. A better understanding is much to be desired when both arsenic and sulfate have disadvantages in terms of health or pollution.

Legal restrictions on the use of arsenic (because it is a poison) are now so severe in several countries that its use as a fining agent has become very rare. This seems rather unfortunate when some more potentially dangerous organic chemicals are readily available. Sulfate is now almost certainly the most popular fining agent and can also assist melting and improve homogeneity. However, it is rather volatile and this may lead to two other serious problems: sulfates may condense on regenerator refractories or attack them as vapors, and concentrations of sulfur-containing gases or particulates may exceed permitted levels in the waste gases emitted from furnaces. The latter problem can usually be solved by using filters or electrostatic precipitators, but these measures add considerably to running costs. Also, sulfate is not particularly effective in all types of glasses, such as borosilicates. Antimony is generally free from the restraints imposed on arsenic but is not always as effective; halides can also produce undesirable chimney emissions. Alternative fining agents would be very welcome.

Many of the empirical observations of fining behavior not discussed here were summarized by Cable (1970).

B. Reboil and Foaming

In various circumstances new bubbles may grow and persist in a glass melt which was previously reasonably well fined, or else bubbles formed early in melting may not collapse as they usually do. The latter can be a very serious defect in a tank furnace heated by flames from above because the foam is an excellent thermal insulator. Foaming is normally a problem, if it occurs, early in melting, and is thus a high-temperature and melting-end problem. Reboil bubbles tend to occur at lower temperatures (1000–1200°C) toward the end of fining and thus are a working-end or forehearth problem which may lead to blisters in the product.

1. Physics

The essential physics of the two processes differs in only one way: foams require some mechanism to stabilize thin liquid films and prevent them from draining until they are so thin that they rupture spontaneously. Reboil may be a problem when there is less than 1% by volume of gas, but foams may easily contain far more gas than liquid. The necessary initial steps in both cases are (1) to make the melt supersaturated with some gas, (2) to nucleate bubbles, and (3) for the nuclei to grow sufficiently to form obvious defects.

Theoretical models for homogeneous nucleation of bubbles are reasonably easy to set up and solve. A useful starting point for the present purpose is a simple model discussed by Fisher (1948) in which he compared the relative ease of forming a bubble or a crack in a glass over a wide range of temperatures. The fracture pressure for bubble nucleation and growth to occur in 1 sec is, according to Fisher's theory,

$$p = -\left[\frac{16\pi}{3} \frac{\sigma^3}{kT \ln(Nk/h) - \Delta F^0}\right]^{1/2} \tag{18}$$

where N, k, h, and T have their usual significance in physics, σ is interfacial tension, and ΔF^0 the activation energy involved in transferring a gas molecule into the bubble.

Nucleation seems to occur more easily than such a model predicts but heterogeneous nuclei are likely to be present: pores and crystals in refractories, platinum particles in the melt, carbon or dust particles in the furnace atmosphere. Whatever the reasons, nucleation can become very easy in practice, despite the fairly high equivalent pressures predicted by homogeneous nucleation theories. Some very interesting data were presented in the original version, but not the published text, of a paper by Faile and Roy (1966). Samples of glass previously equilibrated with high pressures of several gases were reheated in steps, the bulk density being measured each time. All showed a sharp decrease in density because of bubble nucleation at some characteristic temperature. Although the pressures were much lower than those given by Fisher's theory, the temperatures for easy bubble growth were similar. The results suggest that pressures no greater than 50–100 atm may cause nucleation of bubbles at melting temperatures. Fisher's theory was recently used in a different context, to interpret the failure of glass fibers in the transformation range, by Rekhson et al. (1983).

Once a nucleus exists Scriven's model for bubble growth from zero size may be used to estimate its growth [see Eq. (16) above]. By writing down some typical values it is at once possible to see what order of dissolved gas concentration is necessary for the observed phenomena. Foam and reboil bubbles are usually several millimeters in diameter and often grow in times from a few seconds up to about 2 min; take $a = 0.2$ cm, $D = 4 \times 10^{-6}$ cm^2/sec, $t = 100$ sec. This gives $\beta = 5$ and, as the density of the gas in the bubble is of the order of 4×10^{-4} g/cm^3 at high temperature, the actual excess dissolved concentration must be around 2×10^{-3} g/cm^3 or of the order of 0.1 wt. %. Alternatively, one can note that the volume of a foam may easily exceed the volume of the melt from which it formed. If the gas density is again taken to be 4×10^{-4} g/cm^3, the excess dissolved concentration must exceed about 0.02 wt. %. Reboil obviously need not involve such large volumes of gas as foaming.

The effective dissolved concentrations just worked out are high for the solubilities of gases in glass melts. In oxidized glasses only sulfate, oxygen, and water are likely to have sufficiently high solubilities. In strongly reduced glasses sulfate should be replaced by sulfide and nitrogen added to the above list. In some instances halides might also be included.

2. Chemistry

The supersaturations needed to cause reboil and foaming are not likely to be achieved physically in glass manufacture, but reduced pressures have been used in some laboratory studies. When thinking about the formation and behavior of bubbles it is sometimes helpful to consider the increase in external gas pressure that would be necessary to make the existing dissolved concentration the *equilibrium* one.

Glass melts may be, but are not necessarily, supersaturated with some gas when melting is complete. Assume for the moment that the melt is just saturated with some particular gas. The melt may easily be made supersaturated in two ways: by changing the temperature or by altering some chemical balance (e.g., degree of oxidation) so that the equilibrium concentration is changed. The solubility of dissolved gas may have either a positive or a negative temperature coefficient, so cooling does not necessarily involve an increase in solubility (although one would expect this to be true for O_2, CO_2, and 'SO$_3^=$). According to Fisher's theory temperature changes of nearly 1000°C could be involved in some cases of bubble formation, but 200–300°C is more typical; this would hardly change saturation by more than an order of magnitude. A change in oxygen partial pressure (as discussed in Section II.E) is more likely to cause large changes in the apparent pressure needed to keep the existing dissolved gas concentration in equilibrium. The work of Fincham and Richardson (1954) on $SO_4^{2-} \rightleftharpoons S^{2-}$ equilibria in slags shows how such conditions could occur when reducing an oxidized melt containing sulfate, or vice versa. Data for sulfate solubility in Na_2O-SiO_2 melts were reported by Holmquist (1966) and also show how changes in oxidation could affect normal glass melts.

Lack of knowledge of the high-temperature chemistry of glass melts hinders a proper understanding of reboil and foaming, but enough work has been published to furnish matter for the following section.

3. Experimental Studies

Several authors have used reduced pressures to induce reboil or foaming either by reducing pressure at constant temperature, using a gradient furnace at constant pressure, or another equivalent experiment. Jebsen-Marwedel and Dinger (1947) showed that different glasses had different pressure–temperature characteristics for bubble formation and suggested that this might be related to efficiency of fining. This kind of experiment can successfully be used to determine the dissolved gas content of molten metals (e.g., H_2 in Al. Budd *et al.* (1962) used a similar technique to study some reboil problems. Rasul and Cable (1966, 1967) were the first to show that reproducible results can be obtained at constant pressure and that foams can be produced or destroyed by changing only the composition of the atmosphere. This work investigated glass composition (R_2O-SiO_2 systems with R equaling Li, Na, or K), sulfate content, oxygen partial pressure, and water-vapor partial pressure; all had important effects. These observations were extended to $Na_2O-CaO-SiO_2$ glasses by Cable *et al.* (1968a). Foaming appeared to require the presence of sulfate in the glass and to be made easier by a wet atmosphere, but the result that foaming occurred most easily in the most oxidizing atmosphere was rather unexpected. Some excellent work by Hanke and Scholze (1970) suggested that this rather curious result was due to the effect of atmosphere and time on loss of $'SO_3'$ by volatilization before foaming occurred. Hanke's work also provided a good model for the way in which sulfate and hydroxyl interact to influence the solubilities and foaming behavior of each. A later paper by Hanke (1977) showed more clearly how the composition of $O_2-H_2O-N_2$ atmospheres affects foaming and how CeO_2, Sb_2O_3, MnO_2, and As_2O_3 modified the behavior.

Budd *et al.* (1962) made some experiments concerned with reboil in amber glasses and the influence of water. Emer (1964) also studied reboil in amber glasses; his work included analyses of the dissolved $'SO_3'$ and

CO_2 contents by vacuum extraction. Carbon dioxide contents were often higher than those of SO_3.

A very early investigation, chiefly of fining, by Zschimmer et al. (1929) produced very interesting results with respect to the effects of increasing the alumina content of the glass on foaming during melting.

A different kind of reboil problem, involving oxygen bubbles, is best thought of in terms of an electrochemical cell and was studied by Diefenbach-Kaden and Sendt (1964) and Sendt (1965). A very well-planned study of this problem was published by Cowan et al. (1966).

IV. The Homogenizing of Glass Melts

A. Sources of Inhomogeneity

Segregation of various kinds happens easily in the early stages of glass melting. Even if segregation did not occur before and during melting, it would be difficult to produce homogeneous glasses. The last stage of glass melting reactions usually involves dissolving residual sand grains in a melt which is, by then, very close to the final average composition. However, this stage normally involves equilibrium, according to the phase diagram, between liquid and solid at the surfaces of the sand grains.

Suppose one was melting glass of composition 33% Na_2O, 67% SiO_2 at 1300°C; the interfacial composition would be about 16% Na_2O, 84% SiO_2. The production of some cord of composition between about 16 and 33% Na_2O would thus be inevitable. In this case lowering the melting temperature to 1000°C would reduce the range of compositions involved to about 22–33% Na_2O. If a simple Na_2O–CaO–SiO_2 glass, say, 16% Na_2O, 10% CaO, were being melted, the interface composition would not necessarily lie on the tie line joining this composition to pure silica but would probably be close to it. For a melting temperature of 1450°C the interface composition would lie near 10% Na_2O, 7% CaO and production of some bad cord would again be unavoidable. Rosenkrands and Simmingsköld (1962) showed how the range of compositions present in the early stages of melting lead crystal approached the final composition.

In practice refractory corrosion and volatilization may also make significant contributions to production of inhomogeneity during melting, fining, and conditioning of the glass.

B. Diffusion

All glass melts made by normal techniques contain regions of different compositions and properties; the variations are usually outside the limits that can be tolerated and they must be reduced. Only diffusion can reduce

TABLE IV

APPROXIMATE DIMENSIONLESS TIMES TO
REDUCE INHOMOGENEITIES BY FACTORS OF 10
OR 100 BY DIFFUSION ALONE

	Dt/l^2	
	$\Delta c = 0.1\,\Delta c_0$	$\Delta c = 0.01\,\Delta c_0$
Slab	31	2800
Cylinder	2.5	25
Sphere	0.9	4
Sandwich	1.1	2.0

the concentration differences. The standards required vary with application; those set for optical glasses are probably higher than for any other bulk products. Even at high temperature diffusivities for the major constituents in the melt are very small and diffusion will not transport matter over significant distances in reasonable times, as shown below.

Consider the behavior expected of an inhomogeneity represented by a slab of thickness $2l$ lying in a broad expanse of homogeneous material. The concentration difference Δc at the center would decay with time according to

$$\Delta c = \text{erfc}(l/2\sqrt{Dt}); \qquad (19)$$

if a cylindrical inhomogeneity of radius l were involved

$$\Delta c = 1 - \exp(-l^2/4Dt) \qquad (20)$$

or, for a sphere,

$$\Delta c = \text{erf}\left(\frac{l}{2\sqrt{Dt}}\right) - \frac{2}{\sqrt{\pi}}\frac{l}{2\sqrt{Dt}}\exp\left(-\frac{l^2}{4Dt}\right). \qquad (21)$$

For reduction by a factor of 10 or 100 the dimensionless times needed would be, approximately, those given in Table IV.

It is obviously advantageous to distribute a given quantity of material in the most disperse form possible. If layers can be distributed as a multiple-layer sandwich instead of only one there will be the further advantage of reducing the distance over which diffusion must occur. For a regular array of alternate layers all of thickness $2l$ in the center concentration would be

$$\Delta c = \frac{4}{\pi}\sum_{0}^{\infty}\frac{(-1)^n}{2n+1}\exp\left[-(2n+1)^2\frac{\pi^2 Dt}{4l^2}\right], \qquad (22)$$

the results for which are also entered in Table IV above.

For a regular array take $D = 1 \times 10^{-8}$ cm^2/sec, $t = 21{,}600$ sec (6 h). If it is necessary to reduce the initial concentration difference by a factor of 100 the thickest layer that can be effectively dissipated will have $2l = 0.2$ mm, but this is an extremely favorable case. For an isolated slab inhomogeneity the equivalent maximum thickness would be $2l = 15$ μm. It is clear that flow is needed (1) to attenuate and reduce the thickness of individual inhomogeneities, and (2) to redistribute inhomogeneities to reduce the maximum distances over which diffusion will have to transport matter. Melts easily segregate under the influence of gravity, making vertical redistribution very important.

C. Flow in Glass Melts

Velocities of flow in glass tanks and the viscosities of molten glasses are such that Reynolds numbers rarely exceed Re = 10 (based on depth of melt). Flow is thus in the laminar regime and of the simplest and most predictable type. This is very useful in trying to analyze the flow patterns but is a disadvantage for mixing. Turbulent flow or, at least, well-developed eddies would be much better for mixing but cannot usually be achieved in glass melts. It is therefore necessary to consider how laminar flow can attenuate and distribute inhomogeneities.

1. Attenuation of Inhomogeneities by Laminar Flow

The way in which simple shear can attenuate and rotate a layer in a homogeneous liquid has been considered by several authors, notably Mohr (1960), McKelvey (1962), and Cooper (1966a,b). Consider a rectangular element of length a and width b lying at an angle α to the velocity vector U_x; the ends of the element lie at x_1, y_1, and x_2, y_2. The two ends of the element will be moving in the direction of the axis of the element itself at velocities $U_1 \cos \alpha$ and $U_2 \cos \alpha$. If the velocity field is simple shear with

$$dU/dx = 0, \quad dU/dy = G, \qquad (23)$$

and if $y_2 - y_1 = h$, one has

$$(U_2 - U_1) \cos \alpha = Gh \cos \alpha. \qquad (24)$$

From simple geometry $h = a \sin \alpha$, and so

$$\frac{1}{a}\frac{da}{dt} = G \sin \alpha \cos \alpha = \frac{1}{2} G \sin 2\alpha. \qquad (25)$$

If the liquid is incompressible $ab = a_0b_0$, and so $b/b_0 = a_0/a$ and it follows that

$$\frac{b}{b_0} = \frac{1}{(1 + Gt \sin 2\alpha_0 + G^2t^2 \sin^2 \alpha_0)^{1/2}}. \tag{26}$$

Note that reduction in b depends only on the product Gt and not on either G or t independently, also that da/dt is very small for α_0 very near to either 0° or 180° but has its maximum value for $\alpha_0 = 45°$.

Simple shear thus tends to align inhomogeneities parallel to the direction of flow, but as they become aligned more closely the rate of extension becomes very small. It is easy to reduce the thickness of most inhomogeneities by a factor of 10 and some by a factor of 50, but extremely large amounts of shear are needed to achieve a reduction of a factor of 100 in all cases. If so large a reduction is needed the flow system should be changed so that the inhomogeneities once again lie approximately perpendicular to the direction of flow.

There will always be some angles for which any specified amount of shear will compress rather than extend inhomogeneities, but as Gt increases the range of angles for which this is true will become smaller.

2. The Behavior of Inhomogeneous Inclusions

The previous model considered the behavior of an inclusion having exactly the same properties as the liquid surrounding it. Inclusions, of course, need to be attenuated because they differ in properties from the bulk glass and they often differ in viscosity. This is likely to affect the ease with which inclusions can be deformed; intuition suggests that very viscous inclusions will be more difficult to deform.

Consider a fairly typical container glass composition with 14.5% Na_2O, 10.0% CaO, 1.5% Al_2O_3. Reasonable estimates of the interface compositions in contact with silica (sand grains) and their viscosities suggest that between 1300 and 1500°C these interface compositions will be between 2 and 4.5 times more viscous than the bulk melt. Changes in composition caused by volatilization or refractory corrosion could also give viscosities several times greater than that of the original melt.

3. An Infinitely Deformable Inclusion

The approach used here was developed by Eshelby following our discussion of the problem over several years beginning in 1968. Eshelby (1957, 1959) had produced an elegant analysis of the equivalent problem in elasticity, which he easily adapted to the viscous system of interest in glass technology. As a result of these discussions further theoretical investigations were undertaken by the author's colleagues (see Bilby *et al.*, 1975; Howard and Brierley, 1976; Bilby and Kolbuszewski, 1977). This model assumes that there is no interfacial tension which could limit the deformation.

Consider an ellipse of semiaxes a and b; the limiting case of a circle has $a = b$. If one defines the *natural strain* as

$$S_a = \int_{a_0}^{a} \frac{da}{a} = \ln \frac{a}{a_0} \tag{27}$$

and

$$\lambda = (\eta_i - \eta_0)/\eta_0, \tag{28}$$

where η_i is the viscosity of the inclusion and η_0 the viscosity of the surrounding matrix, then, according to Eshelby (1957, 1959), the deformation experienced by an inhomogeneous inclusion ($\eta_i \neq \eta_0$) can be compared directly with that experienced by a homogeneous inclusion ($\eta_i = \eta_0$) when subjected to the same unspecified stress field, and the result is simply

$$S_a + \tfrac{1}{2}\lambda \tanh S_a = S_{aH}, \tag{29}$$

where S_{aH} is the strain for the homogeneous case. Since the material is incompressible one has $a_0 b_0 = ab$ and the decrease in b/b_0 is equal to the increase in a/a_0. The amount of shear needed in any particular homogeneous case can be evaluated from Eq. (26).

Examination shows that viscosity has a large influence on the amount of deformation when the inclusion is originally near to spherical shape but the behavior becomes more and more similar for all viscosities as a/b increases. As might be expected, a low-viscosity inclusion deforms somewhat more easily and a highly viscous one much less readily (see Table V).

Several three-dimensional cases have been examined for ellipsoids in simple velocity fields. The above two-dimensional model is a reasonable approximation for all those that have been investigated. Examples of these were discussed in what may well be thought very tedious detail by Cable (1977). It is clear that compact inclusions distinctly more viscous than the matrix will be difficult to deal with. If inclusions could be formed as long thin layers their viscosity would be less important but it would generally be difficult to form them at right angles to the flow and thus difficult to achieve much further attenuation.

4. Deformation Limited by Surface Tension

If a gas bubble or other fluid has an interfacial tension the deformation will tend to be limited by the surface tension forces. In such a case the

TABLE V

Predicted Amounts of Shear Needed to Achieve $b = 0.10b_0$ for Initially Spherical Inclusions

Viscosity ratio η_i/η_0	$-S_H$	Gt
0	1.83	6.17
0.5	2.07	7.87
1	2.30	10.0
2	2.78	16.1
5	4.26	71.4
10	6.73	840
20	11.61	1.10×10^5

drop will rotate and deform to a limited extent which depends on the *rate* of shear.

Several authors have made mistakes in evaluating this problem. For example, the attractive analysis of Chaffey and Brenner (1967) leads to the obviously incorrect conclusion that the equilibrium deformation of a spherical inclusion increases as the viscosity of the inclusion increases. The analysis of Cox (1969) appears to be quite sound. Cox considers an ellipsoid for which deformation from spherical shape is fairly small and deals with the eccentricity

$$e = (r/a_0) - 1 \tag{30}$$

or the parameter

$$E = \tfrac{1}{2}(e_a - e_b); \tag{31}$$

for fairly small deformations one will find $e_a \simeq -e_b \simeq E$. The equilibrium shape is then given by

$$E = \frac{5[19(\eta_i/\eta_0) + 16]}{4[(\eta_i/\eta_0) + 1][(20/w)^2 + (19\eta_i/\eta_0)^2]^{1/2}} \tag{32}$$

and the equilibrium angle will be

$$\phi_a = \tfrac{1}{4}\pi + \theta \quad \text{(rad)} \tag{33}$$

if $E = F \cos 2\theta$. In these equations

$$w = G\eta_0 a_0/\sigma, \tag{34}$$

where σ is the interfacial tension, and

$$F = G\eta_0 a_0 (19\eta_i + 16\eta_0)/16\sigma(\eta_i + \eta_0). \tag{35}$$

Analysis of time dependence shows an approximately exponential approach to equilibrium shape, with superimposed sinusoidal oscillation, which is generally rather slower to decay than the time to establish steady state flow. The time to establish steady state flow between two flat plates impulsively set in motion is approximately (Schlicting, 1960)

$$t^* = \rho a_0^2 / \eta. \tag{36}$$

In one case that has been examined (for $\eta_i = 10\eta_0$) the time to reach equilibrium was about $t = 30\rho a_0^2/\eta$.

In many cases it seems a reasonable approximation to assume that inclusions in glass melts are deformed without diffusion and that diffusion then occurs without further deformation.

5. Flow in Furnaces

It is not proposed to discuss this topic in detail here, but it should be clear that the glassmaker has sound practical reasons for not wishing to allow unmelted batch to progress too far down the furnace and near to the throat. This would risk having batch stones in the output and also poor fining. It is also clear that complex flows are necessary for reasonably efficient attenuation of inhomogeneities and homogenizing.

The overall throughput of the furnace is obviously due to the continual removal of glass from the forehearths and the supply of fresh batch, at a carefully matched rate, from the batch chargers. If one has a throughput of 200 tonne/day through a melting chamber 7.0 m wide and 1.0 m deep, with an average viscosity of 400 P, the slope of the surface necessary to maintain this flow under the influence of gravity is only about 0.3' in the melting end. It is thus usual to assume that the surface of the bath is exactly horizontal.

Apart from this flow, the batch is heated by flames and largely melted from above or, sometimes, by electrodes immersed in the melt. In either case heat distribution is not uniform and there are losses from the sidewalls and bottom of the tank, so that temperature gradients are sure to occur in the melt. Since the melt expands and contracts as temperature changes, temperature gradients cause density gradients and these, under the influence of gravity, produce convection in the melt. When there is no throughput the thermal convection will still give internal mixing. The glassmaker's problem is how to control both thermal convection and throughput flows to give sufficiently good mixing as well as good thermal efficiency in the furnace, sometimes aided by electric boosting or bubbling. Despite the importance of optimizing these processes, furnace design and operation have tended to be rather by rule of thumb. However, many glass manufacturers and furnace designers now obtain assistance from studies on models or tracer experiments in full-size furnaces. Flow patterns can obviously vary with throughput. Typical container glass tanks do not give glass of the same homogeneity from all forehearths; those nearest the center line often have the shortest residence times and the poorest homogeneity. Better understanding and control of matters like these would be of great value.

6. *Homogenizing of Pot and Laboratory Melts*

Calculations of hydrodynamic boundary thicknesses, for example over flat plates, show that the walls of pots and crucibles must, because of the small size of the vessel, exert a powerful damping effect on thermal convection in such melts. It is thus no surprise to find that the folklore of glassmaking says that it is difficult, perhaps impossible, to make homogeneous glass on a small scale. The design of many full size pot furnaces means that the pots tend to receive a much larger heat flux over the upper surface and the wall to the rear than over the rest of the body; this may encourage thermal convection but still tends to produce glass inferior in quality (cord or ream) to that melted in tanks, and several aspects of pot melting practice reflect this.

It was known and recorded in the middle of the eighteenth century that stirring mechanically with an iron bar could much improve the quality of inferior glass, but it took the skills of Guinand and Fraunhofer very early in the nineteenth century to develop reasonably successful methods of making pot melts of acceptable optical glass. Faraday (1830) was the first to show how to make such glasses on a laboratory scale and to use platinum for containers and for stirrers, but his work had little influence on glassmakers.

Much scientific work in glass research would benefit from the use of homogeneous samples because composition–property relations are rarely linear and inhomogeneous samples may not give the true average value. In other cases, such as optical properties, inhomogeneous samples are incapable of giving precise results. Most of this work needs only small samples, and it would be very wasteful if large melts had to be made whenever homogeneous samples were needed. Trials over the last 20 years or so in the author's department have led to the development of very simple methods of making homogeneous samples of many glasses on a 50–500 g scale, as reported by Cable *et al.* (1983). Further development of stirrer efficiency in such very simple systems has been shown to be possible by the model studies of Jambor Sadeghi (1980) and Wang and

Cable (1985). Such techniques ought to be used much more widely by glass researchers, some of whom seem merely to ignore the difficulty of making homogeneous glasses.

V. Volatilization from Glass Melts

A. Introduction

Many glass melts contain one or more constituents volatile during melting or subsequently from the molten glass. The most obvious volatile constituents are alkalis, boric oxide, lead oxide, halides, sulfur, and selenium. Sometimes the batch material is far more volatile than the final glass and the losses may occur largely during the initial melting reactions; this can be true for boron, lead, and selenium. Loss by evaporation leads to change in glass composition and properties, the continual production of local inhomogeneities, surface scum (e.g., recrystallized silica), attack of furnace superstructure, and corrosion of regenerators. Such phenomena are obviously undesirable.

B. Theory

Our inadequate knowledge of the reactivity of glass melts with gases and vapors suggests that melt and vapor ought to come to equilibrium almost instantaneously and, despite our lack of detailed justification, this must generally be an entirely reasonable assumption. General knowledge of the low diffusion coefficients in glasses (typically $<10^{-10}$ m^2/sec) and in gases (around 5×10^{-5} m^2/sec) has then led many workers to assume that the volatilization must then be controlled by diffusion in the melt. It will be shown here that this apparently reasonable postulate is not generally valid although occasionally appropriate.

If the melt were static and diffusivity independent of concentration, transport within the melt would be governed by Fick's second law,

$$\partial c/\partial t = D\,(\partial^2 c/\partial x^2). \tag{37}$$

As melts are usually open to the atmosphere or swept by rapidly flowing combustion gases, it is generally reasonable to assume that the equilibrium concentration of volatile material in the gas phase is zero, and hence it is assumed that the surface concentration c_i in the melt is instantaneously reduced to zero at time $t = 0$. So long as the sample is thick enough to be considered semi-infinite, the relation between total loss per unit area M and time should be

$$M = (2/\sqrt{\pi})c_0\sqrt{Dt}, \tag{38}$$

where c_0 is the initial uniform concentration of volatile species in the melt. Although some authors such as Oldfield and Wright (1962a) and Matousek and Hlaváč (1971) have shown M versus $\sqrt{t}$ to be a straight line, that line often does not pass sufficiently near the origin for the discrepancy to be accepted merely as due to experimental error; besides, the discrepancies are often systematic, especially in terms of glass composition; the model must therefore be rejected in those cases. This model also makes it difficult to explain why pressure and flow of the gases above the melt can have important effects.

Terai and Ueno (1966) introduced a modified model which successfully described their results for a series of high-lead glasses; Matousek and Hlaváč (1971) have also used this model to interpret data for a lead crystal glass. According to this model, Eq. (38) still applies but a new boundary condition is imposed at the melt–atmosphere interface ($x = 0$). Instead of assuming instantaneous transfer of the volatile species from the surface of the melt to the atmosphere, one writes

$$j = \alpha(c_i - c_\infty) = -D \left.\frac{\partial c}{\partial x}\right|_i \qquad (39)$$

where j is the flux to the atmosphere (grams per square meter second), c_i is the interface concentration ($c_i = c_0$ at $t = 0$, $c_i \to c_\infty$ as $t \to \infty$), c_∞ is the final equilibrium concentration in the melt (usually $c_\infty = 0$), and α is a mass transfer coefficient (meters per second). The surface concentration c_i now falls to zero over a measurable range of times instead of being zero for all times $t > 0$. The required solution to this problem is, according to Carslaw and Jaeger (1959a),

$$M = [(c_0 - c_\infty)/h]\{\exp(h^2 Dt)\,\text{erfc}[h\sqrt{Dt}] - 1 + (2/\sqrt{\pi})h\sqrt{Dt}\} \qquad (40)$$

where $h = \alpha/D$. As two unknowns, h and D or h and $h\sqrt{D}$, are to be evaluated from observations of M and t, the task is best done by computer. The limit of Eq. (40) for $\alpha \gg D$ is Eq. (38).

The depth may not be sufficient to represent a semi-infinite body throughout the whole time of experiment. For a finite melt of depth l, Eq. (39) is replaced by (Carslaw and Jaeger, 1959b)

$$\frac{M}{M_\infty} = 1 - \sum_{n=1}^{\infty} \frac{2L^2}{\beta_n^2(\beta_n^2 + L^2 + L)} \exp\left(-\frac{\beta_n^2 Dt}{l^2}\right), \qquad (41)$$

where M_∞ is the total possible loss of volatile matter, $L = l\alpha/D$, and the β_n are the positive roots of $\beta_n \tan \beta_n = L$. Equation (41) can be used to determine whether deviations from Eq. (39) are due to the finite depth of the melt. The concentration at the melt–gas interface when Eq. (39) applies is

$$\hat{c}_i = (c_i - c_\infty)/(c_0 - c_\infty) = \exp(h^2 Dt)\,\text{erfc}[h\sqrt{Dt}], \tag{42}$$

which approaches zero at sufficiently long times.

Convection of the melt would invalidate Eqs. (38) and (40)–(42). The limiting case, in which stirring was sufficient to keep the liquid of uniform composition, would still be subject to the boundary condition [Eq. (39)], but now

$$c_i = (M_\infty - M)/l, \tag{43}$$

hence the kinetics of volatilization should be described by

$$\ln[(M_\infty - M)/M_\infty] = \alpha t/l. \tag{44}$$

Slower stirring would be expected to produce results lying between the predictions of Eqs. (40) and (44).

One other limit might be noted: the maximum possible rate of evaporation into vacuum which, according to Dushman (1949), is

$$\left[\frac{dM}{dt}\right]_{\max} = 43.7 p \sqrt{\frac{W}{T}}, \tag{45}$$

where p is the equilibrium vapor pressure (atm), W the molecular weight of the evaporating species, and T the absolute temperature.

C. Experimental Data

A good example of careful qualitative work, which can show several important aspects of behavior, is a paper by Kruithof *et al.* (1958). Their observations show how easily convection can influence behavior with some glasses but not with others. Experimenters need to be very careful about this. Even when the surface layer is less dense than the bulk and stable under the influence of gravity, local variations in composition can cause variations in surface tension and small-scale circulations result. The various ways in which surface forces can have a role in glassmaking were reviewed by Hrma (1982).

Oldfield and Wright (1962b) published some very interesting results for borosilicate glasses which showed the problems that can easily arise because of silica scum formation.

Several other workers have shown that results are quite sensitive to conditions outside the melt, which means that the simple diffusion model cannot be correct. For example, Hirayama (1960) reported losses from lead borate and silicate melts at 0.05–0.15 torr, which other data suggest to be about 20 times (for the borate) and 200 times (for the silicate) greater than would have been expected at 1-atm pressure. Data for soda–lime–silica glasses by Cable and Chaudhry (1975) indicated losses in vacuum about 200 times greater than at 1 atm.

Barlow (1965) showed a clear dependence of loss on rate of gas flow over the sample, a result again not capable of explanation by the simple diffusion model. Rate of loss varied by about a factor of 5 over the range of gas flows that he used. Barlow also showed the loss of B_2O_3 to be proportional to the square root of water-vapor partial pressure. This same result was found by Dietzel and Merker (1957) for soda–lime–silica melts. This may be attributed to a reaction of the form

$$B_2O_3(m) + H_2O(g) \longrightarrow 2HBO_2(g) \qquad (46)$$

or

$$Na_2O(m) + H_2O(g) \longrightarrow 2NaOH(g). \qquad (47)$$

Some borosilicate glasses seem not to show this sensitivity to water-vapor partial pressure; other species must be involved in those cases.

Reviews of fairly extensive data for lead glasses and for sodium silicates have been made by Cable *et al.* (1975a,b) and the model described by Eqs. (39)–42) shown to fit the data very well. However, the true significance of α and D is not entirely clear partly because D obviously varies with melt composition and none of the above equations is valid for a concentration-dependent diffusivity, when c_i varies with time. A model investigation by Cable and Cardew (1977) showed that in such a case the best-fit values of α and D could be merely curve fitting parameters which did not represent any appropriate average of the real parameters and this may account for some curious features seen in results for Na_2O–SiO_2 melts.

The constant-property model has been used successfully to describe data for Na_2O–B_2O_3 melts which included both congruently and incongruently evaporating compositions (Cable, 1978). That paper also briefly outlined how the low saturation concentration in the gas phase can be the factor that limits the flux away from the surface of the melt, thus making furnace geometry, flow of the atmosphere, and pressure significant parameters in many cases. The concentration of the volatile constituent in the melt will often be in the range 100–1000 g/liter under conditions where the total gas density may not exceed 5×10^{-4} g/liter; even for a fairly high vapor pressure of 10^{-2} atm the highest possible concentration of volatile material in the vapor would be only about 5×10^{-6} g/liter. Having the diffusivity in the gas phase much greater than in the melt is then not sufficient to make transport in the gas phase trivially easy.

Schaeffer and Sanders (1976) and Sanders *et al.* (1976) explored some of the problems associated with the traditional transpiration technique and were led to design a new kind of experiment. Scholze *et al.* (1983) have recently claimed that losses of fluorine from some glasses of low fluoride content could successfully be interpreted in terms of thermodynamic calculations. Some properly designed Knudsen effusion experiments on soda–lime–silica glasses were reported by Argent *et al.* (1980).

VI. Conclusion

This chapter has briefly reviewed the factors important in melting glasses. It is hoped that readers will now realize that a complex interaction of many physical and chemical processes is involved. Our understanding of nearly all of them is inadequate but sufficient to give some insight into most aspects. Continued study should bring further benefits but appropriate experimental methods are often difficult to devise.

Few books discuss glass melting in detail but there are some not mentioned in the text above. Bosc D'Antic (1780) was a notable pioneer, most of whose essays have been translated by the author. Tooley (1974), has good empirical reviews of many topics, Bezborodov (1968) has an extended interesting discussion of melting reactions with a very extensive bibliography, and Hlaváč (1983) concisely reviews several important topics.

References

Apak, C. (1975). Interactions between bubbles or furnace atmosphere and glass melts. Ph.D. thesis, Sheffield University, Sheffield, England.
Appen, A., and Polyakova, L. B. (1938). *Stekol'naya Prom.* **1**(7), 18–21.
Argent, B. B., Jones, K., and Kirkbride, B. J. (1980). *In* "Industrial Use of Thermochemical Data" (T. I. Barry, ed.), *Spec. Publ.—Chem. Soc.* **34**, Chemical Society, London.
Barlow, D. F. (1965). *Proc. Int. Glass Congr., 7th, 1965, Brussels* Vol. 1, Pap. 10, pp. 1–14.
Barton, J. L. (1982). *Riv. Staz. Sper. Vetro* **12**, 19–31.
Bastick, R. E. (1956). *Symp. Affinage Verre, C. R., 1955, Paris*, pp. 127–138. Union Scientifique Continentale du Verre, Charleroi, Belgium.
Bezborodov. M. (1968). "Synthesis and Structure of Silicate Glasses" [in Russian]. Nauka i Tekhnika, Minsk.
Bilby, B. A., and Kolbuszewski, M. L. (1977). *Proc. R. Soc. London, Ser. A* **355**, 335–353.
Bilby, B. A., Eshelby, J. D., and Kundu, A. K. (1975). *Tectonophysics* **28**, 265–274.
Boffé, M., and Letocart, G. (1962). *Glass Technol.* **3**, 117–123.
Bosc D'Antic, P. (1780). "Collected Works, Containing Many Memoirs on the Art of Glassmaking etc." [in French], 2 vols. Paris.
Budd, S. M. (1966). *Phy. Chem. Glasses* **7**, 210–215.
Budd, S. M., Exeby, V. H., and Kirwan, J. J. (1962). *Glass Technol.* **3**, 124–129.
Bunting, J. A., and Bieler, B. H. (1969). *Am. Ceram. Soc. Bull.* **48**, 781–785.
Buss, W. (1982). *Glastech. Ber.* **35**(4), 167–176.
Cable, M. (1958a). *J. Soc. Glass Technol.* **43**, 20–31.
Cable, M. (1958b). *Symp. Fusion Verre, C. R., 1958, Brussels*, pp. 253–258. Union Scientifique Continentale du Verre, Charleroi, Belgium.

1. PRINCIPLES OF GLASS MELTING 41

Cable, M. (1959). The refining of soda–lime–silica glass in a laboratory furnace. Ph.D. thesis, Sheffield University, Sheffield, England.
Cable, M. (1960). *Glass Technol.* **1**, 144–154.
Cable, M. (1961). *Glass Technol.* **2**, 60–70, 151–158.
Cable, M. (1969). *Glastek. Tidskr.* **24**(6), 147–152.
Cable, M. (1970). *Glastek. Tidskr.* **25**(1), 7–14.
Cable, M. (1977). "The Theory of Homogenizing Glass Melts." Dept. of Ceramics, Glasses, and Polymers, Sheffield University, Sheffield, England.
Cable, M. (1978). *In* "Borate Glasses: Structure, Properties, Applications" (L. D. Pye, V. D. Fréchette, and N. J. Kreidl, eds.), pp. 391–411. Plenum, New York.
Cable, M. (1982). *In* "Molten Salt Technology" (D. G. Lovering, ed.), pp. 223–264. Plenum, New York.
Cable, M., and Cardew, G. E. (1977). *Chem. Eng. Sci.* **32**, 535–541.
Cable, M., and Chaudhry, M. A. (1975). *Glass Technol.* **16**, 125–134.
Cable, M., and Evans, D. J. (1967). *J. Appl. Phys.* **38**, 2899–2906.
Cable, M., and Haroon, M. A. (1970). *Glass Technol.* **11**, 48–53.
Cable, M., and Martlew, D. (1971). *Glass Technol.* **12**, 142–147.
Cable, M., and Martlew, D. (1984). *Glass Technol.* **25**, 24–30, 139–144.
Cable, M., and Naqvi, A. A. (1975). *Glass Technol.* **16**, 2–11.
Cable, M., Rasul, C. G., and Savage, J. (1968a). *Glass Technol.* **9**, 25–31.
Cable, M., Clarke, A. R., and Haroon, M. A. (1968b). *Glass Technol.* **9**, 101–104.
Cable, M., Apak, C., and Chaudhry, M. A. (1975a). *Glastech. Ber.* **48**, 1–11.
Cable, M., Apak, C., and Chaudhry, M. A. (1975b). *Glastech. Ber.* **48**, 127–134.
Cable, M., Smedley, J. W., and Wang, S. S. (1983). *Int. Glass Cong., 13th, 1983, Hamburg, Poster Presentation.* (To be published).
Carslaw, H. S., and Jaeger, J. C. (1959a). "Conduction of Heat in Solids," pp. 70–74. Oxford Univ. Press, London and New York.
Carslaw, H. S., and Jaeger, J. C. (1959b). "Conduction of Heat in Solids," pp. 121–125. Oxford Univ. Press, London and New York.
Chaffey, C. E., and Brenner, H. (1967). *J. Colloid Interface Sci.* **24**, 258–269.
Conroy, A. A., Billings, D. D., and Manring, W. H., and Bauer, W. C. (1963). *Glass Ind.* **44**, 139–143, 175–177.
Cooper, A. R. (1966a). *Glass Technol.* **7**(1), 2–11.
Cooper, A. R. (1966b). *Chem. Eng. Sci.* **21**, 1095–1106.
Cooper, A. R. (1980). *Glass Technol.* **21**, 87–94.
Cooper, A. R., and Kreider, K. G. (1967). *Glass Technol.* **8**, 71–73.
Cowan, J. H., Buehl, W. M., and Hutchins, J. R. (1966). *J. Am. Ceram. Soc.* **49**, 559–562.
Cox, R. G. (1969). *J. Fluid Mech.* **37**(3), 601–623.
Dawson, J. B., and Wilburn, F. W. (1965). *Therm. Anal., Proc. Int. Conf., 1st, 1965, Aberdeen, Scotland,* pp. 273–281. MacMillan, London.
Diefenbach-Kaden, J., and Sendt, A. (1964). *Glastech. Ber.* **37**, 69–72.
Dietzel, A., and Flörke, O. W. (1959). *Glastech. Ber.* **32**(5), 181–185.
Dietzel, A., and Merker, L. (1957). *Glastech. Ber.* **30**, 134–138.
Dubrul, L. (1956). *Symp. Affinage Verre, C. R., 1955, Paris,* pp. 445–451. Union Scientifique Continentale du Verre, Charleroi, Belgium.
Duda, J. L., and Vrentas, J. S. (1969). *AIChE J.* **15**, 351–356.
Duda, J. L., and Vrentas, J. S. (1971). *Int. J. Heat Mass Transfer* **14**, 395–401.
Dushman, S. (1949). "Scientific Foundations of Vacuum Technique." Wiley, New York.
Emer, P. (1964). *Glastech. Ber.* **37**, 59–69.
Epstein, P. S., and Plesset, M. S. (1950). *J. Chem. Phys.* **18**, 1505–1509.

Eshelby, J. D. (1957). *Proc. R. Soc. London, Ser. A* **241,** 276–396.
Eshelby, J. D. (1959). *Proc. R. Soc. London, Ser. A* **252,** 561–569.
Faile, S. P., and Roy, D. M. (1966). *J. Am. Ceram. Soc.* **49,** 638–643.
Faraday, M. (1830). *Philos. Trans. R. Soc. London,* pp. 1–57.
Fincham, C. J. B., and Richardson, F. D. (1954). *Proc. R. Soc. London, Ser. A* **223,** 40–62.
Fisher, J. C. (1948). *J. Appl. Phys.* **15,** 1062–1067.
Fletcher, W. W. (1963). *Glass Technol.* **4,** 152–158.
Frade, J. R. (1983). Diffusion controlled growth and dissolution of gas bubbles. Ph.D. thesis, Sheffield University, Sheffield, England.
Frölich, A. (1946). *Chalmers Tek. Hoegs. Handl.* **49.**
Garstang, A. (1971). *Glass Technol.* **11,** 1–7.
Gehlhoff, G., Kalsing, H., and Thomas, M. (1930). *Glastech. Ber.* **8,** 1–24.
Greene, C. H., and Gaffney, R. (1959). *J. Am. Ceram. Soc.* **42,** 271–275.
Greene, C. H., and Kitano, I. (1959). *Int. Glaskongr. 5th, 1959, Munich. Glastech. Ber.* **32K** pp. 44–48.
Greene, C. H., and Lee, H. A. (1965). *J. Am. Ceram. Soc.* **48,** 528–533.
Greene, C. H., and Platts, D. R. (1969). *J. Am. Ceram. Soc.* **52,** 106–109.
Griffin, C. (1971). Diffusion controlled growth or dissolution of a bubble containing two gases. Ph.D. thesis, Sheffield University, Sheffield, England.
Hanke, K. P. (1977). *Glastech. Ber.* **50,** 144–147.
Hanke, K. P., and Scholze, H. (1970). *Glastech. Ber.* **43,** 475–482.
Harrington, R. V., Hutchins, J. R., and Sherman, R. D. (1962). *In* "Advances in Glass Technology," Vol. 1, pp. 75–92. Plenum, New York.
Hirayama, C. (1960). *J. Am. Ceram. Soc.* **43,** 505–509.
Hlaváč, J. (1983). "The Technology of Glass and Ceramics." Elsevier, Amsterdam.
Hlaváč, J., and Nademlynska, H. (1969). *Glass Technol.* **10,** 54–58.
Holmqvist, S. B. (1966). *J. Am. Ceram. Soc.* **49,** 467–472.
Howard, I. C., and Brierley, P. (1976). *Int. J. Eng. Sci.* **14,** 1151–1159.
Hrma, P. (1982). *Glass Technol.* **23,** 151–155.
Jack, H. R. S., and Jacquest, J. A. T. (1958). *Symp. Fusion Verre, C. R., 1958.*
Jambor Sadeghi, J. (1980). Homogenization of viscous liquids by simple stirrers. Ph.D. thesis, Sheffield University, Sheffield, England.
Jebsen-Marwedel, H. (1936). "Glastechnische Fabrikationsfehler," pp. 64–69. Springer-Verlag, Berlin.
Jebsen-Marwedel, H., and Dinger, K. (1947). *Verres Réfract.* **1,** Aug. 19–22.
Jucha, R. B., McNeil, D. P., and Cole, R. (1982). *J. Am. Ceram. Soc.* **65,** 289–292.
Johnston, W. D. (1965). *J. Am. Ceram. Soc.* **48,** 184–190.
Karlsson, K. (1977). *Glastek. Tidskr.* **32,** 6–10.
Karlsson, K. H., Westerlund, T., and Hatakka, L. (1983). *J. Am. Ceram. Soc.* **66,** 574–579.
Kröger, C. (1948). *Glastech. Ber.* **22,** 86–93.
Kröger, C. (1949). *Glastech. Ber.* **22,** 248–261.
Kröger, C. (1953). *Glastech. Ber.* **26,** 170–174, 202–214.
Kröger, C., and Eligehausen, H. (1959). *Glastech. Ber.* **32,** 362–372.
Kröger, C., and Marwan, F. (1955). *Glastech. Ber.* **28,** 51–57, 89–98.
Kröger, C., and Marwan, F. (1956). *Glastech. Ber.* **29,** 257–289.
Kröger, C., and Marwan, F. (1957). *Glastech. Ber.* **30,** 222–229.
Kröger, C., and Vogel, E. (1955). *Glastech. Ber.* **28,** 426–437, 468–474.
Kröger, C., and Ziegler, G. (1952). *Glastech. Ber.* **25,** 307–324.
Kröger, C., and Ziegler, G. (1953). *Glastech. Ber.* **26,** 346–353.

Kröger, C., and Ziegler, G. (1954). *Glastech. Ber.* **27**, 199–212.
Kröger, C., Janetzko, W., and Kreitlow, G. (1958). *Glastech. Ber.* **31**, 221–228.
Kruithof, A. M., La Grouw, C. M., and de Groot, J. (1958). *Symp. Fusion Verre, C. R., 1958,* pp. 515–527.
Lakatos, T., Johansson, L-G., and Simmingsköld, B. (1972). *Glass Technol.* **13**, 88–95.
Lakatos, T., Johansson, L-G., and Simmingsköld, B. (1975). *Glastek. Tidskr.* **30**(1), 7–8.
Lakatos, T., Johansson, L-G., and Simmingsköld, B. (1976). *Glastek. Tidskr.* **31**(1), 31–35.
Lakatos, T., Johansson, L-G., and Simmingsköld, B. (1977). *Glastek. Tidskr.* **32**(2), 31–35.
Lakatos, T., Johansson, L-G., and Simmingsköld, B. (1978). *Glastek. Tidskr.* **33**(3), 55–59.
Lakatos, T., Johansson, L-G., and Simmingsköld, B. (1981). *Glastek. Tidskr.* **36**(4), 51–55.
Lyle, A. K. (1956). *Proc. Int. Cong. Glass, 4th, Paris,* Paper II-4, p. 93.
Lynn, G. (1932). *Glass Ind.* **13**, 1–4.
McKelvey, J. M. (1962). "Polymer Processing," Chap. 12. Wiley, New York.
Manring, W. H., and Bauer, W. C. (1962). *Glass Ind.* **43**, 379–385, 408.
Matousek, J., and Hlaváč, J. (1971). *Glass Technol.* **12**, 103–106.
Mazurin, O. V., Strel'tsina, M. V., and Shvaiko-Shvaikovskaya, T. P. (1973–1981). "Properties of Glass and Glassforming Melts," Vols. 1–4. Nauka, Leningrad.
Mohr, W. D. (1960). *In* "Processing of Thermoplastic Materials" (E. C. Bernhardt, ed.), Chap. 3. Reinhold, New York.
Morey, G. W. (1954). "The Properties of Glass," 2nd ed. Reinhold, New York.
Mulfinger, H. O. (1974). *Glastek. Tidskr.* **29**, 81–95.
Mulfinger, H. O. (1976). *Glastek. Ber.* **49**, 232–245.
Němec, L. (1974). *Glass Technol.* **15**, 153–161.
Němec, L. (1977). *J. Am. Ceram. Soc.* **69**, 536–440.
Němec, L. (1980). *Glass Technol.* **21**, 134–138, 139–144.
Okamura, T. (1966). *Res. Rep. Asahi Glass Co.* **16**, 1–12.
Oldfield, L. F., and Wright, R. W. (1962a). *In* "Advances in Glass Technology," Vol. 1, pp. 35–51. Plenum, New York.
Oldfield, L. F., and Wright, R. D. (1962b). *Glass Technol.* **3**, 59–68.
Onorato, P. I. K., Weinberg, M. C., and Uhlmann, D. R. (1981). *J. Am. Ceram. Soc.* **64**, 676–682.
Paul, A. (1982). "Chemistry of Glasses." Chapman and Hall, London and New York.
Paul, A., and Douglas, R. W. (1966). *Phys. Chem. Glasses* **7**, 1–13.
Poole, J. P. (1963). *Glass Technol.* **4**, 143–152.
Potts, J. C. (1939). *J. Soc. Glass Technol.* **23**, 129–140.
Potts, J. C., Brookover, G., and Burch, O. G. (1944). *J. Am. Ceram. Soc.* **27**, 225–231.
Preston, E., and Turner, W. E. S. (1940). *J. Soc. Glass Technol.* **24**, 124–138.
Preston, E., and Turner, W. E. S. (1941). *J. Soc. Glass Technol.* **25**, 136–149.
Pugh, A. C. (1968). *Glastek. Tidskr.* **23**, 95–104.
Rasul, C. G., and Cable, M. (1966). *J. Am. Ceram. Soc.* **49**, 568–571.
Rasul, C. G., and Cable, M. (1967). *J. Am. Ceram. Soc.* **50**, 528–531.
Readey, D. W., and Cooper, A. R. (1966). *Chem. Eng. Sci.* **21**, 917–922.
Rekhson, S. M., Grant, M. L., and Peckman, H. F. (1983). *Proc. Int. Glass Congr., 13th. Glastech. Ber.* **56K**(1), 408–413.
Rosenkrands, B., and Simmingsköld, B. (1962). *Glass Technol.* **3**, 46–51.
Rosner, D., and Epstein, M. (1972). *Chem. Eng. Sci.* **27**, 69–88.
Routt, K. R., Cosper, M. B., and Iverson, D. C. (1983). *Proc. Int. Glass Congr., 13th. Glastech. Ber.* **56K**(1), 48–53.
Sanders, D. M., Blackburn, D. H., and Haller, W. (1976). *J. Am. Ceram. Soc.* **59**, 366–368.

Schaeffer, H. A., and Sanders, D. M. (1976). *Glastech. Ber.* **49,** 95–102.
Scherer, G. W., and Schultz, P. C. (1983). *In* "Glass: Science and Technology" (D. R. Uhlmann and N. J. Kreidl, eds.), Vol. 1, Chap. 2, pp. 49–103. Academic Press, New York.
Schlicting, H. (1960). "Boundary Layer Theory" (J. Kestin, trans.), 4th ed., pp. 73–74. McGraw-Hill, New York and London.
Scholze, H. (1965). "Glas: Natur, Struktur and Eigenschaften." Friedrich Vieweg, Braunschweig.
Scholze, H., Tünker, G., and Conradt, R. (1983). *Glastech. Ber.* **56**(6/7), 131–137.
Schultz, P. (1979). *In* "Fiber Optics: Advances in Research and Development" (B. Bendow and S. S. Mitra, eds.), pp. 3–31. Plenum, New York.
Scriven, L. E. (1959). *Chem. Eng. Sci.* **10,** 1–13.
Sendt, A. (1965). *Glastech. Ber.* **38,** 208–212.
Simpson, W., and Myers, D. D. (1978). *Glass Technol.* **19,** 82–85.
Slavyanskii, V. T. (1957). "Gases in Glass" [in Russian], pp. 87–115. Gosudarstvennoe Izdatel'stvo Oboronoi Promishlenosti, Moscow.
Solinov, F. G., and Pankova, N. A. (1965). *Proc. Int. Congr. Glass, 7th, 1965, Brussels,* Vol. 2, Pap. 341.
Swift, H. R. (1947). *J. Am. Ceram. Soc.* **30,** 170–174.
Terai, R., and Ueno, T. (1966). *J. Ceram. Assoc. Jpn.* **74,** 283–294.
Thomasson, C. V., and Wilburn, F. N. (1960). *Phys. Chem. Glasses* **1,** 52–69.
Tooley, F. V. (1974). "Handbook of Glass Manufacture," 2 vols. Books for Industry, New York.
Turkdogan, E. T. (1983). "Physicochemical Properties of Molten Slags and Glasses." Metals Society, London.
Volf, M. B. (1961). "Technical Glasses." Pitman, London.
Wang, S. S., and Cable, M. (1985). In preparation.
Warren, B. E., and Biscoe, J. (1938). *J. Am. Ceram. Soc.* **21,** 259–265.
Weinberg, M. C., and Subramanian, R. S. (1980). *AIChE J.* **27,** 739–748.
West-Oram, F. G. (1979). *Glass Technol.* **20,** 222–243.
Weyl, W. A. (1951). "Coloured Glasses." Soc. Glass Technol., Sheffield.
Wilburn, F. W., and Thomasson, C. V. (1958). *J. Soc. Glass Technol.* **42,** 158–175.
Wilburn, F. W., Metcalfe, S. A., and Warburton, R. S. (1965). *Glass Technol.* **6,** 107–114.
Zachariasen, W. H. (1932). *J. Am. Chem. Soc.* **54,** 3841–3851.
Zelinski, B. J. J., Yinnon, H., and Uhlmann, D. R. (1983). *Glastech. Ber.* **56K**(2), 822–827.
Zschimmer, E., Zimpelmann, E., and Riedel, L. (1929). *Sprechsaal* **59,** 331–333, 353–357, 393–395, 411–413, 422–425.

CHAPTER 2

Flat Glass Manufacturing Processes

W. C. Hynd

PILKINGTON BROTHERS P.L.C.
ST. HELENS, ENGLAND

I. The Development of Flat Glass	46
A. Flat Glass Products	46
B. The Pattern of Flat Glass Development	47
C. The Position Today	49
II. Basic Science of Flat Glass Processes	50
A. Stress and Movement in a Sheet of Glass as It Is Drawn	50
B. Updraw Sheet Processes	54
C. Downdraw Sheet Processes	55
D. Glass Composition	56
III. Rolled Glass Processes	58
A. Short History	58
B. The Table Cast Process	58
C. The Intermittent Double-Roll Process	59
D. The Continuous Double-Roll Process	60
E. Essential Elements of the Continuous Double-Roll Process	62
F. Wired Glass Processes	67
G. The Polished Plate Process	70
IV. Flat Drawn Sheet Processes	70
A. Short History	70
B. Window Glass Manufacture	70
C. The Drawn Cylinder Process	71
D. The Fourcault Updraw Process	71
E. The Colburn Updraw Process	75
F. The PPG Pennvernon Updraw Process	77
G. The Slot Bushing Downdraw Processes	81
H. The Fusion Downdraw Process	81
V. The Float Process	83
A. The Development of the Float Process	83
B. Essential Features of the Float Process: Equilibrium Float	85
C. Theory of the Float Effect	87
D. Float Ribbon Formation	91
E. Chemical Aspects of the Float Process	97
F. The PPG Industries Modified Process	100

VI. Glass for Radiation Control	100
A. New Requirements	100
B. Useful Measures of Performance	101
C. Processes for the Production of Radiation-Control Glass	102
D. The Properties of Radiation-Control Glasses	105
References	106

I. The Development of Flat Glass

A. FLAT GLASS PRODUCTS

The term flat glass covers a wide variety of products ranging from glass less than one-thousandth of an inch in thickness supplied in ribbon form for use in electrical capacitors to the largest sheets of float glass 30 × 15 ft × $\frac{3}{4}$ in. thick used for panoramic windows. The main flat glass products currently marketed are listed in Table I. The main uses of flat glass are as windows and external cladding in buildings; as screens and mirrors in building interiors and furnishings; and as windshields, side lights, and back lights in automobiles. These are mass markets in which the supplier is in competition not only with other glass companies but in some areas with alternative materials or forms of construction. Two main requirements have largely determined the evolution and development of flat glass manufacturing processes. First, the ever-present pressure to reduce manufacturing costs to provide the ultimate user with a cheaper product. Second, the need to produce a better product—one available in a large range of sizes but above all a product of the highest optical quality, free from distortion and local inclusions.

Over the years, the glassmaker has been successful in meeting both requirements. The cost of glass has remained competitive relative to other building materials, and this is one reason for the increased use of glass in houses and commercial buildings seen in recent years. Important though the cost reduction has been, the requirement to produce consistently high quality has been perhaps even more significant in determining the evolutionary path which the development of flat glass processes has followed. Where a new process has become established the product has in almost every case been not only cheaper, but of a consistently higher quality than that which it displaced.

The purpose of this chapter is to describe and analyze the major flat glass processes which are in use today in a way that will be useful and instructive to a practicing glass maker rather than to present a historical account of their evolution. Nevertheless, as a background to this technical study it is useful to set the main pattern of development of flat glass in historical perspective.

TABLE I

FLAT GLASS PRODUCTS

Plain and rough-cast rolled glass
 Rolled sheets with no definite pattern. Includes "rough-cast" and plain surface glass used in window glazing or for horticultural purposes.
Patterned rolled glass
 Rolled sheets with a patterned surface. Used in windows to obscure vision without cutting off light.
Wired rolled glass
 Rolled glass reinforced with steel wire mesh which holds glass together when smashed or damaged by fire. Supplied with either rough-cast or polished surfaces.
Plate glass
 High-quality transparent glass produced by grinding and polishing a rolled sheet. Now obsolete and superseded by float glass except for special compositions.
Drawn sheet glass
 Widely used for windows and low-quality mirrors. Some degree of optical distortion always present, marked in poor samples but only just noticeable in the best sheet glass.
Float glass
 Now widely used for high-quality windows, mirrors, and automotive uses. Virtually free of distortion and available in a large range of sizes and thicknesses.

Speciality Flat Glass

Capacitor glass
 Produced in ribbon form 0.0003 in. thick in special compositions for use in electrical capacitors
Microscope cover glass
 Thin glass (0.003–0.010 in.) produced in special compositions for use in microscopy.
Optical color filters
 Special colors and compositions for various optical and ophthalmic uses.

B. The Pattern of Flat Glass Development

The first relatively clear flat glass was produced in Europe in the early centuries A.D. The earliest form of flat glass to come into general use was broad glass produced by a crude blowing process in which a rough cylinder was blown, opened out, and roughly flattened at the furnace. Color, flatness, and transparency were poor.

Flat glass was also made by a primitive form of the crown process originated in Syria in the seventh century A.D. This process was improved and developed in Normandy in the late sixteenth century to give glass of superior flatness and optical quality to the broad glass currently produced. In the crown process glass melted in a pot was first gathered and blown into a globe. A punty, a small gather on a solid iron rod, was then attached to the globe opposite the blowing iron, which was then cracked off. The globe held on the punty was then reheated and rotated as fast as possible so that centifrugal force acting on the open mouth spun the round globe into an almost flat disk. Although the final disk could be some 40–50 in. in diameter, the largest rectangular pane that could be cut was limited to some 16 × 12 in. and a thickness of 2 mm. In Europe the improved crown process gradually displaced broad glass during the seventeenth century but not until the eighteenth century in Britain.

In the late seventeenth century the blown cylinder process, a much improved form of the broad glass process, was developed in northern France. This process differed from the early broad glass process in that a blown cylinder was allowed to cool before being decapped, split lengthways, and flattened in a separate kiln on a specially prepared table. The blown cylinder process was capable of producing larger and thicker sheets—60 × 40 in. being a typical size—of glass of reasonably good optical quality and flatness, and it began to take over from the crown process in Europe during the eighteenth century, but not until after 1830 in Great Britain. The blown cylinder process was the main source of window glass throughout the nineteenth century in Europe and America.

In parallel with the development of the blowing processes a major step forward in the production of high-quality flat glass was made in France around 1680 A.D. with the introduction of the table cast process. In this process glass from a pot was poured onto a flat cast iron or bronze table and a traveling roller running on guides at the side of the tables passed over it to reduce its thickness and to flatten the sheet. This process not only provided larger sheets of rough glass for glazing than could be obtained from other processes in current use, but, more importantly, provided the raw material for the polished plate process, in which the rough cast sheets were taken and subsequently ground and polished to produce large flat sheets of high-quality glass for use as mirrors or in special glazing applications.

The table cast process, improved and developed in stages, continued to supply the blanks for polished plate glass until the 1920s, when it was superseded either by the discontinuous Bicheroux rolling process or by the continuous-flow rolling process. The polished plate process continued as the source of the highest-quality flat glass until being completely displaced by the float process in the 1960s.

Patterned rolled glass as we know it was first produced in England by improvements in the table cast process introduced around 1850. A discontinuous double-roll process was introduced in England around 1870, but it was not until the 1920s that patterned glass was made by a continuous double-roll process.

The introduction of the float process in 1960 and its subsequent development is probably one of the most far-reaching developments to have taken place in flat glass manufacture.

This summary does not do justice to a fascinating subject; for a further study the reader is referred to some excellent accounts in the literature (Turner, 1930; Barker, 1977; Douglas and Frank 1972; Pilkington, 1976; Scoville, 1948).

C. THE POSITION TODAY

There have been great changes in the structure of the flat glass industry over the past two decades. The major factor underlying these changes was the introduction by Pilkington in 1960 of the float process. By 1980 the float process had completely displaced the polished plate process as the source of supply of the highest-quality flat glass and had become an important source of window glass at the expense of the established flat drawn processes, particularly for the thicker substances (≥ 3 mm) and in market areas where the volume was high enough for advantage to be taken of the float process capacity. These changes are reflected in Table II, which gives data on the relative importance of the different manufacturing processes over the period 1960–1980.

Nevertheless, the updraw processes remain of great importance for the supply of window glass and will continue to be used in areas where they are already established and where the nature and small scale of the market do not justify a changeover to the float process. Developments continue to be made in the flat drawn process, and in recent years important improvements have been made to both Fourcault and Colburn processes.

In recent years developments by Corning Glass have resulted in an improved downdraw process which could become established in certain specialized product areas, particularly for the production of thin flat glass.

TABLE II

RELATIVE MARKET CAPACITY (%) IN FREE WORLD

	1961	1965	1970	1974	1980
Plate	32	15	7	3	0
Sheet	67	74	61	46	24
Float	1	11	31	51	76

II. Basic Science of Flat Glass Processes

The market for flat glass demands a high level of quality. To provide this consistently and efficiently is a demanding task requiring the glassmaker to bring to bear all the art, experience, attention to operating detail, and practical engineering skills at his command. But this alone will not guarantee success. He will fail unless the design of the equipment and the fundamental operation and control of the process take full account of the underlying science and technology. In short the basic chemistry and physics of the process must be right.

The paragraphs which follow seek to set out in a simple manner elements of the underlying science common to all flat glass processes in a way which will help the practical glassmaker to appreciate the design features and to understand the operation of current flat glass processes. Theory that is specific to any one process is dealt with when describing and analyzing that process.

A. Stress and Movement in a Sheet of Glass as It Is Drawn

Nothing is more valuable to anyone charged with operating a flat glass unit than to have a good inherent understanding of the pattern of stresses and stretches present in a sheet as it is drawn. This applies to the float and rolled glass processes as well as to the updraw and downdraw sheet processes and is of the greatest importance in understanding how flatness is promoted and optical distortion minimized. The rigorous and full mathematical treatment is difficult and complex but a useful understanding can be reached by the application of relatively elementary mathematics.

Most glass technologists at some time in their studies will have been familiar with the terms and the equations used in elastic theory (Timoshenko and Goodier, 1951), and this provides a useful starting point for looking at the stresses and movements which take place in a viscous sheet as it is stretched. The equations that apply in elastic theory can be used to describe what is happening at any moment of time to a viscous sheet that is being stretched provided we make the substitution in terms listed below.

Stress	Force per unit area. This concept remains the same for both the elastic and viscous cases.
Strain	Elastic strain, the change in length per unit length, is replaced by the concept of stretch, the change in length per unit length per unit time.
Poisson's ratio	In the elastic case this lies in the range 0.2–0.3 for most elastic solids and is replaced in the viscous case by the constant value 0.5, indicating that there

	is no change in volume when a viscous body deforms under stress.
Young's Modulus	The ratio of stress to strain in the elastic case is replaced by the factor $3n$ in the viscous case, where n is the viscosity.

Having made these substitutions we can use the equations with which we are familiar in elastic theory to study the stresses and movements taking place at any moment in time as a viscous sheet is stretched. There is, of course, one essential difference between the elastic and viscous cases which we must bear in mind. In the elastic case there is no change in configuration with time, while in the viscous case movement continues with time. This complicates the full theoretical treatment but a useful understanding of the key features can be obtained by using the simple equations to look at what is happening at any instant.

1. General Case

Referring to Fig. 1 let us begin by looking at the relationship between stresses and stretches at a point in a viscous sheet of glass which is being stressed. Figure 1 refers to an updraw sheet process but the basic theory applies equally to the horizontal ribbon in the float bath.

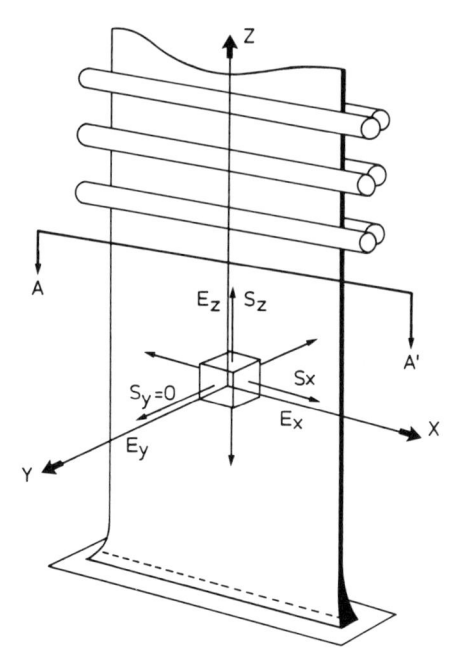

FIG. 1. Stress and stretch in an updraw process (diagrammatic).

We take the z coordinate in the direction of draw, the x coordinate across the width, and the y coordinate through the thickness of the ribbon being drawn. Then using the notation S_x, S_y, S_z for stress in the x, y, and z directions; E_x, E_y, E_z for stretch (rate of strain) in the x, y, and z directions; and n for the viscosity of the glass, we can by analogy with the equations for stretching of an elastic plate write

$$E_x = S_x/3n - S_y/6n - S_z/6n, \quad (1)$$

$$E_y = S_y/3n - S_x/6n - S_z/6n, \quad (2)$$

$$E_z = S_z/3n - S_x/6n - S_y/6n. \quad (3)$$

Since there is no volume change $E_x + E_y + E_z = 0$. For almost every practical case these equations can be simplified since the stress through the thickness of the sheet is always zero, i.e., $S_y = 0$. The exception is the initial stage of float forming, when at the hot end of the bath gravitational effects are significant.

2. Stretching of a Sheet with No Transverse Restraint

Consider first the case in which there is no transverse restraint, i.e., the sheet is free to narrow down in width as it is stretched. This situation applies in certain circumstances in the float process. This implies $S_x = 0$ as well as $S_y = 0$. Then Eqs. (1)–(3) become

$$E_z = S_z/3n, \quad (4)$$

$$E_x = E_y = -S_z/6n. \quad (5)$$

That is, as the sheet is stretched the thickness and the width both decrease at half the rate at which the sheet elongates.

3. Stretching of a Sheet with Transverse Restraint

In most practical sheet forming processes it is desirable to avoid loss in width by providing some means of restraint at the edges of the sheet. In this case $S_y = 0$, $E_x = 0$, and by inserting these values and rearranging Eqs. (1)–(3) we get

$$E_z = S_z/4n, \quad (6)$$

$$E_y = -S_z/4n = -E_z, \quad (7)$$

$$S_x = \tfrac{1}{2}S_z. \quad (8)$$

That is, when the edges are held all the stretch applied results in reduction in thickness, and in addition to the applied longitudinal stress S_z there is a stress S_x of half that value acting transversely.

There are three ways in which we can "hold an edge," i.e., prevent the ribbon from narrowing down as it is drawn. First, in regions where the viscosity is low enough we can wet each edge onto a solid guide and draw the ribbon between the guides, relying on forces of adhesion to overcome the narrowing tendency produced by surface tension and drawing forces. Second, where the viscosity is higher we can hold the edge with rotating grips using pairs of edge rolls as in the sheet process or single top rolls in the float process. Third, we can arrange for there to be a band of colder glass of relatively higher viscosity along each edge of the ribbon so that the drawing tension and curvature in this band can provide a transverse force to resist narrowing.

4. Transverse Variation in Viscosity

In every practical sheet drawing process the glass viscosity varies across the width of the sheet. Sometimes this variation is contrived, as when cold, high-viscosity edges are deliberately created to maintain flatness in the sheet as it is drawn. Sometimes it is inadvertent since there are a number of factors which make it difficult to achieve complete uniformity in viscosity at all points across a sheet. The effects of transverse variation in viscosity can be good or bad; it can be used to promote flatness but may be the underlying cause of thickness variation, lines, and distortion.

We can use our simple equations to examine how transverse variation in viscosity will result in transverse thickness variation when a sheet is stretched while the width is kept constant. The longitudinal stretch E_z is uniform across the sheet, as is also the transverse stress S_x. Let us examine what will happen if owing to variation in temperature or the presence of inhomogeneities the viscosity varies from point to point across the sheet.

From our Eqs. (1) and (3), remembering that $S_y = 0$, we can by simple algebra arrive at the relationship

$$E_x = S_x/4n - \tfrac{1}{2}E_z. \qquad (9)$$

If $S_x = 0$ (as is the case when sheet width is not maintained), E_x is constant across the sheet and there is no problem. If, however, the width is maintained, S_x is positive and since the viscosity is not constant the lateral stretch will vary from point to point across the sheet, depending on the viscosity. Where the viscosity is high the lateral stretch will be low and vice versa. Thus as stretching proceeds streaks of more viscous glass will become relatively thicker and streaks of low-viscosity glass thinner. This effect applies to both large- and small-scale transverse viscosity variations. Large-scale variations in viscosity (often the result of temperature variations) lead to measurable variation in thickness across the sheet, while small-scale variations often caused by inhomogeneity cause lines and other types of optical distortion.

5. *Lines and Distortion in the Drawing Process*

There is always some degree of inhomogeneity in the glass arriving at the drawing process. However carefully raw materials are selected and mixed and however carefully the melting unit is operated to promote homogeneity there will always be some slight variation in composition and hence in viscosity in the glass feeding the drawing machine forehearth. The inhomogeneities that arrive at the forehearth have the form of long extended cords since even inhomogeneities which were approximately equidimensional lumps at the melt end are by the time the forehearth is reached greatly elongated as the result of shear flows which have taken place in the melter.

As the sheet is drawn such an extended cord—assuming that it is of relatively high viscosity—will narrow and thicken, giving rise to a visible line in the drawn sheet. This effect will be present to some degree whenever there is an area of glass of different viscosity, however small, from the main body. The effect will be most marked the bigger the area, the bigger the viscosity difference, and the closer the area is to the surface. Normally the pattern of compositional variation is such that the resulting optical defect is a series of fine lines or bands (1–5 mm).

There is a second important source of variation of viscosity across a sheet. The temperature and hence the viscosity can vary across the ribbon owing to variation in temperature from point to point. This variation in temperature can be present in the glass arriving at the point of draw, or can result from the pattern of air flows in the kiln or nonuniformity of heat extraction across coolers. Again, when a sheet with such temperature variations is drawn the result is to produce bands of uneven thickness across the sheet. In practice the bands arising from nonuniform temperature are wider (10–100 mm) and less sharp than the lines arising from compositional defects. The thinner the final sheet the greater the amount of stretch and the sharper and finer the line and band pattern.

B. UPDRAW SHEET PROCESSES

Let us examine a typical updraw process (see, e.g., Fig. 1) on the basis of the equations and ideas presented above. As a first approximation assume that the width is maintained and that the sheet is cooled as it moves upward but that the temperature is constant across the width.

At any elevation the stretch that takes place at any cross section AA' is determined by the stress and the viscosity of the glass. The stress is substantially equal to the suspended weight of glass divided by the cross section area of the sheet at that point. Other factors do contribute to this stress, for example, any reaction pull at the drawbar, but these are relatively small and can be neglected. The viscosity gradient can be controlled by the placing and size of the water coolers used.

In practice stretching is most rapid at the base of the draw, where the glass is hottest. Clearly the suspended weight and hence the stress increases as we move upward, and unless the glass is cooled to impose a viscosity gradient stretching will become more and more rapid and the sheet will neck in and the draw be lost. It is our ability to impose an appropriate vertical temperature gradient that makes the updraw process practicable. Equally important is the imposition of an appropriate transverse gradient, giving the cold edges that supplement the edge rolls to maintain sheet width. Since sheet width is maintained a transverse tension is set up across the sheet which promotes flatness.

C. Downdraw Sheet Processes

The pattern of longitudinal stress is very different in downdraw processes. The ribbon is anchored to and hangs from the drawing slot, while the speed of draw is controlled at the first pair of rolls. In this case the tensile force is a maximum at the slot where the glass is hottest and decreases from there to the first pair of rolls. The ribbon stretches under the action of these tensile stresses and the viscosity of the ribbon builds up as it moves down from the slot and cools. In the downdraw configuration two alternative conditions can obtain. First, the temperature at the slot and in the ribbon as it moves down may be so low and the corresponding viscosity so high that at the first roll position the ribbon velocity acquired as a result of stretching due to gravitational forces may be below the drawing velocity imposed by the rolls. In this case the rolls provide an additional drawing traction, the sheet is everywhere in tension, the process is in control, and a flat sheet can be produced. If, on the other hand, the temperature at the slot and subsequently is so high and the corresponding viscosity so low that the velocity achieved at the first roll position as a consequence of gravitational stretching is higher than the velocity imposed by the rolls, the sheet goes into compression at some point and buckles. In this case the process is out of control and unworkable.

Two things will at once be clear to the reader. First, there is an upper limit to the temperature at the slot or the draw point at which any given downdraw process can be operated. Second, a downdraw process is much better suited to the production of thin glass than thick. At a given output the roller speed to produce thin glass is higher and so it is much easier to ensure that the ribbon remains in tension. To produce thick glass it is necessary to cool back at the slot both to reduce output and to reduce stretching under gravity; often the cool back has to be so extreme that any imperfections introduced at this passage through the slot fail to "firepolish" out and a dull and liny sheet results. The slot bushing downdraw process is especially suitable for producing ultrathin glass for microscope slides, flake glass, or electrical capacitors.

D. GLASS COMPOSITION

So far in looking at the underlying science of flat glass processes we have examined some aspects of the physics of forming a flat sheet of glass. In practice the chemistry and, in particular, the choice of glass composition is a matter of crucial importance. No matter what forming process we use consistent production of the highest-quality flat glass is only achieved by the glassmaker exercising close control and striking a careful balance in the choice of operating conditions. If an optimum balance is to be achieved every advantage must be taken from the choice of glass composition.

Over the years experience and trial and error have defined and set limits on the range of compositions that can be successfully used on each of the glass processes. In arriving at these compositions there are three main considerations: first, the temperature/viscosity relationship must be suited to the forming process used. Second, to avoid devitrification the liquidus temperature and the rate of crystal growth must be consistent with the temperature regime of the forming process. Third, the product must be accepted on the market place and its durability must be adequate. In practice there are, of course, many other considerations which must be taken into account when arriving at the choice of glass composition, for example, the availability and relative cost of local raw materials are clearly significant. Paramount, however, in the choice of glass composition is that it should be suited to the chosen process and product, and perhaps the single most valuable piece of advice that can be given is that the very greatest caution should be exercised before making any change in an established composition. Before any proposed change is introduced it should be fully researched and pretested as far as possible, but bitter experience shows that even when the greatest care has been taken unanticipated adverse effects on process efficiency or product performance can show up months after the change has been made. Table III sets out a range of glass composition which have been found suitable for each of the flat glass processes in current use together with a typical glass composition for each and a note of the significant physical properties.

Broadly speaking, the Fourcault process demands a low liquidus temperature, and this is achieved at the expense of glass durability. Both the PPG process and the float process can operate with glass compositions

TABLE III
Typical Flat Glass Compositions

	SiO$_2$	Al$_2$O$_3$	Fe$_2$O$_3$	CaO	MgO	Na$_2$O	K$_2$O	SO$_3$	Log$_{10}$ viscosity (P)[a] 2.5	4.0	7.6	13.0	Liquidus T_l (°C)
Rolled glass													
Range	70.3	0.1	0.08	7.9	1.0	12.8	0.03	0.15					
	72.5	1.2	0.27	12.5	1.9	14.2	0.7	0.35					
Typical	70.7	1.1	0.09	11.4	2.2	13.7	0.5	0.25	1271	998	712	560	1000
Fourcault													
Range	71.0	0.6	0.07	6.8	3.6	13.6	0.1	0.18					
	72.3	1.0	0.17	7.8	4.5	15.5	1.0	0.3					
Typical	72.1	0.9	0.13	6.8	4.4	15.2	0.1	0.26	1299	1019	710	546	960
Colburn													
Range	—	—	—	—	—	—	—	—	—	—	—	—	
Typical	72.5	1.2	0.15	8.7	3.7	13.2	0.1	0.29	1323	1038	727	564	1011
PPG Pennvernon													
Range	72.0	0.9	0.09	6.9	3.7	13.0	0.3	0.15					
	73.0	1.7	0.14	8.5	4.0	14.3	0.9	0.3					
Typical	72.7	1.3	0.12	8.0	3.9	13.0	0.6	0.25	1332	1044	727	562	993
Float[b]													
Range	70.9	0.1	0.05	7.3	3.6	13.0	0	0.15					
	73.3	1.7	0.12	9.7	4.3	14.1	1.3	0.3					
Typical high alumina	72.7	1.0	0.11	8.3	3.95	13.0	0.6	0.25	1324	1040	726	561	1000
Typical low alumina	72.7	0.1	0.11	8.8	4.0	14.0	0.03	0.25	1292	1018	716	554	993

[a] Glass temperature in degrees Celcius given below as a function of viscosity.
[b] It appears that in the case of float glass composition a division into two main classes, high alumina and low alumina, has become established.

having a higher liquidus temperature and the inherently high surface quality justifies the use of a more durable composition. The Colburn process operates most successfully with a composition which is relatively high in calcia or alumina.

III. Rolled Glass Processes

A. Short History

1690	Table cast process introduced in France. Used to produce raw blanks for polished plate and for rough glazing glass. Developed in stages over years and continued in use for plate blanks until 1920s.
1847	Table cast process extended in England to produce patterned rolled glass for glazing.
1884–1890	Intermittent double-roll process established in England for production of patterned rolled glass.
1920	Bicheroux improves the double-roll process, making it suitable for large plate blanks.
1920–1923	Continuous double-roll process developed in the United States. By late 1920s the process is established as the main source of plate glass and of patterned and wired rolled glass.

B. The Table Cast Process

The table cast process was introduced in France around 1690. In this process glass melted in a pot is poured or ladled onto an iron casting table. A heavy roller running in cogged guides at each side of the table is then pulled over the pool of glass to form a plate of roughly rectangular shape and of a thickness determined by the separation between roller and table.

The chief use of this product was as blanks for polished plate glass, but it also provided a rough form of glazing glass. The process was improved and extended in stages over the years and continued in use as the main source of blanks for plate until the 1920s. Its use for rolled glazing glass was expanded in England (Chance, 1919) around 1850 by improvements which made it possible to produce flat sheets of wired and patterned glass. By today's standards even the improved process was unsatisfactory. The surface quality and flatness were poor largely because different areas of the sheet were in contact with the casting table for different times. Consequently it was necessary to cast at a much greater thickness than that of the final polished plate and low overall yield and high labor requirements made costs high.

C. The Intermittent Double-Roll Process

In the intermittent double-roll process molten glass from a pot or tank is poured or ladled onto a holding tray to form a pool upstream of a pair of separated water-cooled rolls. Rotation of the rolls draws the glass into and through the roller pass to form a sheet which emerges to be supported on a flat table which moves forward as the sheet is rolled and then to cool and set before being slid into an adjacent kiln or lehr to be annealed.

A double-roll process was described by Bessemer (1846) in a most inventive patent but proved impossible to operate in practice. A later patent by Mason and Conquerer (1884) was acquired by Chance and led to the establishment of an intermittent double-roll process for the production of patterned rolled glass. In the Chance process (Chance, 1919) the glass was ladled from a pot or tank. The process produced much thinner and flatter glass for glazing than the table cast process but bubble and ream quality was not high enough for plate glass.

It proved a difficult problem to improve the quality from the double-roll process sufficiently to make it acceptable for plate glass, and it was not until 1920 that this was achieved by Bicheroux in Herzogenrath, Germany (Wendler, 1927). Bicheroux's main improvement lay in positioning the pouring axis of the pot and in arranging for the holding tray to tilt progressively so that bubbles were not trapped during the pour. Main features of the Bicheroux process are shown in Fig. 2. The process was

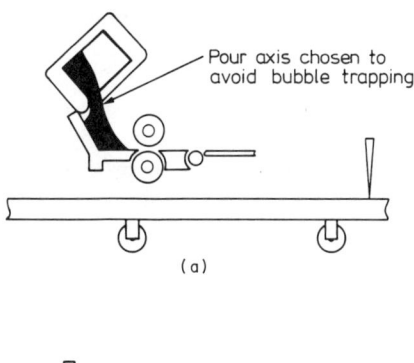

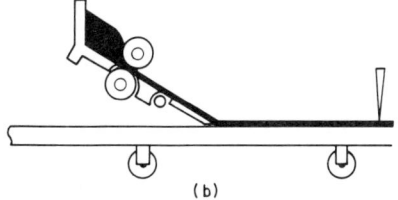

FIG. 2. Bicheroux process: (a) pouring position and (b) rolling positions.

used alongside the continuous double-roll process for the production of plate blanks in the period 1920–1930 and continues in use to this day for the production of colored and special glass in small quantities. Although the surface quality, flatness, and thickness control obtained with Bicheroux was far superior to that possible with the original Table Cast process labor costs remained high and the cost efficiency was markedly lower than obtained with the continuous double-roll process, which from 1930 increasingly took over.

D. THE CONTINUOUS DOUBLE-ROLL PROCESS

There was a growing demand for plate glass, particularly for use in automobiles in the period immediately following the First World War. The Ford Motor Company (Avery, 1930) with great vision applied its engineering and production expertise to develop a continuous double-roll process. In this they had the co-operation of Pilkington (Barker, 1977), who could bring to bear complementary glassmaking experience particularly in the production of high-quality glass from continuous melting furnaces. This development was carried to a successful conclusion and by 1923 the continuous double-roll process was in use at Ford for the manufacture of a narrow (36 in. wide) ribbon for windshield plate. Alternative double-roll processes were developed by other glassmakers and the process was used to produce the raw ribbon for plate glass from the mid 1920s until the plate process was displaced by float in the 1960s. It also became the established process for the production of patterned and wired rolled glass and remains so to this day.

A number of versions of the continuous double-roll process have been developed by different companies, but although there are considerable differences in scale, layout, and engineering the essential technology is common. The chief division lies between processes in which there is a flooded feed to the rolls at the level of the glass in the furnace and processes in which the flow of glass from the furnace is controlled by a tweel and falls into a pool upstream of the rolls.

Two typical processes are examined: the Boudin process, an example of a flood feed process, and the Pilkington flow process, a tweel-fed process typical of the processes used to produce plate glass.

1. *The Boudin Process*

In the Boudin process, originally developed at St. Gobain, France, and shown diagrammatically in Fig. 3, the rolling machine is set so that the roller pass lies some 1–3 in. below glass level; glass is prevented from falling between the forehearth and rolls by a close-fitting shaped refractory termed the "bottom lip." A controlled feed to the rolls is assured by maintaining a steady level of glass in the furnace. The process is not

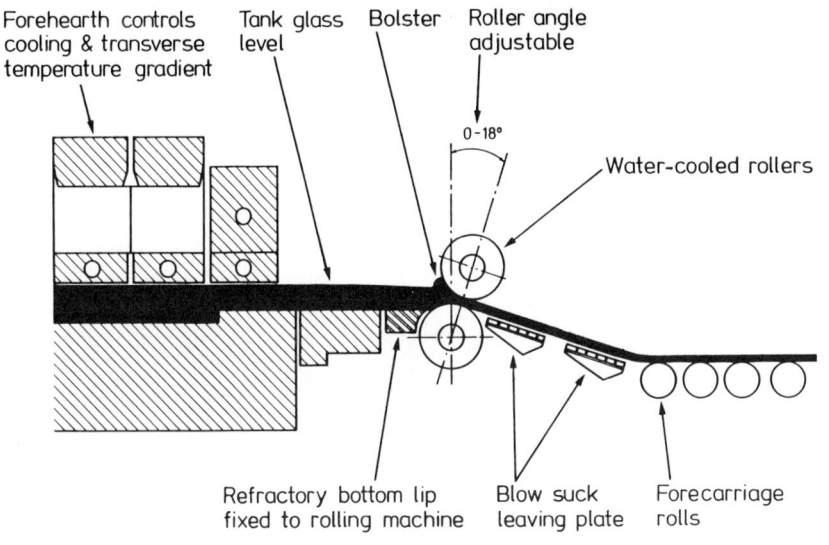

FIG. 3. Boudin continuous double-roll process.

unduly sensitive to furnace level and control to ±1.0 mm is adequate. Nor is the feed to the rolls sensitive to variation in forehearth temperature, although accurate control is important for other reasons.

2. *The Pilkington Flow Process*

In the flow process (Le Mare, 1924) illustrated in Fig. 4 the glass at the end of the forehearth passes over a refractory threshold block and under a

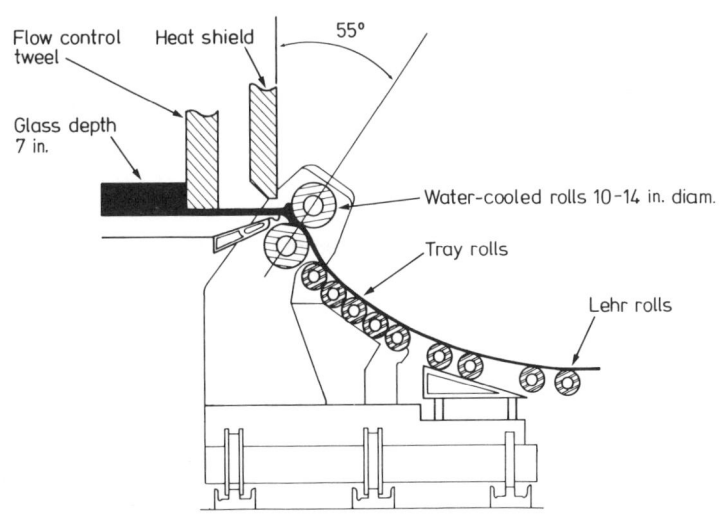

FIG. 4. Pilkington flow process for continuous double rolling.

refractory tweel which can be moved up and down in an exact and controlled manner to regulate the supply of glass falling into the pool upstream of the rolls. To produce a sheet of constant width the amount of glass reaching the rolls must be closely regulated; this demands the closest possible control of the glass temperature and level in the furnace and forehearth. The rolling machine is set so that the roller pass is some 6 in. below glass level, and although the feed arrangement is more complex than in the Boudin machine, it is possible to use larger-diameter rolls and a greater depth of feed glass to make possible higher output and wider ribbons. The capability to cut off the glass supply by lowering the tweel is a very useful aid at startup and during machine changes.

E. Essential Elements of the Continuous Double-Roll Process

The elements of both the Boudin and the flow processes can be considered together.

1. Delivery Forehearth

The forehearth serves to convey glass from the melting furnace and cool it so that it is at the required temperature and viscosity at the pool upstream of the roller pass. In the case of high-output rolling processes very great ingenuity goes into the design of the forehearth, not only to produce the maximum cooling in the shortest distance but also to control the pattern of flow of the glass and to achieve the desired transverse temperature profile. The width of the forehearth is normally only just greater than that of the roller pass and the depth is kept to the minimum that will allow delivery of the required volume of glass without unacceptable draw down. A glass depth of 4–6 in. is usual on a Boudin forehearth, while a forehearth depth of 7–12 in. is usual in the flow process to take account of the higher throughput.

2. The Roller Pass

As the rolls rotate, at nominally the same peripheral speed, contact forces between the roller surface and the glass pulls molten glass into the pass, from which it emerges as a ribbon of thickness determined by the separation between the rolls. In passing through the water-cooled rolls heat is extracted, and so on emerging from the rolls the effective viscosity of the ribbon is high enough to avoid narrowing from surface tension forces at the edges. The newly formed ribbon is then carried forward over driven rolls or on a leaving plate where it is further cooled until it is stiff enough to be supported on the lehr rollers and is then carried forward into the annealing lehr.

The key to successful operation of the rolling process lies in controlling the removal of heat and cooling at every stage. To produce rolled glass of the highest quality demands that the right balance be struck between cooling in the forehearth, cooling upstream of the pass, cooling at the roller pass, and cooling at the leaving plate before the glass enters the lehr. The standard of control called for in operating the rolled process is less severe than that demanded by either the flat drawn sheet processes or the float process. Even so, when working at high output rates there is very little latitude in the choice of operating conditions. The glass composition must be chosen to minimize devitrification at the slow-moving edges of the pool while retaining a steep temperature viscosity gradient to promote true imprint of the pattern.

3. The Rolls

The rolls serve three functions: to form the ribbon, to imprint the chosen pattern, and to remove heat. The roller dimensions and material must be carefully chosen to fulfill each function effectively at the required output rate. In practice it is found that the surface temperature of the roller should lie in the range 400–475°C, at the lower end of the range for high rates of output and vice versa. If the roller surface temperature is too high glass will conform intimately to the roller surface and every grain boundary and surface roughness be reproduced, with the result that the surface finish of the final ribbon will be dull or uneven. In extreme cases the chill imparted may be so inadequate that the ribbon may stick to the rolls or emerge so soft as to be intractable. If the roller surface temperature is too low the pattern imprint may be incomplete. To ensure that they run at the right surface temperature, the rolls are water cooled and the roller material and the external and internal diameters are carefully chosen.

4. Roller Dimensions and Material

For the production of patterned glass, roller diameters in the range 7–14 in. are most common, while diameters of up to 24 in. were used to cope with the wider ribbons and higher outputs required to produce the raw ribbon for the plate process. The internal roll diameter is chosen to achieve the desired roller surface temperature and heat extraction at the output rate targeted. To achieve a sheet of uniform thickness across the ribbon it is sometimes the practice to camber rollers. The most common roll material is selected cast iron, but where a pattern calls for a very high surface finish alloy steel, chromium-plated steel, or aluminum bronze rolls may be used. The casting, machining, and engraving of rolls is normally the province of specialist suppliers. Great care must be taken at all

TABLE IV

Material	Composition (wt. %)									
	C	Si	Mn	S	P	Ni	Cr	Mo	Al	Cu
Cast iron 2	3.85	1.4	0.6	0.4	0.04	2.0	0.5	0.75	—	—
Cast steel 1	0.18	0.35	1.0	0.07	0.05	—	—	—	—	—
Cast steel 2	0.20	0.25	0.45	0.05	0.65	—	12.0	0.5	—	—
Cast steel 3	0.35	2.0	0.80	—	—	14.0	14.0	—	—	—
Aluminum bronze	—	—	1.5	—	—	5.0	—	—	10.0	(Balance)

stages in the casting, annealing, and machining of the rolls to ensure perfect axial symmetry; or otherwise the roll will bend in service. The pattern may be applied in a number of ways, by hammering, rolling, engraving, or chemical etching, but however the pattern is applied it is essential that it be uniform over the whole roller surface. It is also important to avoid "low spots" in a pattern since these will transform into "high spots" in the glass produced, which can lead to serious surface damage when the finished sheet is stacked and transported. Table IV sets out the important physical properties of a number of roller materials in common use.

5. Roller Cooling

Rolls are cooled by passing a continuous and uniform flow of water through rotary glands at the end of each roll. In practice the rate of water flow is kept high and no attempt is made to control the degree of cooling by varying the rate. It is not uncommon to fit a freely rotating finned torpedo inside the roller bore to increase the water velocity over the roller surface. It is of prime importance that the inner roll surface be absolutely clean and free from rust and deposits, which, if present, invariably lead to nonuniform heat extraction and bending of the roll, which, in turn, causes periodic variation in thickness along the length of the ribbon. It can often be advantageous to acid-clean the bore of a roll immediately before putting it into service with a warm 10% solution of sulfuric acid.

The heat transfer mechanism at the water–roller interface is complex but in practice either of two conditions is likely to obtain. If a torpedo is used and water flow velocity is high (>20 liters/cm^2 min) the heat transfer coefficient will be high, and provided the input water temperature is below 50°C it can be assumed that the roller bore will be at the same temper-

PROPERTIES OF ROLLER MATERIALS

Fe	Density	Expansion coefficient (°C × 10⁻⁶)	Thermal conductivity (CSG units)	Specific heat (CSG units)	Brinell hardness
(Balance)	7.3	11	0.12	0.17	250
(Balance)	7.84	12.7	0.106	0.117	180
(Balance)	7.73	10.5	0.061	0.13–0.17	250
(Balance)	8.00	16.5	0.043	0.121	220
5.0	7.58	17.6	0.10	0.104	240

ature as the cooling water. At lower water velocities the point at which evaporative boiling occurs is reached at the water–roll interface, and in this condition the roller bore temperature can be taken as 110°C.

The main factors which determine the roller surface temperature and the heat extraction at the pass are the following:

(1) the roller material, in particular the thermal conductivity,
(2) the roller dimensions—the outer roll diameter is normally determined by other requirements of the rolling machine so that in practice the roller bore diameter is the important variable, and
(3) the chief operational variables: temperature of incoming glass, length of the arc of contact with glass on top and bottom rolls, ribbon thickness, and rolling speed.

The curves of Fig. 5 indicate how roller surface temperature and heat extraction are affected by choice of material and bore diameter.

It will be apparent that as the glass passes through the rolls it is chilled and the chilled layer extends into the thickness of the forming ribbon. When operating within the range of acceptable roller surface temperatures this chilled region extends 1–1.5 mm into the glass even at the highest rolling speeds. Although by optimizing rolling conditions it is possible to roll a high-quality ribbon of 2 mm thickness, the presence of this chilled region makes it difficult to produce rolled glass on a routine basis at a thickness less than 3 mm. At lower thicknesses the chilled region extends right across the glass thickness and this stiff glass is crushed or pressure vented as it passes through the rolls. Even at thicknesses greater than 3 mm crushing and pressure venting can occur if the heat extraction is too great for the rolling speed in use.

Roller Data	OD (cm)	Bore (cm)	Outer temp.(°C)	Bore temp.(°C)
Curves A ——	26.7	Varies	450°C	40°C
Curves B ----	26.7	15.2	Varies	40°C

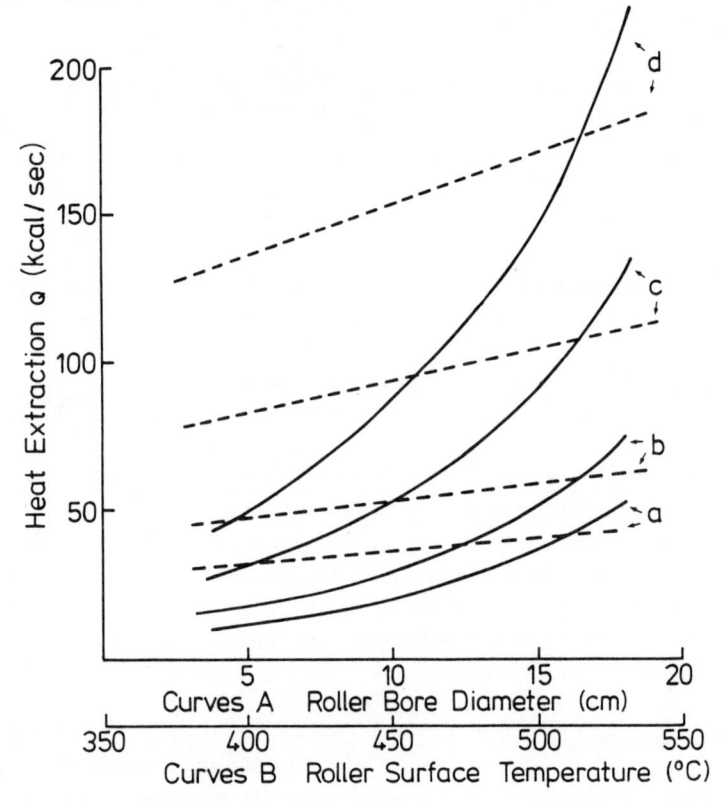

FIG. 5. Heat extraction from rolls as a function of roller material, roller bore, and outer-surface temperature. [a, hot-rolled steel (1); b, hot-rolled steel (2); c, cast iron; d, aluminum bronze.]

6. After-Pass Cooling: The Leaving Plate

At low rolling speeds the ribbon as it emerges from the roller pass may be cool enough and stiff enough to be carried on rollers and go forward to enter the lehr. But when making patterned glass at high outputs the ribbon emerging from the pass is too soft to be handled on rollers. In this case the ribbon immediately after the pass is supported on a leaving plate. Originally in the Boudin process this was done by sliding the ribbon over an air-cooled ridged cast iron leaving plate. As rolling speeds increased this simple leaving plate became inadequate. First it was replaced by simple air support conveyor but today it is usual to use a "blow suck" air support tray of the type developed in the gas hearth process for bending and tempering.

There may be some slight distortion in the pattern as the ribbon emerges from the pass. To correct this it is usual to impose a positive pull and stretch by setting the lehr speed at 8–15% in excess of the rolling speed.

7. Rolling Machines

The machines used for rolling patterned glass range from lightweight units with 5-in.-diam rollers producing a 1-m-wide ribbon at speeds of 30 in./min to much heavier machines with 9–16 in. diam rollers producing 2-m-wide ribbon at rolling speeds of up to 400 in./min. for 3-mm glass. The roll separation may be fixed or the top roll may be loaded to float above a stop position. In theory this latter arrangement makes some allowance for the safe passage of cold glass or other foreign bodies between roller passes, but even so damage to rolls almost invariably occurs. The machine may provide a means to change the angle of the rolls from a position where the top roll lies vertically above the bottom roll, giving a horizontal pass, to a position where the roll axis lies at 45° to the vertical. One of the most important considerations of machine design lies in the facility with which a pattern roll can be changed. The range of patterns is so wide that frequent changes must be made to meet market demands and to balance stocks, and any steps which reduce the time lost in changeover are worthwhile. Many devices have been tried but the most practical and widely adopted is for two rolled-plate machines to be set up side by side on a switch rail system which will allow one or the other to occupy the rolling position. With such an arrangement, the new pattern rolls can be mounted in the spare machine ready to be pushed into place when the changeover is required. It is common to use some form of variable-speed dc drive on the rolling machine and to tie this drive in electrically to the line drive so as to facilitate creation of the positive speed differential needed to give the required degree of stretching to the sheet after it leaves the roller pass.

F. WIRED GLASS PROCESSES

The Boudin process and the Pilkington flow process have both been modified and extended to produce wire-reinforced glass.

1. Pond/Boudin Process

The wired development of the Boudin process was carried out by Pond working with Lewis at the Blue Ridge Glass Company in Tennessee

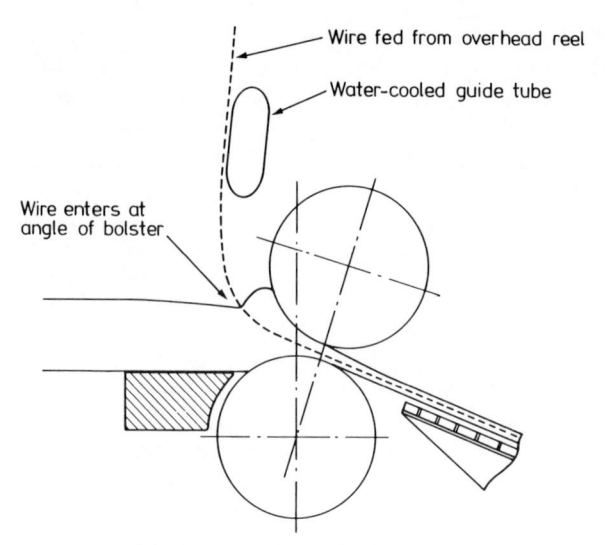

FIG. 6. Wire feed on Pond/Boudin wired glass process.

in the late 1920s (Pond and Lewis, 1929). This process is widely used today, though varying in detail from company to company.

By means of a standard Boudin machine setup wire mesh from a roll suspended above the machine is fed down onto the ribbon upstream of the roller pass and enters the pool at the base of the so-called bolster. In the version depicted in Fig. 6 the wire mesh is loosely fed and is dragged into the glass over the water-cooled bar by the pull of the ribbon. In other arrangements provision is made for positive drive of the mesh between knurled rolls. The process produces a good quality product at low rolling speeds (50 in./min), but becomes more difficult to operate at higher speeds (70 in./min), with an increasing tendency for the wire mesh to lie off center and to be displaced toward the top surface of the ribbon. At higher speeds the glass emerging from the pass is soft and falls away locally between the mesh, giving the sheet a puddled or baskety fault.

2. The Pilkington Double-Pass Wired Process

The development of the flow process to produce wired glass was made in St. Helens in the period 1934–1936.

Figure 7 shows the double-pass machine in cross section. The first roller pass is fed by a short forehearth direct from the furnace, and at this pass a 3-mm sheet is rolled. At this sheet passes forward to the second roller pass, the wire mesh—preheated—is laid down and enters at the

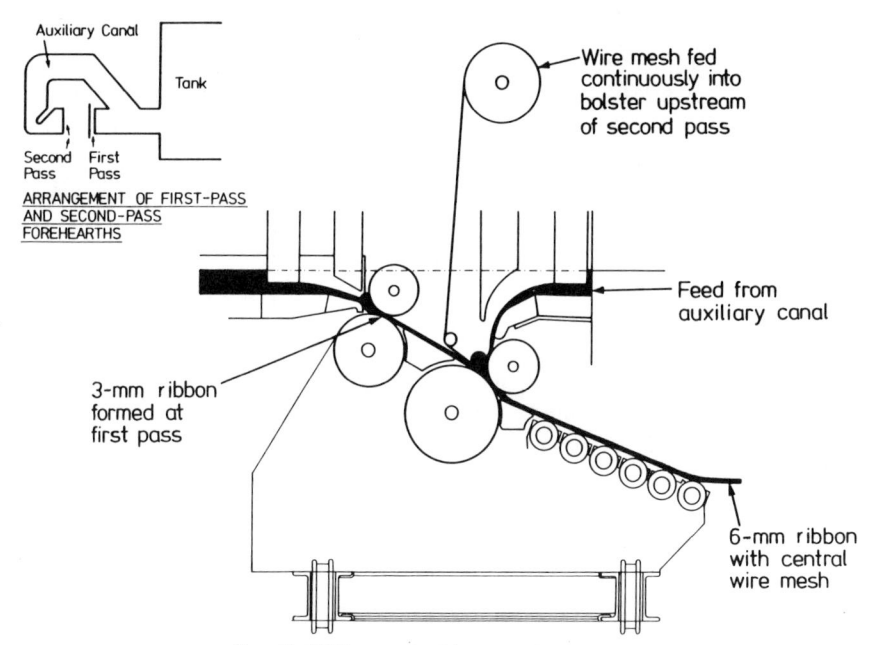

FIG. 7. Pilkington double-pass wire process.

second pass. At the second pass additional glass is fed in and the final sheet 6 mm thick emerges with the wire mesh in an exactly central position. The feed of glass to the second pass is by a long J-shaped forehearth set to one side of the first forehearth.

The double-pass process has considerable advantages over the single-pass process. Rolling speeds are considerably higher and wider ribbons can be provided. The quality of the product is also superior. Not only is the wire mesh always exactly central but the wire can be freed of bubble or disfiguring halo. As a consequence ribbon from the double-pass process is more suitable for subsequent conversion into polished wired glass.

3. The Wire Mesh

For both processes the specification and control of the wire mesh is of great importance if good quality wired glass is to be produced. The original mild steel billets must be free from entrapped slag or other inclusions, smooth well-polished dies must be used to draw the wire, and the finished mesh must be clean and free from contaminants. When producing Georgian (square mesh) wire if the wire mesh is cold when it enters the glass, subsequent expansion may be localized to produce "feather" distortion of the mesh. This fault can be prevented by preheating the mesh before the point of entry. To promote preheating and to prevent widespread oxidation, the mesh may be given a protective surface treatment (Pilkington, 1965). There is no requirement—as in conventional glass/metal seals—to match the expansion of the glass and the wire. The bond between the glass and the wire set up at the rolling stage is a weak one and as the ribbon cools in the lehr the wire breaks away, so when the ribbon is cold the wire mesh lies free and there is a minute air space between the wire and the glass.

G. THE POLISHED PLATE PROCESS

The development of the polished plate process was a major breakthrough in glassmaking and over a long period of years provided the source of the highest quality of flat glass. As finally developed using continuous rolling and on-line twin grinding and polishing the process was the epitome of engineering applied to glassmaking, and the development of the process makes a fascinating study. In the past decade the process has been superseded by others and a detailed description, therefore, lies outside the purpose of this chapter. There are, however, some excellent accounts in the literature to which the reader is referred (Vincent, 1960; McGrath and Frost, 1961).

IV. Flat Drawn Sheet Processes

A. SHORT HISTORY

100–1600 A.D.	Flat glass produced in Europe mainly by broad glass process.
~1560	Improved crown process developed in France.
~1700	Blown cylinder process developed in France.
1895–1910	Drawn cylinder process invented by Lubbers and developed in the United States.
1902–1915	Fourcault process invented and developed in Belgium.
1906–1915	Colburn process invented in the United States and commercially developed after formation of Libbey Owens Company in 1915
~1918–1925	PPG Pennvernon process developed in the United States.

B. WINDOW GLASS MANUFACTURE

The early history of window glass manufacture has been outlined in Section I.B. By the year 1900 virtually all window glass was produced by means of the blown cylinder process supplemented by a small quantity of plate glass for certain special requirements. Until that time much of the glass was pot melted, but by the early years of this century continuous melting units capable of supplying glass of adequate quality were becoming available, and both the potential and the need for a low-cost means of mechanical production of sheet glass were widely recognized. The hand-blown sheet process made heavy demands on skilled labor which both in Europe and America was becoming increasingly organized and expensive. This stimulated invention on both sides of the Atlantic, leading first to the invention and introduction of the drawn cylinder process followed by the three flat drawn processes.

C. The Drawn Cylinder Process

The first mechanized process to come into commercial use was the drawn cylinder process invented by John H. Lubbers around the turn of the century and from 1903 taken forward to a commercial scale by the American Window Glass Company. In this process glass from a melting tank was ladled into a special draw pot into which dipped a flanged metal disk held on the end of a vertically mounted blow pipe. As this disk bait was slowly raised between guiding shafts a cylinder of glass was drawn up, its diameter being kept uniform by a controlled feed of compressed air through the blow pipe. In this way a cylinder 25 in. in diameter and 25 ft long could be drawn. This cylinder was then removed mechanically, split, cut into smaller lengths, and flattened.

Although worked commercially in the United States, Canada, and Britain in the period 1905–1928 this process did not meet commercial requirements: it remained an intermittent process, labor requirements were high, and the need to flatten inevitably led to some degradation of the surface. The quality was barely sufficient to match that of the best blown cylinder glass.

D. Fourcault Updraw Process

The Fourcault process was conceived by E. Fourcault around the turn of the century and developed at various plants in Belgium, only becoming widely established on a production basis in the years following the First World War.

1. Equipment and Operation

Figure 8 shows in section a typical layout of a Fourcault canal and drawing kiln. Depending on the output required, from one to nine such machines may be fed from the conditioning chamber of a continuous melting unit, the disposition of the canals being chosen to achieve as uniform a temperature as possible at the point of entry.

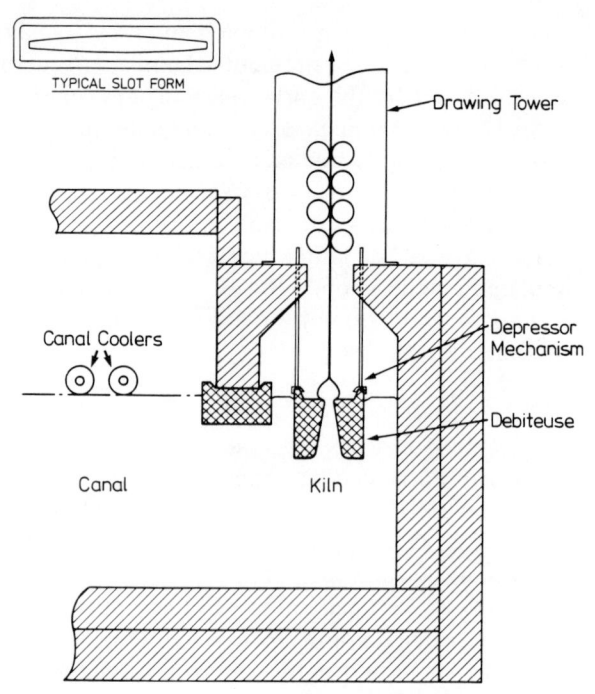

FIG. 8. Fourcault flat drawn process.

Manufacture by the Fourcault process is cyclical. A cycle begins with the heating of the drawing kiln by auxiliary burners to the point where the glass in the kiln is fluid and any pockets of devitrification have been melted out. The drawing process is initiated by launching the clay debiteuse into the drawing kiln. The kiln is then allowed to cool back and at the appropriate temperature the debiteuse is pressed down (normally by between 1 and 2 in.) below its normal buoyancy level and is then held down at each end by the depressor mechanism. Glass begins to flow up through the slot of the debiteuse and is then captured on the end of a steel mesh bait reversed down from the drawing tower which is then slowly pulled up until the sheet is gripped by the lower rollers of the drawing tower. The sheet as it forms on emerging from the slot of the debiteuse is cooled by the water coolers. The draw thus established, the kiln is then sealed up and every effort made to stabilize drawing conditions.

The rate of draw and the cooling provided are such that the temperature of the formed sheet is low enough as it leaves the kiln to be gripped by the bottom rolls of the drawing tower without surface damage. The sheet then proceeds up the tower, being annealed and further cooled until at the top it reaches a temperature at which it can be cut off.

2. Control Variables

The output rate is primarily determined by the volume of glass flowing up through the debiteuse as a consequence of the depression. The amount of flow is a function of the width of the slot, the amount of depression, and the temperature and hence the viscosity of the glass at the slot, and is affected only to a minor degree by the rate of draw. The size and position of the coolers must be matched to the debiteuse setting.

The thickness of the sheet is largely determined by the rate of drawing. An analysis of the interplay of these variables has been given by Goerk (1962). The coolers have two important roles to play in the drawing process. First, they impose the vertical temperature gradient on the sheet as it forms to give the desired pattern of stretch in the direction of draw. Second, they are an important element in the chain of devices used to impose the required transverse temperature distribution across the ribbon as a means of promoting flatness and uniformity of thickness across the ribbon.

Control of the transverse temperature distribution begins as the glass moves down the canal. The forward flow is faster at the canal center than at the edges, and so the glass near the edges is exposed to the cooling effect of the canal for longer than the glass at the center. The overall degree of cooling in the canal may be increased and the natural cold edge/warm center pattern intensified by the insertion of uniform-diameter water pipes across the canal above the glass level. Alternatively, by using nonuniform water pipes with the center third of greater diameter it is possible to increase the cooling in the middle of the glass going forward to the drawing kiln.

Natural cooling is the main control used to achieve the desired temperature distribution in the glass flowing up the slot. Once the sheet is formed the pattern of cooling is then largely governed by the size and shape of the water coolers.

The primary means to achieve uniformity of thickness across the sheet lies in the choice of the contour of the debiteuse slot. Figure 8 shows a typical contour. In practice the choice of debiteuse width and contour is arrived at from experience.

To maintain stable drawing conditions within the kiln the level of glass in the furnace and the temperature of glass fed to the canal must be kept constant. In practice variation of not more than ± 0.5 mm in level and not more than $\pm 1°C$ in working end temperatures are good targets.

The greatest care must be taken to avoid in-leakage of air or other drafts both at the drawing kiln, where they can affect the thickness distribution and surface quality, and in the drawing tower, where they can lead to breakage.

At the lower range of drawing speeds the Fourcault process will operate without edge rolls, relying on the "cold edge" effect to maintain sheet width, but at higher draw speeds there is a clear advantage in using knurled edge rolls to assist width maintenance.

3. Quality Control in the Fourcault Process

In a well-operated, modern Fourcault plant it is possible to produce a quality of glass sufficient at best for use in car windshields, but in the majority of Fourcault plants around the world the glass produced is of inferior quality, the sheets showing heavy draw lines, distortion, and variation in thickness. As is the case in any other flat drawn process the surface quality obtained with the Fourcault process is susceptible to the presence of inhomogeneity in the input glass from the furnace, but the Fourcault process contains a second and more serious source of lines. The glass passing up the clay debiteuse is in contact with the clay on both back and front faces and picks up a very thin film of dissolved clay products. Only in exceptional circumstances will such a thin film stretch uniformly and give a line-free sheet as the sheet is formed. In practice the surface of the debiteuse is slightly rough and the parting line (the line where the glass sheet leaves contact with the debiteuse) is always slightly uneven. As a consequence the thin surface film is uneven in thickness and this gives the classic condition (Section II) for line formation on stretching the sheet. The problem of lines is kept to a minimum by the choice of the debiteuse clay, by polishing the inside of the debiteuse slot to the highest possible standard, and by controlling the temperature condition of the slot to induce exactly the desired shape of "onion" at the initial point of draw.

It has already been pointed out that manufacture in the Fourcault process is cyclical, and it is not possible to maintain completely stable conditions throughout a cycle. As the machine life proceeds there is a gradual increase of devitrification on the inside surfaces of the drawing kiln, particularly at the corners, which modifies the pattern of flow to the debiteuse. At the same time there is a growth of devitrification on the slot surface of the debiteuse, particularly along the parting line, which increases the propensity to form lines. Toward the end of the cycle this increased devitrification can be great enough to affect the pattern of flow through the slot and make edge control difficult.

As a consequence of these progressive changes the quality varies throughout the cycle. In general quality is poor for some 10–20 h after startup during the period when the draw is being settled down. Quality then rises to a peak once steady running conditions have been established and continues so for some 3–6 days, after which time the gradual increase of devitrification in the corners of the kiln and along the parting line leads to gradual deterioration and increased formation of lines. Eventually after some 180–240 h the deterioration in quality reduces the good yield to a point at which it is necessary to knock down the sheet and start the cycle once again. The cycle times quoted are typical of many Fourcault operations. In recent years, by careful choice of glass composition and the use of a special form of debiteuse cycle, times have been considerably lengthened (Takahashi and Ichinose, 1980).

E. THE COLBURN UPDRAW PROCESS

The process was invented by Irving W. Colburn in the United States around 1906 but only brought to commercial success after the formation of the Libbey Owens Company in 1916.

1. Equipment and Operation

It is usual for one or two Colburn forehearths to be taken from the conditioning zone of a conventional flat glass melter, the glass flowing into each forehearth under an immersed C-shaped skim bar. A typical arrangement of forehearth and drawing kiln is shown in Fig. 9. The glass flows into the drawing chamber, where it is contained in a single-piece refractory bowl 1 m long and from 1 to 2 m wide. The glass is 150–200 mm deep; it is maintained at the required temperature and devitrification is prevented by heating the outside of the bowl by a gas- or electrically heated chamber.

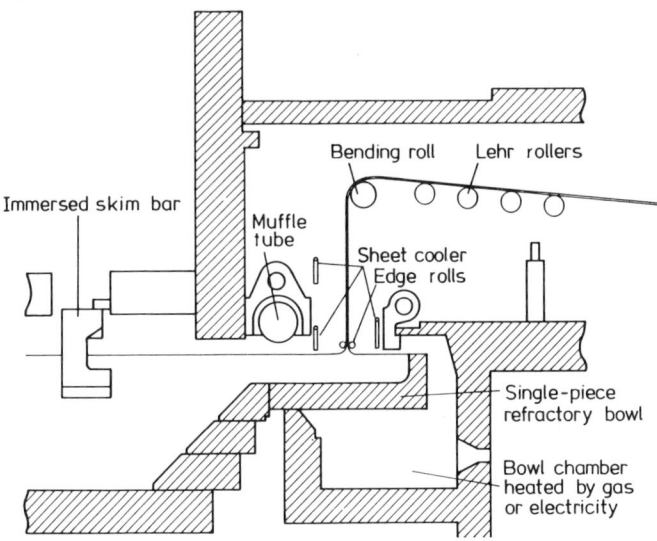

FIG. 9. Colburn Libbey–Owens flat drawn process.

The unique feature of the Colburn process is the horizontal bending roll, a highly polished nickel alloy roll 200 mm in diameter and some 70 cm above glass level around which the sheet bends before entering the horizontal lehr. The bending roll is cooled by air blown down the bore.

The draw equipment consists of front and back coolers, a second vertical cooler on the front side, and a horizontal roll cooler. The sheet is gripped at each edge by a driven knurled roll and there are edge burners and a full-width horizontal muffle to reheat the formed sheet so that it is sufficiently plastic to bend freely around the guide roll, which is maintained at between 700 and 800°C without undue marking.

The draw is initiated by reversing a flexible comb-shaped bait once the edge rolls, top coolers, and burners have been set in place. Once the sheet is established the main coolers are put in place and the kiln closed up.

2. Control Variables

Control of the operation is more critical than with other updraw processes, the control of temperature depending less on natural heat loss in the forehearth and kiln and more on the influence of the applied burners and coolers. The primary control of output (per unit width) is the degree of cooling which can be imposed on the sheet, but this is limited by the requirement that the sheet be sufficiently hot and flexible to bend over the guide roll. Draw speed is the prime determinant of sheet thickness. Draw rates are high, and to hold the edge it is important to carefully control both longitudinal and transverse temperatures.

3. Glass Quality, Advantages, and Disadvantages

Advantages offered by the Colburn process are the high output, up to 70 tonnes/day per unit, the ability to make sheets up to 4.2 m in width and to produce glass of thickness ranging from 0.6 to 20 mm. Since the process is horizontal there is not the same restriction on lehr length as with the updraw processes, and in general Colburn sheet can be annealed to a lower stress level and better cutability than with Fourcault or PPG sheet. Against this must be set the complex startup and operating procedures and the problem of manufacturing and fitting—hot—the massive refractory bowl.

The Colburn process is at a disadvantage in respect of glass quality. The complexity of control makes it difficult to achieve consistently good control of transverse thickness and distortion. The drawing force is transmitted from the lehr by pull around the bending roller and the contact inevitably leads to some degree of "orange peel" surface defect. The process is as sensitive as the PPG process to the standard of chemical homogeneity of the glass supplied.

4. Glaverbel Modification

In the 1960s the Glaverbel Company introduced a modification to the process by using a draw bath some 90 cm deep in which a refractory draw bar in cross section approximately 45 cm wide and 23 cm deep is immersed below the surface to stabilize the draw. One result of this deepening is to set up a system of return flow which stabilizes the temperatures in the draw kiln without the need for externally applied heat.

F. THE PPG PENNVERNON UPDRAW PROCESS

Great credit must be given to the pioneers Fourcault and Colburn and their associates. Not only did they have to master the then novel technology of the flat drawn process but they had to develop many essential engineering elements—the melting furnace and forehearth, the drawing kiln and coolers, and in the case of Fourcault the vertical drawing tower. But in the final analysis both processes were left with fundamental weaknesses which prevented the attainment of the highest quality. In the case of Fourcault the weakness lay in the cyclical nature of the process and tendency to produce lines as a result of drawing through a slot. In the Colburn process contact with the bending roll inevitably tended to impair the surface and the means of control were complex.

The development by the Pittsburgh Plate Glass Company of the Pennvernon flat drawn process and its introduction in 1926, while drawing on much of the pioneering work of Fourcault, represented a deeper technical appreciation of the fundamental requirements affecting sheet quality and a more subtle and practical approach to the means of control.

1. Equipment and Operation

A typical arrangement of delivery forehearth and kiln is shown in Fig. 10. Differences exist from company to company and the arrangement shown represents one European practice. The key points to note are the following:

(1) The adoption of a full-depth kiln in which there is a considerable volume of return flow by which the operating temperature is maintained without the need of extra heat and which holds devitrification to a minimum.

(2) The shape and disposition of the top clay work. The skim bar and the ell blocks have functions far beyond the mere covering of the kiln.

(3) The drawbar, which serves to divide the flow of the glass to the draw and to anchor the point of draw, which would otherwise be unstable under the temperature regime obtaining in the kiln. There is no upward flow of glass through the drawbar slot.

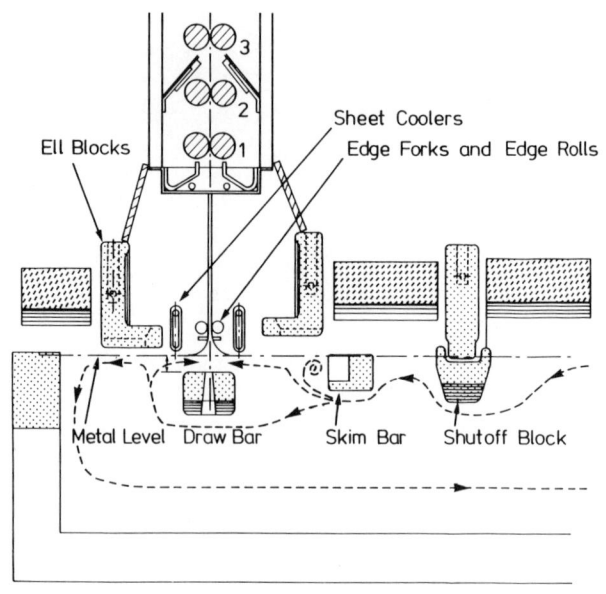

FIG. 10. PPG Pennvernon flat drawn process.

The coolers, edge forks, edge rolls, and kiln seals serve much the same purpose as in the earlier flat drawn processes. Because drawing speeds are higher the distance from glass level to the first pair of rolls is in the region of 46 in. and in normal operation after startup the first grip is made by the third pair of rolls 60 in. above glass level. Tower height varies from 19 to 27 rolls high in more recent constructions.

The Pennvernon process is cyclical, machine life being limited to around 1000 h by buildup of devitrification, particularly between the kiln and the ends of the drawbar. The production cycle is started by thorough heating of the kiln to remelt any devitrification, a period of cooling back, followed by the introduction of edge rolls and coolers at which stage the draw is initiated by using a metal bait. Marketable sheet can be produced within 5–10 h of initiation, but it usually takes from 20–30 h for glass and kiln temperatures to stabilize and for a stable edge configuration to be attained. Unlike the Fourcault process the sheet quality does not deteriorate with time and glass of high quality can be made right to the end of the machine life—the end coming when the buildup of devitrification interferes with the flow of glass forming the edge.

2. Operational Variables

The output from a Pennvernon machine is controlled by the degree of cooling that can be imposed starting right back in the conditioning end of

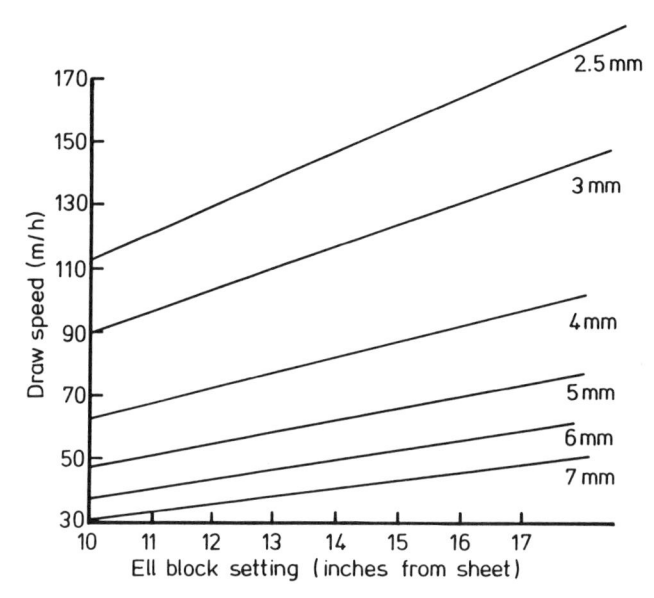

FIG. 11. PPG Pennvernon process draw speed as a function of sheet thickness and ell block setting.

the melter through the forehearth to the separation of the ell blocks on either side of the draw, and by the size and disposition of the water coolers. The greater the cooling the higher the output. Draw speed has only a secondary effect on machine output. For a given machine setup output increases as the 0.20 power of draw speed. Sheet thickness is primarily determined by draw speed. Figure 11 gives typical levels for output, draw, speed, and thickness for a range of ell block settings.

The edge is held and sheet width maintained by a number of interdependent factors. First, a natural cold edge/hot center condition is established in the flow to the line of draw. Second, at the root of the draw the sheet is run through edge forks and the edges are then gripped by knurled edge rolls driven at the appropriate speed. A controlled indraft through each side of the kiln serves to further cool and stabilize the edges.

It is difficult to make sheet less than 2.0 mm thick by the PPG process and tower rolls have to be in prime condition if pressure breakage is to be avoided when making thin glass. In a well-designed unit uniformity of thickness across the ribbon is normally good but it is possible to correct local heavy spots by hanging pads on the face of the cooler at the appropriate point. Excessive use of pads is bad practice.

Successful operation of the Pennvernon process demands care, skill, and discipline from line managers, machine operators, and tower attendants. The necessary standards are only attained with sustained effort, but once mastered the process itself has built-in features which contribute to the production of glass of high quality and which have led to its acknowledged superiority over other flat drawn processes.

3. Quality

The pattern of glass flow (Fig. 10) within the forehearth and drawing kiln is crucial in achieving the high distortion quality of which the PPG process is capable. The flow pattern is designed to ensure that none of the glass forming either the front or the back surface of the sheet has become contaminated by passing over refractory. The front half of the thickness of the sheet is formed by glass which has flowed under and around the drawbar and the back half by glass flowing directly from the forehearth. By controlled cooling at the kiln front the front flow is caused to divide so that "new glass" goes forward to form the front surface of the sheet. It is the function of the skim bar and the kiln geometry as a whole to promote a division and spread of flow so that the glass moving forward to form the back surface of the sheet has minimal refractory contamination.

The PPG process is just as sensitive as other flat drawn processes to body inhomogeneity in the feed glass. Because, as discussed in the previous paragraph, one important source of lines has been eliminated it becomes all the more important to supply the drawing unit with glass of consistently high homogeneity. To produce the highest quality of sheet the glass may be stirred and the disposition of the canals around the conditioner specially arranged to promote a laminar disposition of any ream remaining in the sheet.

Distortion can also arise from lack of uniformity of cooling within the kiln at and just above the meniscus. Any nonuniformity in the cooling produced by the water coolers results in sheet distortion, and in practice every effort must be made to eliminate variation by such means as continuously depositing a film of carbon black on the surface of the coolers to ensure uniform emissivity or by acid-cleaning the inside of the coolers to remove any deposits which could impair the water cooling effect. Even with perfectly uniform coolers cold air "waterfalling" down the face of the cooler may produce distortion, and a number of devices have been developed to minimize this effect.

In seeking the highest quality and yield it is also important to keep breakage in the tower to a minimum since debris from falling glass may adhere to the surface at the meniscus. To minimize breakage tower doors must be close fitting and free of indraft, the rolls must be true and their surface rough enough to provide good traction, and the tower drive must be stable and vibration free.

G. The Slot Bushing Downdraw Processes

Reference has already been made (Section II.C) to the advantages and disadvantages of downdraw processes and their special suitability for the production of thin glass. A great variety of downdraw processes have been described in the patent literature and a number of these have actually become established commercially to provide speciality glass such as that used for electrical capacitors, microscope cover slips, and optical filter glasses. Almost all these processes are based on drawing glass through an accurately dimensioned slot in a platinum bushing. A variety of means are used to maintain the edge and prevent narrowing of the ribbon. Figure 12 shows the outline of a typical slot bushing process.

While these processes serve the special requirements they have been designed to meet, because the glass is drawn through a slot none are capable of producing glass of the very highest distortion quality.

H. The Fusion Downdraw Process

This fundamental drawback of slot bushing processes was overcome in a very elegant and effective way by Corning Glass Works in their

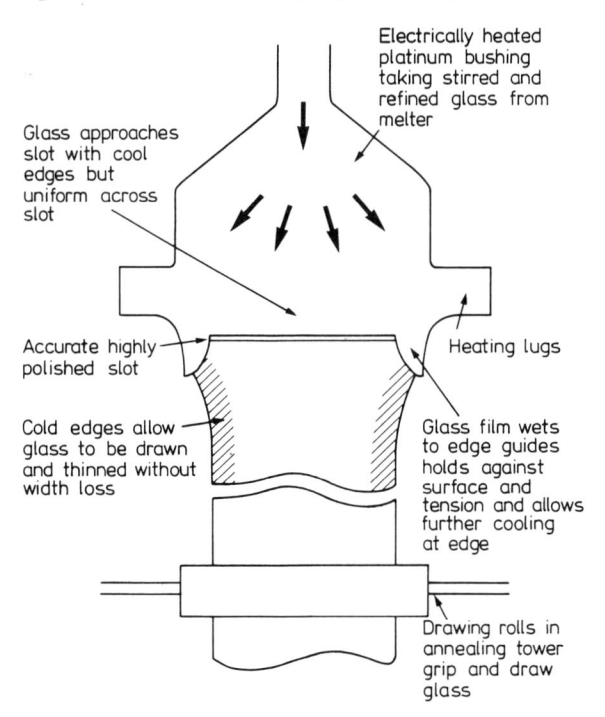

Fig. 12. Elements of a slot bushing downdraw process.

invention of the fusion process. Instead of feeding through a platinum slot, glass is supplied by the overflow from a trough, the front and back flows recombining to form a sheet with uncontaminated virgin glass forming both surfaces.

1. Equipment and Operation

The sketch shown in Figure 13 illustrates the basic concept of the process. Well-stirred hot glass is delivered through suitable conduit tubes to one end of a rectangular trough. Typically the viscosity of the glass would be 40,000 P at this point. The upper edges or weirs of the trough are slightly inclined downward from the inlet end and co-operate with the upwardly inclined trough bottom in such a way that the pressure drop is linear with distance along the trough. The glass overflows the weirs evenly along the full length of the trough, runs down the sides, and the two streams rejoin or fuse together at the root or apex of the troughlike bar. The viscosity of the glass at this location is around 300,000 P. Small metal rolls grip the edges of the sheet just below this apex to prevent excessive narrowing of the sheet as it is stretched. Further cooling as the sheet proceeds downward permits pulling rolls to be engaged on the sheet without causing any damage to the sheet surface. A vertical annealer with suitable pulling rolls conveys the glass ribbon to the cutoff station, where it is cut to length.

The process can operate over a wide range of outputs—glass flow rates from several hundred to several thousand pounds per hour have been reported. The process is particularly well suited for producing thin (≤2-mm) sheet, but thicknesses of 0.6–11 mm have been produced. Usable sheet width is 60 in. after edge trim.

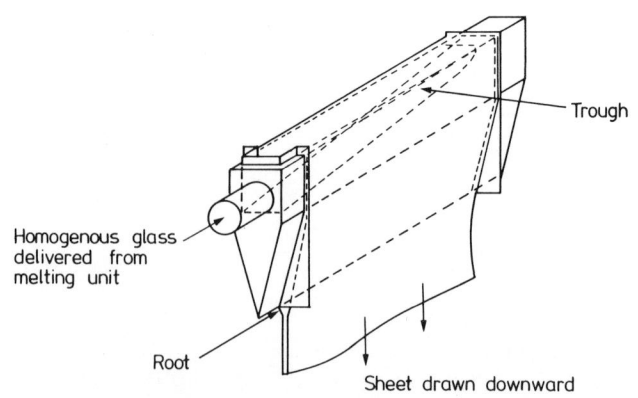

FIG. 13. The Corning fusion downdraw process.

2. Advantages

(1) The glass surface is untouched by anything except air until it is sufficiently hard to resist marking. This yields a smooth fire-finished surface.

(2) The rate of glass flow is determined by the impedance of the delivery tubes and trough and is not influenced in any way by the speed of the pulling rolls. If the sheet breaks between the trough root and the first set of pulling rolls, the flow continues and gravity conveniently supplies glass to the pulling rolls to restart the draw. This is particularly advantageous when drawing thin sheet. It also provides for unusual thermal stability in the glass melting furnace and the glass delivery system.

(3) Well-stirred glass can be delivered to the trough by passing it through a narrow forehearth specifically designed for efficient glass mixing. This is a major factor in eliminating striae and lines.

(4) The process can be attached to a variety of glass melting and delivery systems which may be necessary for a particular glass composition. This coupled with the fact that there is no contact or chemical contamination of the soft sheet surface during forming makes the process applicable to a wide range of glass compositions. Fusion drawn sheet has been produced from hard borosilicate, hard chemically strengthenable glass, and soft photochromic glass.

3. Quality

Provided every care is taken in delivering glass of the highest homogeneity and in establishing and maintaining uniformity of flow and temperature along the length of the trough the process is capable of producing sheet of good optical quality. Thin sheet produced by the fusion process has been tested as being suitable for use for laminated auto windshields and represents a very high quality for a drawn sheet product.

Compared with the older established flat glass processes the fusion process is technically complex and calls for a high level of sophistication in control but it would appear that in specific product areas it can produce excellent glass.

V. The Float Process

A. THE DEVELOPMENT OF THE FLOAT PROCESS

By the early 1950s a pattern of flat glass manufacture had become established. Developments in melting technology and furnace design and in the understanding and control of the drawing process had brought the flat drawn processes and, in particular, the PPG process, to a point where they were capable of supplying glass of consistently high quality adequate to meet the standard required for window glass and by selection some of the less demanding applications previously supplied by plate glass. But this deeper technical understanding had also made it clear that to improve the distortion quality from the flat drawn process to a point where it equalled the quality of plate would require exceptionally high standards of glass homogeneity and of temperature uniformity in the drawing kiln. The technical demands and the cost of meeting them were seen to be so great as to set a practical limit on the quality from the updraw processes.

Corresponding developments (all directed to reducing the cost of production) had taken place in the plate process. By the early 1950s the most modern plate plants, operating with improved furnaces and rolling techniques and using continuous twin grinding and polishing, were approaching a plateau of technical and engineering efficiency from which any major advances seemed likely to call for radical and costly process changes. Over a range of thicknesses from 3 to 25 mm the quality of plate glass was adequate to meet all market requirements but the cost of production was undesirably high. This arose from the very high capital cost of the plate line, the intrinsic inefficiency of the plate process, in which at least 20% of the glass melted had to be ground away, and the considerable labor demand.

The potential for a process which could produce glass of the same quality as the plate process at a cost of production competitive with that from the sheet process was widely recognized. In hindsight one can identify two other requirements to be met by the new process. First, the new process should have the capability to exploit the potential advances arising from the use of computers, and second, it should take account of the rising trend in labor costs and so make the minimum demand for operating labor.

A number of ingenious approaches were explored. One process which was examined comprised rolling a ribbon of glass through porous steel rolls with high-pressure steam fed to the roller bores to prevent direct contact between the glass and the steel of the rolls. Another approach considered was the possibility of forming a ribbon of glass between rapidly vibrating platens. None of these approaches showed signs of being successful in producing glass which would be of comparable quality to plate glass.

The answer was provided by the invention of the float process (Pilkington and Bickerstaffe, 1953) and its subsequent refinement under the leadership of L. A. B. Pilkington (later to become Sir Alastair Pilkington), leading to the announcement of its commercial success in 1959. An account of this development has been given by Sir Alastair

(Pilkington, 1969). Great improvements and extensions have been made to the process since it was first announced and today the float process meets all the requirements recognized at the outset. (1) The process is capable of producing flat glass of the highest quality in thicknesses ranging from 2 to 25 mm and in ribbon widths greater than 3 m. (2) There is no perceived limit on the output from a float unit. At present plant output ranges from 1000 tonnes per week up to 5000 tonnes per week. (3) The process is inherently efficient. The width of the ribbon can be readily altered to match the dimensions of the product required. The only waste glass is a narrow strip cut from either side of the ribbon. (4) The process is not cyclical. Under good conditions the float bath can start up at the beginning of a furnace campaign and continue in operation with only minor on-line repairs throughout the whole of the campaign of the melting furnace. (5) By producing a horizontal ribbon the process lends itself to achieving a high standard of annealing and to subsequent mechanized and automatic cutting at the lehr end. (6) The technology of the float process is well understood and the process operates under much surer control than is the case for most glass manufacture. The whole line from batch mixing to final cutting lends itself to co-ordination and control by computer. (7) In a modern float process the operating labor requirement is low. It is not uncommon to walk the whole length of a float line and see no labor activity other than the one or two men required to monitor the melting, forming, and annealing processes. (8) The process lends itself to the on-line addition of surface treatment processes to produce solar control and other special glasses.

It is the consequence of these advantages that some one hundred plants have been built all over the world and the penetration into the world glass market by the float process shown in the Table II has been achieved.

B. Essential Features of the Float Process:
Equilibrium Float

The glass is supplied from a continuous melting furnace, designed and operated to provide the output required at the highest possible standard for bubble, solid inclusions, and homogeneity. Glass passes from the furnace conditioner to the float bath via a forehearth where the flow is controlled by passage under a movable refractory tweel to the spout lip, from which it flows forward and falls freely onto the molten tin contained in the bath.

The float bath (Fig. 14a) is a large unit made up of a refractory-lined steel container holding a pool of molten tin and enclosed above in a refractory and steel casing to contain the nitrogen/hydrogen atmosphere

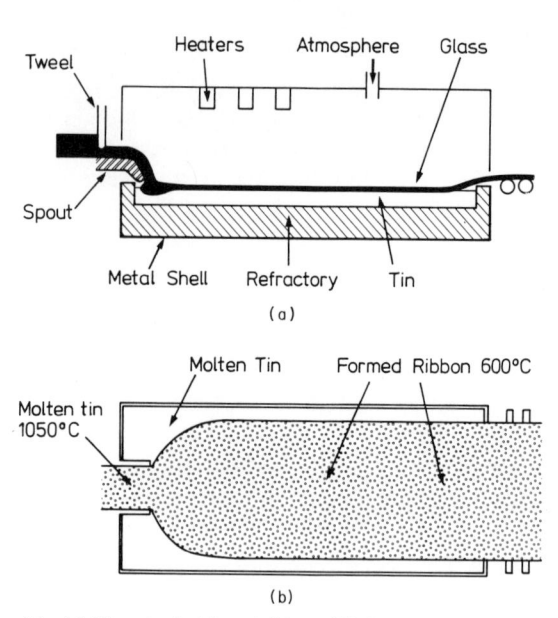

FIG. 14. (a) Float bath (a) and (b) equilibrium mode of operation.

which is necessary to prevent oxidation of the molten tin. Provision is made within the bath for localized overhead heating and for the insertion of water coolers, guides and barriers, edge rolls and top rolls, and other devices used to heat or cool the ribbon or to control the ribbon position as it passes down the bath. The float bath leads directly via a special takeout section into a horizontal annealing lehr where the sheet is carried forward on rolls to emerge annealed and cooled ready for cutting on-line into the final product.

In the original and simplest form of the float process, equilibrium float as illustrated in Fig. 14b, the stream of glass flowing onto the molten tin spreads out to form a wide flat pool of glass of "equilibrium" thickness (approximately 7 mm) which is then continuously transported down the bath as a ribbon of glass by the pull of the lehr. The ribbon is cooled as it passes down the bath so that at the exit end it has reached a temperature at which it can be carried on the lehr rolls without suffering surface damage. The glass is already perfectly flat and uniform in thickness in the pool at the hot end of the bath and the width to which the pool spreads and the subsequent width of the ribbon is determined by the relationship between the quantity of glass flowing down the spout and the lehr speed. It is usual to operate at a lehr speed at which the pool spreads to within a few inches of the sides of the bath since this gives the maximum time to establish uniform thickness and flatness.

An important feature of the process lies in the detailed design of the spout and the point of entry to the bath at the so-called wet back area. The disposition of refractories is such that the thin skin of glass which has flowed over refractory in its passage down the forehearth and over the spout is caused to flow preferentially outward and finishes up confined to the extreme outer border of the ribbon.

Today the float process is seldom, if ever, operated in the equilibrium mode described above, but the description is included to help the reader to achieve a basic understanding of the essential features of the float process. Clearly the equilibrium process, limited to producing glass 7 mm thick, is not versatile enough to meet real market needs, and a number of variants of the float process have been developed which today allow all thicknesses of glass from 2 to 25 mm to be produced efficiently and to provide the high quality demanded by the market.

To help the reader to understand these variants in detail it will be useful to give some account of the science and technology that underlie the float process.

C. Theory of the Float Effect

In this section, a short account is given of the theory underlying the float effect whereby forces of gravity and surface tension operating on the glass floating on the bath of molten tin combine to produce a sheet of glass which is perfectly flat and of a specific and uniform thickness—the so-called equilibrium thickness.

1. Equilibrium Thickness

When oil is poured onto water it does not spread indefinitely but forms a pool of uniform thickness. If more oil is added the pool is enlarged but the thickness remains the same. This is a familiar example of a nonspreading liquid system. Molten soda–lime–silica glass floating on clean molten tin, under an inert or reducing atmosphere, is also a nonspreading system. This is illustrated in Fig. 15, which shows a vertical cross section normal to the length of a floating ribbon of glass of thickness T. The important physical properties are ρ_g, the density of the molten glass, typically 2350 kg/m^3 at forming temperature; ρ_t, the density of the molten tin, 6500 kg/m^3, at forming temperature; S_{gt}, the surface tension at the glass/molten tin interface, ~0.5 N/m; S_{ga}, the surface tension at the molten glass/atmosphere surface, ~0.35 N/m; and S_{ta}, the surface tension at the molten tin/atmosphere surface, ~0.5 N/m.

The requirement that the system be nonspreading means that the sum $S_{ga} + S_{gt}$ is greater than S_{ta}; thus as a result of surface tensions there is an inwardly directed force $S_{ga} + S_{gt} - S_{ta}$ along each unit length of ribbon

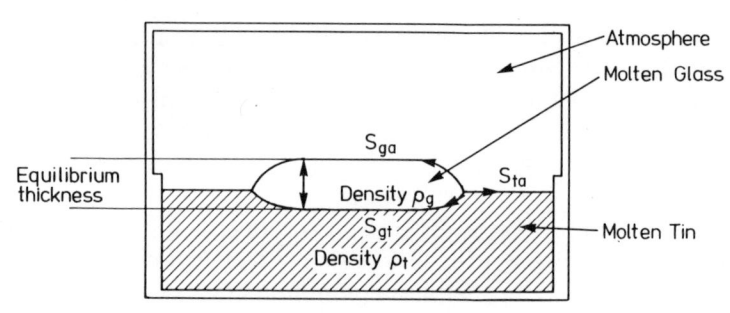

FIG. 15. Equilibrium thickness.

edge. Since the glass is of lower density than molten tin, the layer of glass floats above the level of the molten tin as shown in Fig. 15 and gravitational forces act to spread the glass; it can easily be shown that the net outward gravitational force per unit length of ribbon edge is AT^2, where A is a constant given by the relation

$$A = g(\rho_g/2)(1 - \rho_g/\rho_t).$$

The ribbon spreads out and thins to the point where the surface tension and gravitational forces are in equilibrium at the so called equilibrium thickness T_e given by

$$T_e = \sqrt{(S_{ga} + S_{gt} - S_{ta})/A}.$$

For soda–lime–silica glass floating on clean molten tin under a nitrogen/hydrogen atmosphere T_e is close to 7 mm. The equilibrium thickness is relatively insensitive to changes in temperature or to small changes in composition of glass or metal. Hence the need for special variants of the process to produce glass of other thicknesses.

2. Gravity and Surface Tension Forces and Uniform Thickness

The forces of gravity and surface tension which determine equilibrium thickness also combine effectively to promote the flatness and uniformity of thickness of the floating ribbon of glass so that the final product emerges from the bath free from optical distortion. In fact by the time equilibrium thickness has been reached, the upper and lower surfaces of the ribbon remote from the edges are perfectly flat and parallel. In practice, the time required for this "float processing" to be completed depends on the viscosity of the glass and for float composition at 1050°C (10^4 P) the time is approximately 1 min.

2. FLAT GLASS MANUFACTURING PROCESSES

Some quantitative understanding of this timescale can be gained by considering how a floating ribbon, initially of thickness T greater than equilibrium T_e by a small amount a, reaches equilibrium thickness as a result of the combined influence of gravitational and surface tension forces.

We have

$$T = T_e + a, \quad a \ll T_e.$$

We have already seen that the net outward gravitational force per unit length of ribbon edge is AT^2.

$$AT^2 = A(T_e + a)^2 = AT_e^2 + 2AaT_e.$$

Again we have already seen that the inward force per unit length due to surface tension is equal to AT_e^2. So in our case we are left with a net outward force per unit length of ribbon of $2AaT_e$. This corresponds to an extensive stress across the ribbon of $2Aa$ acting to reduce the thickness.

We can now gain some idea of the rate at which the thickness correction takes place by making use of the simple theory of viscous deformation given in Section II. Using the same symbols as Section II assume that we are at a part of the ribbon where longitudinal stretch is small ($E_z = 0$). There is no stress across the thickness of the ribbon ($S_y = 0$) and the lateral stress $S_x = 2Aa$.

Putting these values into the general Eqs. (1)–(3) we get

$$E_y = -Aa/2n.$$

Now E_y represents the time rate of change or ribbon thickness $(1/T)(dT/dt)$. Substituting $T = T_e + a$ we get

$$(1/T_e)(da/dt).$$

On integration we obtain the relation

$$a = a_0 e^{-t/\tau},$$

where a_0 represents the initial value of a at time zero and τ is a time constant with the value $2n/AT_e$.

The equation shows that the ribbon thickness approaches exponentially toward the equilibrium thickness T_e. For float composition at 1050°C the values of n and A are such that τ is approximately 40 sec, which is in good agreement with practical experience.

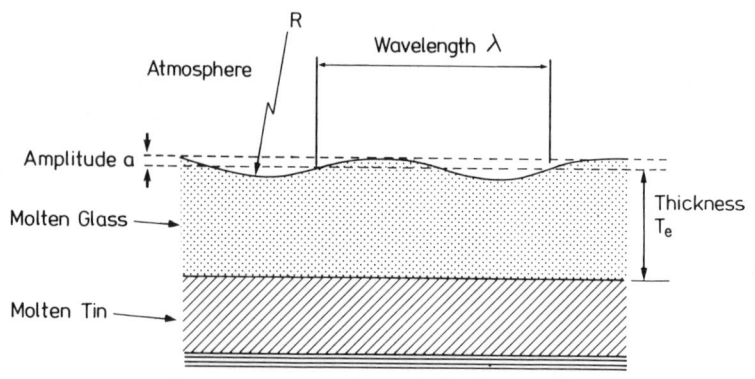

FIG. 16. Idealized distortion of a float ribbon.

3. Gravity and Surface Tension Forces and Flatness

Consider a floating ribbon of equilibrium thickness having an idealized corrugation* of its upper surface as pictured in Fig. 16. The amplitude a of the corrugation is assumed to be small relative to the wavelength λ and the thickness T_e. Two systems of forces are acting to remove the corrugation and to flatten the ribbon, surface tension, and gravity.

Looking first at the action of the surface tension forces. From geometry we see that $R = \lambda^2/32a$. From elementary surface tension theory we can see that there is an increase in pressure beneath the convex cyclindrical corrugation of magnitude S_{ga}/R and a decrease in pressure of the same magnitude beneath the adjacent concave corrugation. Thus due to surface tension, there is a differential hydrostatic pressure of magnitude $2S_{ga}/R$ or $64aS_{ga}/\lambda^2$ acting to move glass from the thicker to the thinner regions of the ribbon.

Now look at the gravitational forces which are also acting. Between the thicker and thinner parts of the ribbon there is a differential hydrostatic head of $2ag\rho_g$ acting to equalize thickness.

Thus the combined surface tension and gravitational forces acting to equalize thickness amount to

$$((64S_{ga}/\lambda^2) + 2g\rho_g)a.$$

* Such an example of is not representative of actual practice since distortions in top and bottom surfaces are not independent. The first consequence of introducing a positive distortion on the upper surface is a simple vertical flow which reduces the top surface distortion by creating a corresponding negative distortion on the bottom surface. The initial redistribution between top and bottom surfaces takes place very quickly and then is followed by horizontal viscous flow to make the thickness uniform and both top and bottom surfaces plane. In looking at the forces which promote this viscous flow it is simpler and entirely adequate to consider the single-surface distortion case.

These two components are equal for a corrugation of wavelength approximately 22 mm.

This implies that for corrugations of wavelength greater than 22 mm gravity is the important corrective agent, while for shorter-wavelength corrugations surface tension becomes of overriding importance.

4. Optical Distortion

Optical distortion whether in reflection or transmission is almost entirely the result of waviness of the surface. In transmission both surfaces play a part and distortion can be associated with variation in thickness. In our idealized examples (Fig. 16) it will be seen that the thicker parts of the ribbon form convex cylindrical lenses of positive power $(\mu - 1)/R$ and the thinner parts concave cylindrical lenses of corresponding negative power; viewing an object through these contiguous lens elements creates the distortion. Our analysis shows that the corrective action is directly proportional to the lens power; i.e., the greater the defect the greater the corrective force. This is a remarkably useful property and accounts for the great success of the float process in producing the highest quality of glass.

D. Float Ribbon Formation

In the previous section we saw how the float effect operates to produce a pool of glass which is completely flat and uniform in thickness. We go on to examine what happens as the glass is then transported down the bath by the traction applied by the lehr. What are the forces involved, how do they interrelate, and how can conditions be controlled to produce a final ribbon of the exact thickness and width required by the market?

A number of mathematical models (Narayanaswamy, 1977, 1981) have been developed which have been used to derive the stress at any point in the ribbon and the corresponding longitudinal stretch, lateral contraction, and change of thickness and to show how these factors vary with output, ribbon speed, and applied cooling gradient. For a nonspecialist the mathematics is complex and the numerical solution calls for considerable computer facilities. The technologist operating a float plant can, however, gain a very useful understanding by applying the simple concepts set out below.

It is useful to begin by listing the controls available to the operator to aid in the production of a ribbon of the required thickness and width. They are (1) output—volume rate of glass flow through the bath, (2) lehr speed; (3) control of the viscosity gradient down the length of the bath by the use of coolers or heaters, and (4) the use of various devices in contact with the ribbon to provide mechanical restraint—examples are edge rolls and top rolls. Other aspects which affect ribbon formation including the bath dimensions are fixed by the bath design and are generally not available as operating variables.

1. Production of Ribbon of Specified Thickness and Width

The glass flowing down the spout onto the bath of molten tin spreads out to form a pool of near equilibrium thickness and is then carried down the bath as a ribbon by the traction of the lehr. At the hot end of the bath the viscosity is low enough for the ribbon to respond to any tensile tractive force which may be present by undergoing longitudinal stretch, lateral contraction, and thinning as described in Section II. At the cold end of the bath the viscosity is so high that no significant stretch or thinning takes place and the ribbon moves forward with no change in form or dimensions.

By setting the lehr speed the operator determines only the motion of the ribbon as it leaves the bath. The level and distribution of the tractive force upstream depends on the conditions within the bath. How the motion imparted to the ribbon at the exit end is transmitted back up the bath is determined by the pattern of stretching of the ribbon. Where the ribbon is stiff and there is no stretching the ribbon velocity remains constant. In regions where stretching is taking place the ribbon velocity decreases as we move back up the bath. The pattern of stretching is determined by two factors: (1) the change in viscosity along the bath and (2) the level of tractive force at various points along the bath length.

The operator can within limits impose the viscosity gradient of his choice by introducing overhead water coolers or by activating electric heating panels provided at various points in the bath roof. By establishing a viscosity gradient, the operator has some control over the pattern of tensile stress along the length of the ribbon. Tensile stress in the ribbon in response to the motion imparted by the lehr is generated by the reaction to (1) inertial forces due to the acceleration of the ribbon, (2) tin drag as the ribbon imparts motion to the tin, and (3) reactive forces generated at points of physical restraint—for example, at the spout area where the ribbon is in contact with bath-end refractories or where as in variant 2 the ribbon is gripped by edge rolls.

2. Ribbon Formation at Low Output: Equilibrium Float

At low outputs—1000 tonnes/week—and low ribbon speeds the reactions generated by ribbon acceleration and tin drag are so low as to be negligible. The only significant reaction force is provided by the viscous attachment to the bath end wall refractories. This reactive force is low and only of the same order as the gravitational and surface tension forces. Because there are no acceleration or tin drag reactions the tensile force is uniform down the length of the bath and stretching will therefore be concentrated at the hot end where the viscosity is lowest. We have already seen how at this part of the bath the surface tension/gravitational forces operate very effectively to maintain equilibrium thickness, so in this situation if the ribbon stretches this results only in narrowing of the ribbon with no reduction in thickness.

To reduce the ribbon thickness at such low outputs it is necessary to introduce a physical restraint to provide a reaction to the stretching forces at a point in the bath where the viscosity is so high that the surface tension/gravity equilibrating effect is ineffective. It is also important that these tensile forces not be transmitted back up the bath to the region where the ribbon viscosity is low. How this is achieved in practice is illustrated diagramatically in Fig. 17 and described below as variant 1.

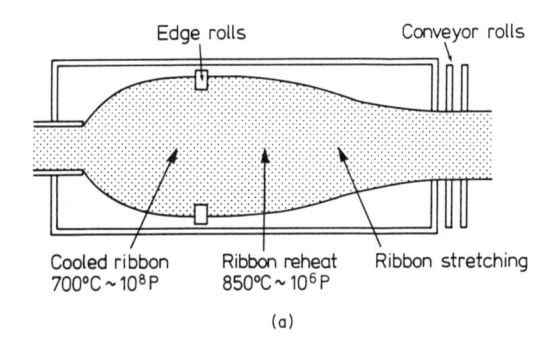

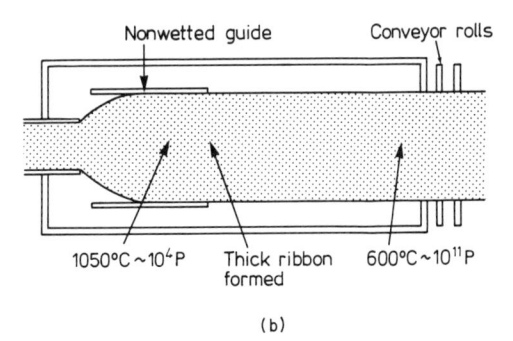

FIG. 17. (a) Float variant 1, the manufacture of thin float at low load; (b) float variant 4, the manufacture of thick float.

3. Variant 1

This is the modification which first enabled high-quality thin (3-mm) float glass to be produced. The main elements of this process are the following:

(1) Glass flows onto the bath and spreads to form an equilibrium ribbon of nearly the full width of the bath.

(2) As the equilibrium ribbon is moved forward, overhead water coolers are used to cool the ribbon so that by the time it reaches the edge roll position the viscosity is in the region of 10^8 P.

(3) At this position the edges of the ribbon are gripped by knurled edge rolls rotating at a speed appropriate to maintain the desired ribbon width upstream. The cold glass lying between the edge rolls effectively forms a stiff beam capable of providing the reaction to downstream stretching forces.

(4) Downstream of the edge rolls the ribbon is reheated by means of overhead electric heaters to lower the viscosity to a point where the ribbon can be stretched by the tractive force provided by the lehr to produce a sheet of the thickness desired.

(5) No attempt is made during stretching to maintain ribbon width and so the distortion quality is not sensitive to glass homogeneity to anything like the same extent as in the flat drawn processes. After stretching the ribbon is cooled in the normal way and passes into the lehr.

One advantage of variant 1 is that it can be operated effectively over a wide range of outputs and can be used to produce thin glass of very high distortion quality. The procedure is somewhat complex and the use of edge rolls to grip a relatively cold ribbon demands great care in operation. The cost of providing for reheating is not negligible.

4. Ribbon Formation at Higher Outputs

As output increases the inertial reaction generated by ribbon acceleration increases. This reactive force is not uniform along the length of the bath but builds up where the ribbon is stretching and increases as we move down the bath. Figure 18a, which refers to variant 2, illustrates how this reactive force increases.

Of greater significance is the increase in the reactive force provided by the tin drag, which increases as the square of ribbon speed. Again this is a force which increases as we pass down the bath as indicated in Fig. 18a. By choosing an output which is sufficiently high and by setting up the correct pattern of viscosities we can generate tensile stresses of sufficient magnitude to stretch the ribbon at viscosities for which the surface ten-

2. FLAT GLASS MANUFACTURING PROCESSES

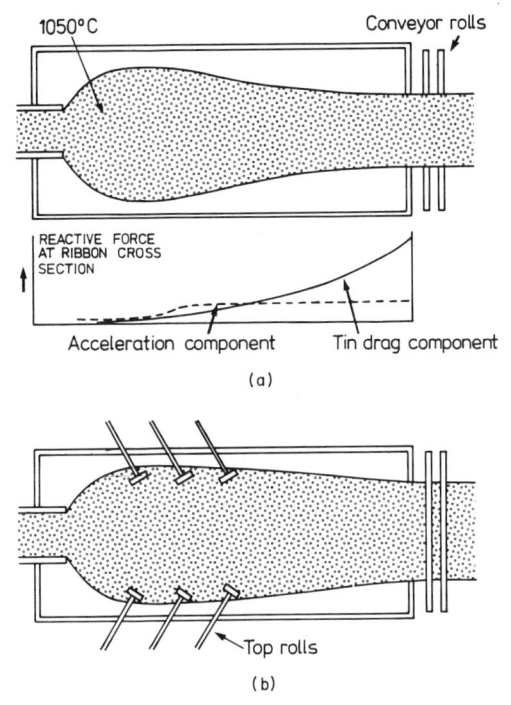

FIG. 18. (a) Float variant 2, the manufacture of thin float at high load; (b) float variant 3, the manufacture of a wide ribbon at high load.

sion/gravity thickness equilibrative effect is negligible. Under these conditions the effect of stretching is to reduce ribbon thickness and ribbon width equally, and this procedure forms the basis of variant 2. By introducing top rolls it is possible to reduce the narrowing of the ribbon and to concentrate the effect of stretching on thickness reduction, and this is the basis of Variant 3.

5. Variant 2

At outputs above 3000 tonnes/week it becomes possible to generate sufficient reactive force from ribbon acceleration and tin drag to create tensile forces of sufficient magnitude to stretch and thin the glass in a region where the viscosity is high enough for the float equilibrating effect to be negligible and to avoid transmitting undesirably high tensile force back to the hot end of the bath without the need to introduce any physical restraints such as edge rolls. This is the basis for variant 2, which is illustrated diagrammatically in Fig. 18 and comprises the following elements:

(1) In variant 2 the operator is setting out to produce a ribbon of required thickness and width. The starting point is to fix the plant output at an appropriate value.

(2) The operator then decides on the sizing and disposition of the water coolers to procure the appropriate temperature gradient down the bath.

(3) Assuming that the correction dispositions have been made by setting the appropriate lehr speed, the operator then produces glass of target thickness and width.

(4) Clearly, considerable experience is called on to determine appropriate settings, which tend to be specific to the design of bath used. The target thickness is determined by lehr speed, and if then the ribbon width is off target or there are any other difficulties, adjustments can be made to the number, shape, and placing of the coolers.

Variant 2 has important advantages. There is no additional edge loss from roller grip, and provided care is taken in placing the water coolers the distortion quality is high. Once established the process is simple and operationally reliable and is particularly suited to the production of glass 4, 5, and 6 mm thick. In any given bath it is not always possible to find settings to produce all the widths of ribbon called for over a range of outputs. For example, at the lower end of the output range there can be difficulty in producing thin glass in sufficient ribbon width and at the upper end of the output range there can be problems in producing a narrow ribbon of 6-mm float.

6. Variant 3

As the float process developed and the output range extended increasing emphasis came to be placed on high cutting efficiency, particularly in close matching of ribbon width to market requirements. In these circumstances benefit could be seen in introducing an additional control on ribbon width, and this led to the development of variant 3.

Variant 3 is an extension of variant 2; it is illustrated diagrammatically in Fig. 18b. An array of driven toothed disks called top rolls is introduced on either side of the ribbon over the region where stretching takes place. The teeth imprint and grip the ribbon edge, and by inclining the axis of rotation of the top rolls to a greater or lesser degree it is possible to restrain the ribbon from narrowing as it is stretched. It is clearly important to have very close control over the angling and the speed of rotation of each top roll. This introduction provides in effect an independent control over ribbon width, and if properly used serves to overcome the difficulties that can arise with variant 2. Once established variant 3 is operationally reliable. The main disadvantage over variant 2 arises from the additional edge loss produced by the top-roll imprint.

7. Variant 4

Variant 4 differs from the earlier variants in that it is directed to the production of glass of greater than equilibrium thickness. By its introduction it became possible to produce float glass of the highest quality up to a thickness of 25 mm. The layout and equipment used in this variant of the float process are illustrated in Fig. 17b. Glass from the forehearth flows as usual down the spout onto the tin bath and then spreads freely until coming in contact with a line of nonwetting guides introduced along either side of the bath. The temperature is maintained at approximately 1050°C for a time sufficient for the float effect to produce a ribbon of high surface quality and of uniform thickness which, under these circumstances, is determined by the mass flow, the distance between the guides, and the ribbon speed. The guides are set accurately parallel to give the width of ribbon required and are long enough to allow the ribbon to be cooled sufficiently that on emerging from the guides the thickness and width are stably maintained. In spite of the apparent complexity associated with the introduction of the guides the process in normal operation is very stable and trouble-free.

E. CHEMICAL ASPECTS OF THE FLOAT PROCESS

So far we have directed our discussion to the physics of the float process, showing how high-quality glass of the required width and thickness is produced. The chemistry of the process is no less important, and any failure to meet the exacting standards demanded inevitably results in the production of reject glass.

1. The Properties of Tin

Molten tin is the metal chosen to support the glass ribbon in the float process. Pure tin provides a combination of properties uniquely suited to the process. Despite extensive search no other pure metal or alloy with suitable properties has been identified and the use of tin is unlikely to be superseded. There are potential problems with the use of tin. For example, at the temperature of the float bath it oxidizes rapidly in the presence of oxygen, and impurities in the tin can have serious adverse effects. The relevant properties of tin are summarized in Table V.

2. Preventing Oxidation

To prevent oxidation of the tin, a protective atmosphere of nitrogen and hydrogen is provided inside the bath, which is carefully sealed to

TABLE V
Properties of Tin

Property	Value
Density	6500 kg/m^3 (at 1050°C)
Melting point	232°C
Boiling point	2623°C
Vapor pressure	Between 1×10^{-3} and 1×10^{-4} mm Hg (at 1027°C)
Surface tension	
Tin–atmosphere	0.5 N/m (at 1050°C)
Tin–glass	0.5 N/m (at 1050°C)

prevent ingress of atmospheric oxygen. The protective atmosphere is fed to the bath at a number of locations in the bath roof and the bath pressure is maintained positive with respect to atmosphere.

3. Oxygen and Sulfur as Impurities in the Tin

At the temperatures of the float bath there is little interaction between pure tin and glass, and the quantity of condensable vapor emitted is negligible. The situation is very different if the tin is even slightly contaminated with sulfur or oxygen. The increase in vapor pressure resulting from contamination of tin by either element at the level of 10 ppm is shown in Table VI. Where the impurity is present stannous oxide and/or stannous sulfide are emitted into the bath atmosphere and condense to form a lower-melting-point tin-rich compound on the cooler parts of the bath roof space which may build up and fall back in particulate form onto the top surface of the ribbon to produce the defect known as "tin speck." The specks though small are surrounded by a distortion halo and can cause rejection of areas of the ribbon so affected. The contamination must be kept to a minimum, but even so some degree of condensation is inevitable in the roof of a long-running float bath, which can lead to outbreaks of tin

TABLE VI
Oxygen and Sulfur Contamination

Impurity in tin	Composition of vapor	Tin content in vapor (mg/m^3)
Pure tin	Tin	0.3
Oxygen (10 ppm)	Stannous oxide	3
Sulfur (10 ppm)	Stannous sulfide	100

speck when the bath roof temperature is changed. To keep the problem within tolerable bounds and to avoid rejection resulting from tin speck it may prove necessary at intervals to clean the bath roof.

4. Other Problems

Two other adverse effects can arise from the presence of oxygen as an impurity in the tin. First, because of increased interaction the undersurface of the glass ribbon may take up tin to form a thin surface layer rich in stannous tin. The effect on the glass sheet is not immediately apparent, but if the sheet is subsequently reheated for bending or tempering the stannous tin is oxidized. The change from stannous to stannic tin is accompanied by an increase in volume and the tin-rich layer expands to give a minutely wrinkled surface which shows as a faint bluish haze and can force rejection. This defect is known as bloom and is avoided by ensuring that oxygen contamination is kept sufficiently low.

The second difficulty arising from oxygen contamination is a consequence of the change in the solubility of oxygen in tin with temperature shown in Table VII. The drag of the ribbon as it moves down the bath causes a general circulation of tin within the bath, and tin from the hot end is transported down to the cold end. At the hot end the tin takes oxygen into solution and this oxygen may be precipitated out as stannic oxide dross at the cold end. If extensive such precipitation can be troublesome and the dross may accumulate to become trapped and caught up on the underside of the ribbon. This can in itself cause rejection of the glass, but in addition small lumps of dross may adhere to a lehr roller and cause persistent trouble by imprinting onto the ribbon as dross marks for a long period of time.

Contamination cannot be entirely avoided. Oxygen contamination arises not only from air ingress into the bath as a consequence of deficient sealing but may be introduced as a low-level impurity either in elemental form or as water in the nitrogen/hydrogen atmosphere. The glass entering the bath contains some available oxygen and sulfur which may be evolved as water vapor or hydrogen sulfide from the top surface of the ribbon or dissolve directly into the molten tin from the bottom surface. The effect of

TABLE VII

SOLUBILITY OF OXYGEN IN TIN

Temperature (°C)	600	800	1000
Solubility (ppm by weight)	5.4	95	630

sulfur as an impurity is confined to its volatilization as stannous sulfide and subsequent condensation as a source of tin speck.

In the early days of float production there was concern that in spite of the best control measures that could be applied impurities would build up to unacceptable levels in the course of a long campaign. Fortunately experience has belied these fears, and, provided a strict control is kept on the purity of the atmosphere supply and on the standard of the bath seal, the rejection of glass for bloom, tin dross, and top speck can be kept to an entirely acceptable low level.

The tin remains in the bath throughout a campaign and it is normally not important to exercise special control of the impurities introduced in the tin when it is fed into the bath at the start of a campaign. Commercial tin is supplied to a high level of purity and any contaminants introduced are normally eliminated in the course of the first few hours of the campaign.

F. The PPG Industries Modified Process

In 1975 PPG Industries announced a modified arrangement of the float process which has been described and analyzed by McCauley (1980). McCauley describes the main structural differences from the Pilkington process as the following:

(1) Replacement of the free-fall spout and the flow separation produced at the wet back area by a full-width lipstone dipping into the bath tin.

(2) In the Pilkington process the glass at the hot end of the bath flows unhindered to form a flat, uniformly thick pool which extends close to but away from the bath sides. In the PPG process the glass is immediately in contact with special guides which form the sides of the bath at the hot end.

One operational difference is that in the PPG modification the final width remains constant for all glass thickness whereas in the original process the ribbon can be adjusted to meet final product requirements and hence minimize cutting waste.

VI. Glass for Radiation Control

A. New Requirements

Historically the main function of glazing was to admit light and keep out weather. In recent years a number of new factors have arisen which extend these requirements:

(1) a rising insistence on amenity and comfort in the home, the workplace, and the automobile;
(2) increased use of air conditioning in buildings and autos;
(3) a steep rise in the relative cost of energy;
(4) a general increase in the glazed area employed; and
(5) a recognition of the important contribution that glazing can make to the architecture and appearance of a building.

The most important new requirements that arise are outlined in the sections that follow.

1. Solar Control

In the sunnier countries where air conditioning is common in buildings and automobiles glazing should have the ability to exclude the sun's heat (infrared radiation) while admitting as much of the sun's visible light as is desired. In some cases the highest possible visible light transmission is sought; in other cases a tinted glass is specified.

2. Minimizing Heat Loss at the Window

In cold climates it is important that expensively generated room heat not be lost through windows by conduction and radiation. The air space in double glazing is effective in reducing loss by simple conduction, but in clear glass units there remains some heat loss by radiation across the air space from the inner to the outer pane. This loss can be significantly reduced by providing a coating on the inside of the outer pane which has a high reflectivity—low emissivity—for long-wavelength infrared radiation while retaining a high transparency for shorter-wavelength visible light.

3. Aesthetic Appeal

Glass is a main feature of many modern buildings which otherwise have little other architectural appeal. The use of tinted glass or glass with a colored or metallic reflective coating can transform the outward appearance of a building. Giving such architectural prominence to the glazing carries associated demands. The glass must be flat and free of distortion when glazed and must be uniform in color and reflectivity.

B. Useful Measures of Performance

In this short account it is not possible to treat in depth the underlying physics of radiation-control glasses. The reader is referred to excellent accounts in the literature (Bamford, 1982; Gillery, 1982).

It is useful to note two measures that are commonly used to specify the performance of radiation-control glass.

1. Shading Coefficient

This measure is used to specify solar control performance and is defined as the ratio of the solar heat admitted by any given window glazed with a solar-control glass to that admitted by an equivalent window glazed with clear 3-mm glass.

2. U and K Values

These measures are both used to specify the heat insulation performance of a window. Both measure the total heat transfer across a window from inside to out under defined conditions of wind velocity. In the United States the U value is used the units being British thermal units per square foot per degree fahrenheit. In Europe the K value is used, the units being watts per square meter per degree centigrade. The American and European standards for wind velocity differ, but a rough conversion between the two units is given if K value is taken as five times the U value.

C. Processes for the Production of Radiation-Control Glass

1. Body-Tinted Glass

The first approach to radiation-control glass lay in the production of a green-body-tinted glass containing 0.5 wt. % iron oxide. Typically such a glass at 6-mm thickness would transmit 74% of the sun's light and 47% directly of the sun's total energy. These green-body-tinted glasses have been followed by a range of heat-absorbing glass in gray, bronze, neutral blue, etc., produced by the addition of varying amounts of nickel oxide, cobalt oxide, chromium, and selenium to the basic iron oxide colorant. Typical properties are shown in Table VIII.

Tinted glass is normally used as single glazing and in this form it has some disadvantage in that the glass itself warms up as it absorbs the sun's rays and some secondary heat is therefore reradiated to the room. Notwithstanding, heat-absorbing body tints continue to retain an important and valuable part of the radiation-control glass market and can be the base glasses used in coated products.

2. Coating by Wet Chemical Processing

The silvering process for the production of mirrors is the prototype for wet chemical processing. Thin, highly reflective, partially transparent metallic films can be produced on glass by using technology and equipment similar to those used in silvering to reduce solutions of metallic salts and deposit them on the glass. Films of gold, copper, and cobalt are commonly produced in this way. Although simple in concept and equipment the films produced by the wet process are not sufficiently hard or scratch resistant to be used on exterior surfaces and are normally employed on the inner surface of a double glazing unit.

3. Pyrolytic Methods

For many years tinted glass has been produced by spraying organometallic compounds in air onto hot glass at around 500°C. The organo compound breaks down to deposit a hard adherent film of metallic oxide on the glass surface. Coatings of iron, cobalt, nickel, chromium, titanium, and vanadium oxides or a combination of these have been produced. Pyrolytic processes can be low in cost and lend themselves to on-line production of coated glass, particularly in the float process. Uniformity and durability can be a problem. With oxide coatings the main reflection is in the visible spectral region and not in the near-infrared so desirable for radiation-control glass. The properties of pyrolytic coatings are also listed in Table VIII.

Pilkington has developed a variant of the normal pyrolytic process to produce a hard and highly reflective coating of elemental silicon on glass within the confines of the float bath. Silane gas (SiH_4) is passed over the ribbon while it is still in the reducing atmosphere of the float bath. The silane gas is reduced to deposit a resistant uniform film of silicon onto the upper surface of the float ribbon. The film has a semimirror appearance and is reflective to infrared radiation.

4. Vacuum Coating Technique

A wide range of vacuum deposition processes has been applied to produce thin films on glass. These include vacuum evaporation, vacuum sputtering, and the new highly efficient magnetically enhanced vacuum sputtering. The equipment ranges from relatively simple jobbing vacuum units for small-scale production to large, complex, and costly units capable of coating the largest sizes of float glass on a continuous basis. After much development the vacuum coating technique has been improved to the stage where it can produce a wide range of radiation-control products. The range covers combined metal–dielectric coatings of various light transmission/solar heat transmission properties and color, films of selected oxides or fluorides which combine high visible transmission with low emissivity in the infrared, and conducting films which allow the glass to be heated electrically. Vacuum-deposited films can be made of high uniformity but are generally too soft to be used externally in glazing. Their high infrared reflection and low emissivity properties make them particularly suitable for use as the inner surface of a double glazed unit.

5. Production of Thin Films in the Float Process

The support of the glass on molten tin in the float process provides a ready-made opportunity for the production of a colored film by ion exchange. In the electrofloat process a fixed copper bar electrode situated halfway down the bath is wetted onto the ribbon across the full width by using a 2% copper–lead alloy. When a suitable dc circuit current is passed from the bar electrode (anode) to the bath tin (cathode) ions from the lead alloy pass into the top surface of the ribbon. As the ribbon passes down the bath these ions interact with the bath atmosphere, are reduced, and agglomerate to form color centers in the ribbon and produce a bronze tint

TABLE VIII

Process	Single (SG) or double[a] (DG) glazing	Glass	Light transmittance (%)	Light reflectance (%)
Body tint	SG	6-mm Antisun Green[b]	75	6
	SG	6-mm Antisun Bronze[b]	50	5
	SG	6-mm Antisun Gray[b]	41	5
	DG	6-mm Antisun Green[b]	65	10
	DG	6-mm Antisun Bronze[b]	44	7
	DG	6-mm Antisun Gray[b]	36	6
Wet process	DG	6-mm PPG Solarban 550-20 Clear	20	18
Pyrolitic	SG	6-mm Reflectafloat[b]	33	43
	DG	6-mm Reflectafloat[b]	29	43
	SG	6-mm Glaverbel Stopsol	42	32
	DG	6-mm Glaverbel Stopsol	38	34
Vacuum coating	SG	6-mm Suncool Silver[b] 20/34	20	23
	SG	6-mm Suncool Bronze[b] 10/24	10	19
	SG	6-mm Suncool Blue[b] 30/39	30	16
	DG	6-mm Suncool Silver[b] 17/25	17	23
	DG	6-mm Suncool Bronze[b] 9/16	9	19
	DG	6-mm Suncool Blue[b] 26/29	26	17
Electro-float	SG	6-mm Spectrafloat	51	10
	DG	6-mm Spectrafloat	44	12

[a] Double glazing properties are for the glasses glazed with 6-mm clear float glass as the inner pane and an air space of 12mm.

[b] Pilkington Products.

which has considerable metallic reflection. The electrofloat film is completely integral with the glass and is highly abrasion resistant and suitable for use in external glazing. By varying the conditions other colors than bronze can be produced. The original electrofloat process has been modified and developed by using a shaped electrode and a pulsed dc current to produce a brightly colored repetitive pattern on the glass.

D. The Properties of Radiation-Control Glasses

Table VIII sets out a summary of the properties of a selected range of radiation-control glasses. There is a much wider range on the market and

PROPERTIES OF RADIATION-CONTROL GLASS

Direct solar heat transmittance (%)	Solar heat reflectance (%)	Solar heat absorption (%)	Total solar heat transmittance (%)	Shading coefficient	K value (W/m² K)
46	5	49	61	0.70	5.4
44	5	51	60	0.69	5.4
44	5	51	60	0.69	5.4
36	6	58	49	0.56	2.9
34	7	59	47	0.54	2.9
34	7	59	47	0.54	2.9
15	15	70	27	0.31	2.6
43	28	29	52	0.60	5.4
34	29	37	43	0.50	2.9
50	26	24	56	0.64	5.4
42	29	29	49	0.56	2.9
16	18	66	34	0.39	4.4
6	21	73	24	0.28	4.3
21	18	61	39	0.45	4.6
13	18	69	25	0.29	2.6
5	21	74	16	0.19	2.4
17	18	65	29	0.34	2.7
54	10	36	65	0.75	5.4
42	12	46	54	0.62	2.9

there is every probability that new and improved types will be developed in the future. Some of these glasses are used singly glazed, while others are used as part of a double glazed unit.

Acknowledgments

Acknowledgment is made to the Director of Research and staff in Pilkington, particularly H. Charnock and D. Gelder for help and criticism in preparing the manuscript, and to Corning Glass Works for information on their fusion process, with which the author has no first-hand association.

References

Avery, C. W. (1930). *Glass Ind.* **11**(4), 75–78.
Bamford, C. R. (1982). *J. Non-Cryst. Solids* **47**, 1–20.
Barker, T. C. (1977). "The Glassmakers," pp. 274–278. Weidenfeld & Nicholson, London.
Bessemer, H. (1846). British patent 11,317.
Chance, J. F. (1919). "History of Chance Brothers." Spottiswoode, Ballantyne, London.
Douglas, R. W., and Frank, S. (1972). "A History of Glassmaking." G. T. Foulis, London.
Gillery, F. H. (1982). *J. Non-Cryst. Solids* **47**, 21–27.
Goerk, H. (1962). *Inf. Glastech.* **5**(1), 1–8.
Le Mare, E. B. (1924). British patent 216,586.
McCauley, R. A. (1980). *Glass. Ind.* **61**(4), 18–23.
McGrath, R., and Frost, A. C. (1961). "Glass in Architecture and Decoration," pp. 52–56. Architectural Press, London.
Mason, F., and Conqueror, J. (1884). British patent 13,119.
Narayanaswamy, O. S. (1977). *J. Am. Ceram. Soc.* **60**, 1–5.
Narayanaswamy, O. S. (1981). *J. Am. Ceram. Soc.* **64**, 666–667.
Pilkington, L. A. B. (1965). British patent 216,586.
Pilkington, L. A. B. (1969). *Proc. R. Soc. London, Ser.* **314**, 1–25.
Pilkington, L. A. B. (Sir Alastair) (1976). *Glass Technol.* **17**, 182–193.
Pilkington, L. A. B., and Bickerstaffe, K. (1953). British patent 769,692.
Pond, L. M., and Lewis, J. H. (1929). U.S. patent 1,824,365.
Scoville, W. C. (1948). "Revolution in Glassmaking." Harvard Univ. Press, Cambridge, Massachusetts.
Takashashi, S., and Ichinose, M. (1980). *Glass Ind.* **61**(4), 18–23.
Timoshenko, S., and Goodier, J. N. (1951). "Theory of Elasticity," pp. 1–28. McGraw-Hill, New York.
Turner, W. E. S. (1930). *Proc. Inst. Mech. Eng.* **119**, 1077–1127.
Vincent, G. L. (1960). *Ceram. Ind.* **75**(3), 100–105.
Wendler, A. (1927). *Glass Ind.* **8**(10), 231–238.

CHAPTER 3

Container Manufacture

*George W. Keller**

TECHNICAL DEVELOPMENT DEPARTMENT
GLASS CONTAINERS CORPORATION
FULLERTON, CALIFORNIA

I. History	107
II. Furnaces	114
A. Furnace Operation	117
B. Refractories	119
III. Forming Machines	125
IV. Raw Material and Glass Analytical Procedures	127
A. Glass Analyses	128
V. Coatings	129
A. Hot-End Coating	129
B. Cold-End Coating	131
VI. Pollution Control and Its Effect on Industry	133
A. Air Pollution	133
B. Water Pollution	135
C. Hazardous and Toxic Wastes	135
D. Solid Wastes	136
E. Conclusion	136
References	136

I. History

Glass and glass products, as we know them today, have become an essential part of present-day living. However, it may not be generally realized that glass and glass products have been in continuous development since 1500 B.C. A brief summary of glass and glass manufacturing prior to our primary discussion will afford a better appreciation of today's glass container manufacture.

Artifacts, retrieved over the years, indicate that the first glasses used by man were the result of natural phenomena and were fashioned into

* Present address: Container General Corporation, Chattanooga, Tennessee.

tools and ornaments. One of the two sources of this type of glass was volcanic eruption, which, depending upon the cooling cycle of the molten rock, would form glass or crystallize into a rock mass. The second source was lightning, which, striking sandy soils, would form fused-silica glass, imperfectly melted and of varying lengths and shapes.

It was during this time frame (between 2000 and 500 B.C.) that the first hollow glass objects were made by man. This primitive process involved the dipping, or coating, of a sand, or soft clay core, attached to a rod, into a crucible of molten glass. The core material was removed, leaving a hollow container. The size of these containers was such that they were used solely for perfume and cosmetics.

The historical archives are devoid of information pertaining to glass, or its manufacture, from 500 B.C. to the fourth century B.C., when technologies for the production of beads, hollow ware, and flat ware were established. Glass globes, used for water storage in the home, were also made during this period, and these containers replaced the ceramic jars that had previously been used for this purpose.

The discovery of the blow pipe around 100 B.C. allowed the oil and wine industry to replace ceramic jars with glass for storing and transporting these products. The blow pipe also allowed the discovery of off-hand blowing and the use of cup molds for the purpose of forming ware.

The demand for blown containers continued to grow during the rapid expansion of the Roman empire, primarily because of the wide variety of designs and sizes of glass bottles. The ware produced was utilitarian in nature, the cost reasonable.

The majority of the containers produced up to, and during, this period had round bottoms that required the use of metal and wooden stands for stability. The development of metal and wooden molds enabled artisans to produce bottles with flat bottoms which were free-standing.

The period from the fifth to the fifteenth century A.D. has been recognized as the Venetian or artistic era of glass. It was during this time that high degrees of manufacturing skill and glass formulation were developed. Better understanding and improved manufacturing skills of pressing, drawing, and blowing marked the beginning of today's mechanization of glass manufacture. With the exception of chemical etching, all of the decorative processes now known were used, or developed, during this period.

The Venetian industries reached their peak in the fifteenth and sixteenth centuries, and the processes developed during that era were eventually carried into Northern Europe. Glass products were used for both their artistic and utilitarian properties, and had become the accepted material for fine tableware for the wealthy. The common bottle was not in widespread use in Europe, as it was confined to the packaging of medicines and cosmetics. With the development of civilization in Egypt came the early use of glass for magnification and burning. Primitive scientific developments such as a simple microscope and sealed thermometer provided the impetus and wherewithal for the development of additional scientific instruments which would explain, to a greater degree, the mystique of glass and glass manufacture. Only within the last century have the projects of European scholars, followed by American scientists, explained many of the complexities of glass, i.e., their chemical and atomic structure.

Over the centuries, thousands of glass formulations have been developed utilizing oxides, carbonates sulfates, and nitrates of silicon, sodium, potassium, calcium, magnesium, alumina, and boron. Since the early 1900s, rare earths have been developed to modify glass compositions for the control of the working range.

Present-day container manufacture is the direct result of several major factors. The mechanical revolution, beginning about 1800, produced molds in which threads could be incorporated on the finish of a jar or bottle. When this had been accomplished, airtight closures were developed for the purpose of sealing glass containers. Subsequently, in the late 1800s and early 1900s the development of semiautomatic and automatic machines, using the press and blow, and blow and blow processes, enabled greater productivity. These developments resulted in the availability of a greater variety of shapes and sizes of containers, which encouraged more demand for the product. During this time, bottles were utilized for the packaging of mineral water, soda water, and milk, expanding from there to the bottling of beer and wines on a commercial basis.

Sieman's development of the regenerative-type furnace in the late 1800s allowed manufacturers to produce tons of melted glass per day rather than being limited to the production of pounds of glass per day, melted in crucibles (day tanks). The glass industry, as a whole, has slowly evolved from very primitive artisanal operations, through periods of basic understanding, refinement, and development, to reach today's state of the art.

Today the glass container industry, in the United States, encompasses 28 companies operating over 108 plants in 30 states, producing a combined glass tonnage in the range of 13.1 million tons. Approximately 70% of the production is flint glass, 20% amber, 7% green, and the remainder miscellaneous colors.

The following discussion is based on the published standard definition of commercial glass products as perceived by the American Society for Testing Material, through its Committee C-14.

Glass—An inorganic product of fusion which has cooled to a rigid condition without crystallizing.

(a) Glass is typically hard and brittle and has a conchordal fracture. It may be colorless or colored, transparent or opaque. Masses or bodies of glass may be colored, translucent, or opaque, by the presence of dissolved, amorphous, crystalline material.

(b) When a specific kind of glass is indicated, such descriptive terms as flint glass, barium glass, and window glass should be used following the basic definition, but the qualifying term is to be used as understood by trade custom.

(c) Objects made of glass are loosely and popularly referred to as glass, such as glass for a tumbler, a barometer, a window, a magnifier, or a mirror.

It should be noted however, that it is an accepted fact that not every noncrystalline product of fusion is a commercial glass.

A typical glass container formula consists of silica sand, alumina, high-calcium limestone, and/or dolomite, soda ash, and sulfate. The formula of any commercial container glass must meet the criteria of the forming process while at the same time satisfying the requirement for the end use of the finished product.

The oxides and their functions in such formulations are used as follows: (1) glass-forming oxides—silicon dioxide and boric oxide make up the bulk of a container composition which gives the glass its characteristic ability to become stable in the vitreous state; (2) glass modifiers—soda oxide and potassium oxide function as fluxing agents in the glass melt and facilitate the melting of silica (SiO_2) at lower operating temperatures; (3) stabilizing oxides—alumina, calcia, and magnesia stabilize the sodium silicate and give the glassy phase the ability to resist moisture attack. They are also used to control the viscosity of the melt.

The following is a range for a typical formulation used by the container industry:

SiO_2	69–73%
Al_2O_3	1.0–4.0%
Na_2O	12–15%
K_2O	Trace–1%
CaO	10–12%
MgO	Trace–4%
SO_3	1%

Today's glassmaking process begins with the development of a series of chemical and physical specifications for the various raw materials. Chemical specifications for major oxides of a given raw material are generally considered to be working ranges but not absolute limits (see Table I). The range of each oxide may vary depending upon the working properties of the glass to be melted, the nature of the material deposits, and the supplier's processing capabilities.

Modern mining techniques, incorporated with improved raw material processing methods (flotation and electromagnet systems, etc.), have al-

TABLE I
Typical Chemical Specifications for Glass Raw Materials[a]

High-silica sand	SiO_2	99.5% min.
	Fe_2O_3, TiO_2	0.03% max.
	Al_2O_3, CaO, MgO, Na_2O, K_2O	±0.05%
	Cr_2O_3	0.001% max.
Alumina sources	SiO_2, CaO, Na_2O, K_2O	1.0% range
	Al_2O_3	18–23% min., ±0.05%
	Fe_2O_3	0.10% max.
Feldspathic sand	SiO_2	±0.5%
	Fe_2O_3	0.05% max.
	Al_2O_3, Na_2O, K_2O	0.5% range
Natural carbonates	SiO_2, Al_2O_3	±0.5%
(limestone and dolomite)	Fe_2O_3	0.10% max.
	CaO, MgO	±0.5%
Soda ash	Na_2O	58.0% min.
	NaCl	0.03% max.
	Na_2SO_4	0.20% max.
Sulfate sources		
Saltcake	Na_2SO_4	99.0% min.
	Fe_2O_3	0.15% max.
	NaCl	0.5% max.
Gypsum	$CaSO_4 \cdot 2H_2O$	95.0% min.
	Fe_2O_3	0.5% max.
Barytes	$BaSO_4$	93.0% min.
	Fe_2O_3	0.5% max.
	SiO_2	5.0% max.
Cullet (soda–lime)	SiO_2	66–75%
	Al_2O_3	1–7%
	CaO + MgO	9–13%
	Na_2O	12–16%

[a] From Kephart and DeNapoli (1981).

lowed suppliers to meet container manufacturers' specifications for both chemical and physical properties on a consistent basis. At the same time, the supplier has improved yield and efficiency.

Consistency in the supplier's product, both in chemical and physical properties, has in turn offered container manufacturers the opportunity to modify their batch formulations and improve both melting efficiencies and energy savings.

Should the chemical consistency or the physical sizing of the raw materials fail to be maintained, the whole melting process, as well as glass properties, would be adversely affected. The working properties of glass, i.e., density and viscosity, are very sensitive to these changes and could result in quality problems in the finished product.

For continued assurance of product quality, the suppliers are required to submit physical and chemical analyses of their product on a monthly basis. Also, every plant is required to submit a random sample of each material to the control lab for analysis.

Minor contaminants are critical and are limited by specification. For instance, iron and chrome, in small amounts, impart a green color to the glass, and refractory heavy minerals, i.e., zircon, sillimanite, kyanite, do not melt, and if larger than 60 mesh will produce stones in the finished ware.

The ideal physical size of raw materials used in container manufacture varies with material and deposit. However, the accepted limits fall between U.S. Standard 20 mesh and 140 mesh, with a trace of product on 200 mesh screen.

The beginning of any glass manufacturing operation is in the batch house. Any errors made in the unloading of raw materials, in the weighing of the batch materials, or to a lesser degree, in the mixing operation, will assuredly result in a glass quality problem.

As the demand for glass container production increased, the post–World War II batch houses were designed to automate those areas involving unloading, weighing, and mixing raw materials, thereby improving the efficiency and reproducibility of the operation.

The configuration of these batch houses has necessarily varied. Of primary consideration, when planning a new batch house, are the number, type, and quantity of raw materials to be handled and stored. Also of prime importance is whether the material will be delivered by truck, rail, or a combination of the two. Once these details have been resolved, the reality of space constraints and economics will determine the size and shape of the structure.

In most instances, the batch houses are multistory structures with the raw material storage located above the weighing devices. The raw materials are discharged into the mixer, with gravity transporting the raw materials from storage to the mixing operation. Cement, steel, and tile are the standard materials used in the construction of modern batch houses.

The raw materials are delivered to the plants by either truck or rail and before being unloaded are verified as the material specified in the bill of lading. A visual inspection is then made to determine if there has been any in-transit contamination. Then, as the materials are unloaded, representative samples of the material are taken and checked for physical sizing. These samples are then retained for approximately one week and are used as a material check should a glass upset occur. Upon completion of these precautionary steps, the material is ready to be transferred into the storage system.

3. CONTAINER MANUFACTURE

The majority of the unloading systems are programmed, or electronically keyed, so that each function is verified before the next function can be operated. For instance, the distributor head, located on top of the storage silo, will confirm through a series of limit switches that the distributor head has reached a predetermined position; then the bucket elevator can be energized, followed by the unloading conveyor, etc. Conversely, the system is also programmed so that the unloading conveyor will stop first, then the elevator, etc., to ensure that the system is purged after the material has been unloaded. It is with this type of control equipment that unloading contamination and the discharging of material into the wrong storage bins has been minimized.

As the raw material is drawn from either the rail car or truck, it enters an unloading hopper which, in turn, empties into a screw conveyor or onto a belt conveyor. During this period of unloading the dry materials, precautions have been taken to minimize ambient dust by the use of enclosed conveyors and employment of dust collection systems. A bucket elevator then receives the material from the unloading conveyor and it is lifted to the top of the batch house to be discharged into the storage silos. The materials are then in position to be weighed into batch formulations.

Predetermined batch formulas are programmed into a variety of systems allowing the batch material to be weighed in a sequential manner. In a single weigh hopper configuration, a single material is weighed and when the scale tolerance is met it proceeds with the weighing of the next material. In those instances in which a single weigh hopper is located beneath each raw material bin, the materials are weighed simultaneously, with the weighing of each material checked to assure correct weight. In this process, the weighed batch material is discharged into an enclosed conveyor, starting with the silica sand followed by soda ash, limestone, etc. By loading the belt conveyor in this manner, a layering effect is obtained, improving mixing. Most systems are designed so that when a misweighing of material occurs, the entire system will stop and an alarm sound. Only when the alarm condition has been corrected and/or accepted can the operation continue.

Upon completion of the weighing operation, the material is discharged into a hopper over a batch mixer or onto a belt that transports the material into the batch mixer. Although various types of mixers are presently used in batch systems, the majority are of the pan variety. Once the material has been received in the mixer, water and wetting agents are added to the batch as it is being mixed. Wetting the batch material at this point minimizes material segregation during handling and eliminates dusting of the material being delivered to storage behind the furnace. Caustic soda has also been used as a wetting agent, performing the same function as the water. Another advantage of wetting the batch is the phenomenon of partially dissolved alkali etching the sand grains, improving melting capabilities. As the wet batch is charged into the furnace it frits over quickly, thereby minimizing the amount of raw material that can normally be carried over into the checker area of the furnace via products of combustion as they enter the furnace. Again, caustic will perform the same function.

The mixed batch is either delivered to storage bins located behind the furnace or to individual batch cans that are subsequently delivered to the furnace. Both the bins and the delivery systems are designed to provide a steady flow of mixed batch material to the furnace with a minimum of segregation. High–low indicators located in the storage bins are employed to keep the bins full at all times, thereby minimizing the distance drop of raw material into a bin. The low bin indicators are utilized to alarm and then automatically start the batching system so that an optimum of mixed batch is maintained at the furnace at all times.

Once the material has been mixed and delivered to the furnace, it is ready to be charged into the furnace via a number of systems. Over the years, sideport furnaces have been equipped with a series of four or five screws, positioned in the back wall of the furnace. There have been furnaces designed with two pusher-type systems using two doghouses, also located in the back wall of the furnace. A commercial charging system allows charging into a triangular doghouse located on the center line of the furnace, utilizing two pusher-type chargers. Blanket chargers have been used in conjunction with an open doghouse charging the materials beneath a shadow wall. All of these batch charging systems are controlled by a glass level indicator located on the refiner end of the furnace. The glass level is automatically maintained as the indicator calls for the charger to start or stop, depending upon the measured level.

The endport furnaces utilize pusher-type chargers located on one or both sides of the furnace. In most cases, electric furnaces are charged with a belt conveyor that has been engineered to move both perpendicular and parallel to the melter and completely covers the melter with a uniform batch pattern. It is important that all charging systems be designed so that they are capable of producing and maintaining a uniform batch pattern over long periods of time.

II. Furnaces

The original regenerative-type furnace was designed by Sir Karl Wilhelm Siemans and patented for use in the steel industry. It was adapted for use in the glass industry in the late 1800s. With the exception of a few minor design modifications, the furnace remains relatively unchanged. Regenerative-type furnaces are the most commonly used in the glass industry except for the all-electric furnaces that have been built in low-cost electric power areas. The all-electric melting furnaces are in limited use in the glass container industry for several reasons: (1) their melting capacity is restricted to approximately 100 tons/day; (2) their melting cost per ton of glass is at least double that of the conventional furnace—this expense will vary with the cost of power; (3) although less of a problem today, the change in furnace pull is limited in any given 24-h period. These furnaces are used more effectively in the melting of special, i.e., lead, borosilicate, and low-alkali glasses.

The glass container industry uses two types of regenerative furnace. One is described as a sideport furnace, the second as an endport furnace. The furnace shown in Fig. 1 is a sideport furnace, having four ports located on either side of the melting chamber. However, the number of ports can and does vary from three to seven depending upon the physical size of the furnace. The endport furnace (Fig. 2) allows only two ports, located on the backwall of the furnace.

The placement of the regenerators depends on port location. A sideport furnace requires them to be located on either side of the melter whereas the endport furnace requires that they be located on the backwall.

Figure 2, showing an endport furnace, illustrates a single-pass regenerator; however, there are a number of endport furnaces in operation today using double- and triple-pass regenerators. In the case of a single-pass

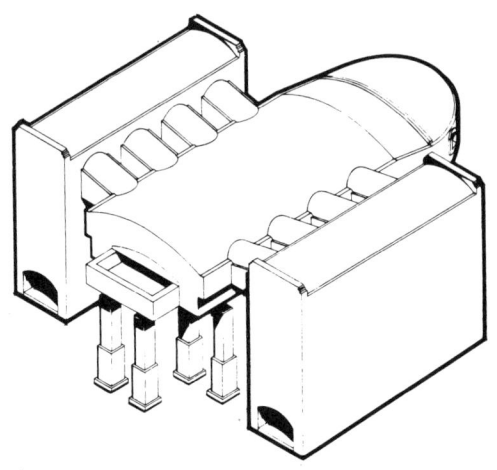

FIG. 1. Typical side-fired box-type regenerator glass furnace. (From Kaiser Refractories, Kaiser Center, Oakland, California. Reprinted with permission.)

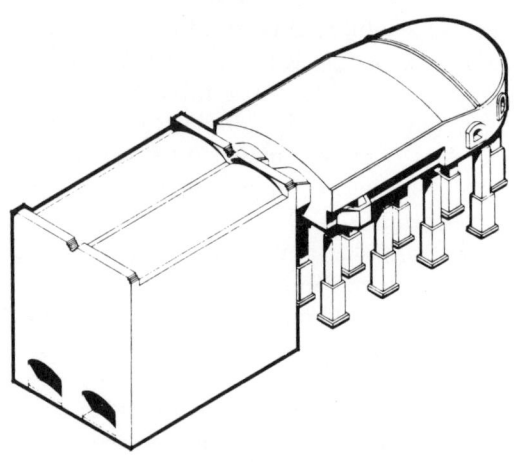

FIG. 2. Typical end-fired box-type regenerator glass furnace. (From Kaiser Refractories, Kaiser Center, Oakland, California. Reprinted with permission.)

regenerator, the products of combustion enter at the top of the checkers, flow downward, and exit through the canal to the stack. In multiple-pass regenerators, the flow is again downward but rather than exiting to the stack, the flow is redirected upward then downward to the stack.

Another heat recovery system that is sometimes used in conjunction with primary regenerators, is that of secondary regenerators. These regenerators are smaller than the primary structures and are located apart from the primary regenerator. The secondary regenerator is located over the canal and receives the flow of hot gases from the canal through the bottom of the structure. The gases then flow upward through the reversing valve and forced-draft stack located on top of the regenerator.

Even though there are noticeable differences in design, physical size, and operation, both the sideport and endport furnaces utilize regenerators for fuel economy; a melter for converting raw material into the liquid phase, employing time and temperature; and a refiner where the hot molten glass from the melter is thermally conditioned and the gaseous inclusion reabsorbed into the glass as it cools, before the glass enters the forehearth.

A submerged throat is used in both furnaces to convey the molten glass from the melter to the refiner. These areas of the furnace (melter and refiner) are separated by a series of refractories called the bridgewall.

Forehearths are shallow refractory channels, equipped with firing and cooling capabilities, through which the glass flows to the forming machines. The forming operation is almost entirely dependent upon thermal control. Since the forehearth affords the last opportunity to impart thermal control, its design and operation are paramount to this phase of the operation.

The combustion process for the two types of furnaces are similar. Combustion air enters the furnace through a reversing valve into a canal connecting the reversal valve with the bottom of the regenerator. The air then passes up through preheated brick (checker) into the ports, where gas is introduced under controlled conditions. The product of combustion continues to flow over the glass melt and exits through the ports into the regenerator located on the opposite side of the furnace. The hot gases then flow downward through the checkers into the canal through the reversing valve, and exit in one of several ways, through a natural draft stack, Venturi stack, or hot fan. This cycle, firing from one side of the furnace to the other, is automatically controlled on a time basis and varies from 15 to 30 min, depending on the combustion air preheat temperatures. The average air preheat temperature varies between 1600 and 2300°F, depending on the size and design of the individual furnace. It is an accepted fact that a well-designed regenerative-type furnace will require approximately 35% less fuel than a furnace with no regenerators.

The melting area of a sideport furnace varies from 500 to 1200 ft^2 and is capable of melting up to 345 tons of glass per day, using the design criteria of 3.5 ft^2 per ton of glass melted. Endport furnaces utilize smaller melting chambers, varying from 350 to 600 ft^2, with the capability of melting up to 175 tons of glass per day, using the same design criteria.

The majority of today's furnaces are designed to use natural gas as the primary fuel, with low-sulfur number 6 or number 2 oil as the alternative fuel. Recent shortages of natural gas prompted many glass manufacturers to provide for the use of propane to fire forehearths and lehrs while firing oil in the furnace.

Both the sideport and endport furnaces can and do utilize electric boost for increased production and thermal control of the melt. The size of the furnace and the tonnage requirement over and above the melting capacity of the conventionally fired furnace dictate the size of the boost system to be used. Boost systems vary in size from 500 to 3000 kV A. The electrodes used are 2 in. in diameter, made of molybdenum, and inserted through water-cooled holders in the sidewall of the melter. They are located 14–15 in. above the melter bottom and fire from one side of the furnace to the other.

A. Furnace Operation

The operation of any given furnace will differ slightly from another in a number of particulars. However, there are a number of operating practices that are always followed for efficient operation, regardless of the variables.

Time and temperature are the basic requirements for melting glass. The longer the time the lower the temperature required, and vice versa. Beyond that is the need to establish proper thermal currents created by convection within the molten glass. These thermal currents can be the result of furnace design, firing systems, glass composition, and the given pull of the furnace; however, they can be controlled.

The thermal currents are the result of temperature gradients between the bridgewall and the charging end. The gradient curve on the sideport furnace will differ with the number of ports involved. However, for good operation, the backwall temperature (2700–2750°F) will always be lower than the bridgewall temperature (2790–2850°F). For an endport furnace the same is true; however, due to the firing and exiting of gases through one port the temperature differential is less than in a sideport furnace. It is very important that the same gradient from the charging end to the bridgewall be the same on both firing sides of the furnace. With imbalanced gradients there would be an uneven batch pattern which could ultimately result in a batch stone problem.

The thermal gradients are set up originally by using an optical pyrometer, sighting through the ports in a sideport furnace or through observation holes in the breastwall of an endport furnace. Because of the rapid cooling of the furnace when fires are turned off, only one reading per reversal can be obtained. Once the gradient has been established, crown thermocouples, bottom thermocouples, and bridgewall radiamatics will help maintain the gradient on a day to day basis.

Proper combustion is a must in maintaining optimum thermal gradients; therefore, a program for taking O_2 analyses on a routine basis should be established so that the fuel-to-air ratio set point is as accurate as possible.

Several factors contribute to the melting capabilities of today's furnaces. One is better control of raw materials and mixing operations; another is the programmable controls now used in the batch house operation, which allows a more uniform batch to be delivered to the furnace, the results being improved melting at lower temperatures and greater fuel reduction. Improved furnace design, i.e., port design, adjustable burners, deeper furnaces, and added insulation, have contributed to better operation. Analog instrumentation and recordings have allowed the operator to follow and make adjustments to the operation when a trend has been determined.

Production tonnage is now being scheduled to optimize furnace melting capabilities. This scheduling has improved both furnace and forehearth operation by minimizing the thermal forays created by load variations. In the future programmable controls for both the furnace and forehearth operations will bring great benefits. The pulls on both the furnace and individual forehearths are such that it is next to impossible for an individual to monitor and make timely adjustments to the control of the operation. A digital control system has the ability to continuously monitor a greater number of sensors programmed to sound an alarm when certain limits have been reached, or exceeded, thereby warning the operator of impending problems. The operator can then react to correct the situation.

These new controllers will allow the use of control loops which can work separately, or in a cascading fashion, whereby the output of one loop can be fed into another. This is ideal for the sensing of the temperatures throughout the process, with the final control being at the forehearth and feeders.

B. REFRACTORIES

Both sideport and endport furnaces use the same type of refractories in comparable parts of the furnace. The following areas and types of refractories are commonly used.

1. Melter

   a. Bottom. Normal practice employs laminated construction 15–27 in. thick, consisting of one or more courses of 6-in. dry-pressed clay block against bottom steel plus several layers of 3-in.-thick high-fired superduty brick plus ½-in. zircon ram mix plus 3-in. zircon brick plus either 3-in. Zirmul or 3-in. AZS pavers. Expansion may be provided for by slip joints in top paving courses.

   b. Sidewalls. Ten-inch AZS regular cast blocks (34% zirconia) with improved material, e.g., void-free 41% zirconia for high-wear areas such as electrode blocks, throats, and doghouse corners.

   Insulation practices vary with furnace depth, pull rate, and type glass being melted. Materials used range from 1 in. to several layers of 28-lb density ceramic fiberboard applied directly to sidewall blocks to a sandwich structure of 2-in.-thick alumina insulating block backed up with 1–3 in. of ceramic fiberboard.

   c. Doghouse. Various batch charger configurations dictate doghouse designs ranging from blanket chargers with suspended backwalls, Emhart chargers with triangular doghouses, and Gana-type chargers with semirectangular doghouses.

   Glass contact refractories are normally 10-in., 34% AZS with 41% corner blocks and void-free casting. Two-piece corner blocks are common with 41% VF in the top half and 34% RC underneath for cost savings.

The same insulation practices as in melter sidewalls are generally used, but with vertical joints exposed to prevent joint leakage in case of block movement during heatup.

d. Tuckstones. Tuckstone thickness is usually 6–8 in., with 34% AZS, normally cast on the cold face. Depending upon tuckstone length, this may require a deep- or end-cast block to keep casting scar away from sidewall or breastwall joints.

Solid-cast blocks are used in high-wear areas, such as opposite the doghouse in endport furnaces, where batch piles frequently impinge on the wall area.

e. Bridgecovers. Normal thickness is 8 in. of regular cast blocks in one piece spanning from melter to refiner on straight bridgewalls; on tapered bridgewalls, two-piece construction is required.

A solid support, e.g., piers of hard brick at joints (with insulating brick in between), is usually necessary to prevent bridgecover cracking and potential shadow wall tilting.

f. Breastwall. Breastwall construction on both endport and sideport furnaces is similar, using 9–12-in.-thick 41% AZS material, insulated with firebrick and/or ceramic fiberboard.

A major design consideration is to ensure adequate steel support for the breastwall blocks. Sidewall blocks wear thin, thus providing no structural support. Proper design ensures breastwall stability, with no dependence upon sidewalls.

Double steel angle support is common practice for the tuckstone and breastwall, ensuring that the load of breastwall rests securely on steel, and is not balanced between steel and sidewall blocks, thus avoiding breastwall failure when sidewalls wear out.

g. Backwall. Melter backwall designs vary in sideports with the type of doghouse used, and in endports, with port design.

Endports require special attention to guard against loss of support, either from sidewalls tilting during heatup and moving the backwall away from its support steel or from sidewall wear later in a campaign, resulting in inadequate steel support. This problem is the result of moving sidewall blocks toward buckstays to increase melter area, resulting in clearances between block cold face and buckstays of 6–8 in. To allow space for overcoating, support steel must project out no more than 3–4 in. If any insulation is used between backwall blocks and buckstays, the backwall block weight may be shifted too far forward to allow stability when sidewall wear occurs.

An additional support course, notched for a large support angle, may be used above the tuckstone; a joint between tuckstones and sidewall blocks and careful attention to paving expansion (i.e., expansion in top paving courses just in front of sidewall blocks) are added precautions against backwall structural problems.

On heavily pulled endport tanks, the normal backwall construction is of 9–15-in. AZS with varying thicknesses of insulation, normally fiberboard.

On sideports with triangular or Gana-type chargers, AZS is normally used around the doghouse opening for mantle blocks, jack arches, corner tuckstones, etc. The remainder of the backwall may be partial AZS, silica, or a veneer of zircon and silica, or superduty.

h. Crown. Melter crowns generally are of a bonded construction of 12–13½-in. thick silica brick (thicknesses of 15–18 in. may be used in larger tanks) plus additional layers of various insulating materials. The most common insulating practice is to use two courses of silica insulating brick covered with approximately 1–2 in. of lightweight insulating castable. The use of a ceramic fiber blanket over a buffer course of silica insulating brick is also quite prevalent.

2. Refiner

a. Bottom and Paving. Bottom construction is generally the same as melter construction except that zircon or fused Alpha Beta alumina is frequently used as a top paver.

b. Sidewalls. Sidewalls are generally 8–10 in. thick with 2–3 in. of insulation, such as Insulflux, Zedsulite, or ceramic fiberboard, etc.

Refiner depths are usually 36–42 in.

c. Forehearth Entrances. Standard one-piece U-shaped channels are most commonly used because three-piece construction is more susceptible to joint attack. Oxidized or conditioned AZS material is now used interchangeably with fused Alpha Beta alumina.

d. Breastwall. A typical refiner breastwall construction is of 12–12½-in. silica, with the use of insulation dependent on refiner operating temperature, furnace pull rate, and forehearth length.

e. Shadow Walk. Construction techniques vary greatly from company to company and furnace to furnace. Silica brick 12–15 in. thick is commonly used in gas- and/or oil-fired sideports. Zircon or bonded AZS brick can be used in thicknesses of 12–15 in.; occasionally fused Alpha Beta is used. Some furnaces are constructed with the shadow walls projecting to within 6 in. of the melter crown, others with the shadow wall projecting through the crown to separate the melter and refiner atmosphere. In these cases, a refiner stack is combined with a level bay for the purpose of locating the level probe out of the forehearths.

f. Refiner Crown. Configuration may be semicircular, rectangular, or separate rectangular. The materials used are the same as for the melter crown, but with less insulation to facilitate heat losses where high temperatures are a problem.

g. Throat and Sump. Throats are susceptible to high wear and are fairly inaccessible for repairs; therefore, the best AZS material available is generally used. Solid cast 41% AZS is usually used for cover blocks and facers, particularly on the melter side. Refiner facers and sump blocks may be downgraded to regular cast 34% AZS.

Chrome-bearing fused cast material is commonly used in throat and sump areas of furnaces melting colored glass and certain types of fiberglass. Many combinations of materials are used in throat bottoms; these include 12-in.-thick clay flux, 27-in.-thick laminated insulating block clay, fiberbrick paved with 3-in. zircon, bonded AZS, or DC6 pavers.

3. Ports

Port neck designs are rather basic, the size determined by fuel flow rate, burner block location, and type of fuel being used. The length of the port neck is determined by the melter–regenerator relationship.

A recent trend in the industry regarding endports has been to angle the ports toward the melter center line to reduce flame impingement on the breastwalls and, in some cases, provide more uniform exhaust air distribution in the regenerators.

a. Bottom. The bottoms usually consist of 3-in. DCL tile, or bonded AZS paving, over two courses of superduty brick, over ceramic fiberboard. Hard brick piers support the port wall loads.

b. Sidewalls. Sidewalls are predominantly built with 6–9-in. 34% AZS, with 2–3-in. ceramic fiberboard, or $4\frac{1}{2}$ in. of insulating fiberbrick.

c. Regenerator Entrance. The regenerator entrance is normally built with all 34% AZS, but can be 50% direct-bonded basic or bonded AZS.

d. Burner Blocks. In initial installations burner blocks are normally made of 34% AZS. For hot replacement, different materials may be used with varying degrees of success, such as bonded AZS, unfired bonded AZS grain, unfired sillimanite, fused silica, or fused Alpha Beta alumina.

4. Regenerator

a. Walls. Wall thickness usually averages 18 in. plus varying thicknesses of mineral fiberboard, or a sprayed-on lightweight insulating mix, or a combination of both. A typical construction would be 9 in. of 50% MgO direct basic bonded or 40% MgO brick plus $4\frac{1}{2}$-in. 2300 insulating fiberbrick with 3-in. mineral board, and $\frac{1}{2}$-in. finishing cement or a hard-coat castable. A continuous row of 18-in. tie bricks are generally placed every 5 ft for six courses, including $\frac{3}{16}$-in. steel tie plates, which are anchored back to the buckstays with a sliding connection.

Forty-percent MgO bricks are not recommended for use where temperatures exceed a mean of 2300°F because of the danger of subsidence. Fifty-percent direct-bonded MgO, or comparable higher-strength brick, is used at temperatures above 2300°F. It is common practice to spray or trowel a thin layer of insulating castable, or cement, over the mineral fiberboard after heatup, for protection against air infiltration. Another common practice is to use a combination of basic key brick and cardboard to controlling thermal expansion and hot-face growth due to chemical alteration of the brick surface.

b. Rider Arches. Rider arches are normally 6–9 in. wide, using 12- or $12\frac{1}{2}$-in.-long bricks in series. The arches are laid up with air-set mortar and extra-strength castable to minimize cuts over the arch brick during the leveling course installation. Normally high-fired superduty brick is used in rider arches, with the occasional use of high-alumina brick. Direct-bonded basic brick is currently being applied on some furnaces with secondary and multipass checkers where higher than normal flue temperatures exist.

c. Crown. The normal practice in crown construction is to use 12 or $13\frac{1}{2}$ in. of 50% direct-bonded MgO brick with as much insulation as is cost effective. There are temperature limitations when using 40% MgO brick due to the danger of subsidence above the 2300°F mean temperature in the basic brick. There have been many cases of 40% MgO crowns slumping because of too much insulation.

The usual practice in insulating the crown is to use two 3-in. courses of insulating fiberbrick, 2 or $2\frac{1}{2}$ in. of lightweight insulating castable, and $\frac{1}{2}$ in. of hard-coat castable (to act as a seal to prevent batch dust penetration). Some crowns are then covered with aluminum or stainless foil, as added protection against batch dust and/or penetration. With proper batch wetting and attention to operating practices, a direct-bonded basic crown is expected to last through several furnace campaigns.

d. Checkers. Refractory developments in the past 30 years have generated a shift from all-clay checker brick, through a combination of top basic and clay brick, to the present use of all-basic checkers. Improved refractories coupled with batch wetting practices has resulted in longer checker life and greater regenerator efficiency. Clay checkers, by the nature of their composition and glassy bond, presented a wet surface which was susceptible to attracting, and holding, solid batch carryover prior to extensive batch wetting practices. Basic checkers are dry by nature and have less affinity for the lesser amounts of carryover present in today's operating furnaces.

A typical basic checker setting for a container furnace is comprised of five to seven courses of 95% MgO unburned brick, over fifteen courses of 90% MgO unburned brick, over ten courses of 90% MgO burned brick, and over fifteen courses of 50% MgO direct-bonded brick. The total number of courses, and the distribution by composition, varies with the regenerator height and temperature profiles. An open basket-weave setting utilizing 12 or $13\frac{1}{2} \times 4\frac{1}{2} \times 2\frac{1}{2}$-brick is predominantly used; however, there are many variations, from 9-in. open batt settings to $18 \times 6 \times 3$-in. four-point settings. The brick supporting the checkers above the rider arches, such as rider tile, support course, and transition course, traditionally have been high-fired superduty brick. The improvement in checker life in recent years has been prompted by improved refractories capable of withstanding thermal cycling and occasional sodium sulfate buildup, prevalent at the base of the checkers. Chrome alumina, 50% direct-bonded MgO, bonded AZS, and other materials are being tried with varying degrees of success as an improvement over the high-fired superduty brick now being used.

e. Secondary Regenerators. Many furnaces today are equipped with a forced-draft stack and reversing system, mounted above a second, but smaller, regenerator system used for additional heat recovery. Exhaust gases and combustion air pass over a second checker setting which alternately absorbs and emits heat to and from the waste gas and combustion air. The system is believed to reduce fuel consumption by an additional 10–14% in the average furnace. The bricks normally used in the checker area are a combination of first-quality brick over superduty brick. The walls of the regenerator are normally constructed of 9-in. first-quality brick with $4\frac{1}{2}$ in. of insulating firebrick, plus 3 in. of mineral fiberboard.

f. Forehearths. The trend from narrow and deep forehearths has gradually changed to ever increasing widths (from 16, 26, 36, and 48 in.) and generally a 6-in. glass depth. The need for thermally conditioned glass at the higher pull rates is responsible for these changes. Modular superstructure refractories are a recent development, simplifying the refractory shapes and eliminating, or reducing, the multitude of sizes. Refractories normally used in forehearths vary from channels of oxidized, or conditioned, fused AZS, to fused Alpha Beta alumina and bonded AZS. The superstructure is comprised of cast sillimanite or mullite, and various types of 2300°F insulating materials.

III. Forming Machines

The first fully automated glass-container-forming machine was developed in 1903 by M. J. Owens. The Owens machine was a large unit, 14–15 ft in diameter, cam operated, made up of a central nonrotating section and a rotating section. The rotating section, depending on its size, was equipped with 10–15 molds. The machine picked up the glass to be formed from a large shallow pot 10–13 ft in diameter with 3–6 in. of glass depth. As the molds, attached to hangers on the rotating section, passed over the edge of the pot holding the molten glass they dipped just below the surface of the glass, then a vacuum was applied to the mold and the glass entered the mold cavity. As the nose of the mold emerged from the glass, a knife severed the excess glass. As the finish was being formed a plunger made a small indention in the finish. This series of functions completed the forming of a parison. The parison was then transferred to a blow mold, where it was blown into its final shape. The bottle was then removed from the machine, and the process started over. The Owens machine produced, for the first time, bottles which were uniform in size, weight, contour, capacity, and finish dimensions.

In 1922 Hartford-Empire developed a gob feeder with a reciprocating plunger, thereby allowing the corresponding development of automatic gob fed machines, i.e., O'Neill, Lynch, Miller, Roirant, Knox WD, and the Hartford IS (individual section) machine.

The majority of the containers manufactured today are being produced on IS machines (Fig. 3). Individual sections (IS) means that the machine is made up of independent sections each of which produces a finished product. These machines are comprised of six, eight, or ten sections lined up side by side. As each section is a straight-line operation, any one, or more than one, section can be immobilized for repair or replacement without interfering with the rest of the machine operation. Each individual section is air operated and controlled by a timing drum driven by an electric motor through a variable-speed reduction unit which also operates the feeder mechanism. Each section of an IS machine can be equipped, moldwise, to handle single-, double-, triple-, or quadruple-gob operation.

FIG. 3. IS (individual section) machine. [From Grif Duncan, *GCC Forming Handbook,* Glass Containers Corporation, Fullerton, California (1967).]

The IS machine has the capability of making glassware by two distinct methods, the "blow and blow" process (Fig. 4) and the "press and blow" process (Fig. 5). In the blow and blow process, air is blown into the blank mold when the parison is partially formed. The parison is then transferred to the blow mold, where air is again used to blow the parison into its final shape. In press and blow operation the glass is dropped into the blank mold, then a plunger is inserted, forming the parison, then the parison is transferred to the blow mold, where air is used to blow it into its final shape.

The quadruple ten-section IS machine is capable of producing as many as 400 bottles per minute.

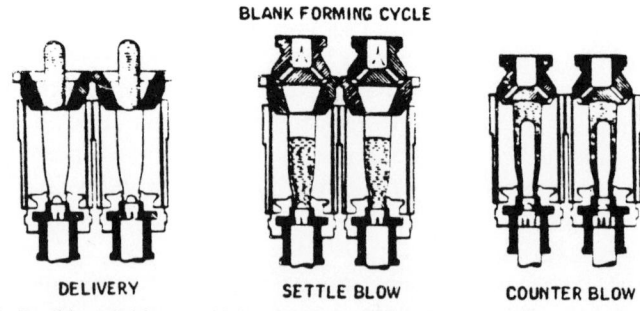

BLANK FORMING CYCLE

DELIVERY   SETTLE BLOW   COUNTER BLOW

FIG. 4. Double-gob blow and blow process. [From Grif Duncan, *GCC Forming Handbook,* Glass Containers Corporation, Fullerton, California (1967).]

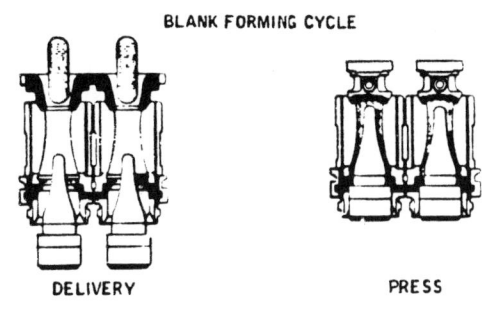

FIG. 5. Double-gob press and blow process. [From Grif Duncan, *GCC Forming Handbook*, Glass Containers Corporation, Fullerton, California (1967).]

IV. Raw Material and Glass Analytical Procedures

The raw material samples are collected at the unloading station and forwarded to the central laboratory for physical and chemical analyses. Unless a particular problem is suspected with an individual sample, complete analyses are routinely performed on composite samples.

Analyses of raw materials are warranted because of their affect on the melting and forming operations. The assurance of acceptable quality is attained through the employment of a strong analytical program.

Physical testing for particle size distribution and purity involves mechanical screening, flotation, and magnetic separation processes. Magnetic separation is a check for contaminate-bearing ferrous metals. Flotation, using heavy liquids, is a method of testing for refractory-type contaminates.

Sieve analysis is probably the most frequently employed quality check of raw materials. Simplified screening procedures could be a function of the material unloading practices where only two or three screens would be necessary to determine if the material sizing is within acceptable limits. Thorough sieve analyses are normally performed in an area of the central laboratory and the procedures should follow ASTM or vendor guidelines.

The following equipment is required for raw material particle size determination: a mechanical sieve machine, such as Roto-Tap, manufactured by the W. S. Tyler Company; a set of U.S. Standard or Tyler equivalent screens, timer, scales, or balance, and a cleaning brush. A typical set of screens would include U.S. Standards No. 20, 30, 40, 70, 100, 140, and 200, lid and pan.

Composition analysis of raw materials is somewhat more sophisticated and the frequency of this analytical method should be based on experience with particular materials. The composition of raw materials can be determined by wet chemical methods, by x-ray fluorescence, or by atomic absorption.

Raw materials of known composition are weighed, processed, and charged into the furnace where melting occurs and the glass is prepared. It is important to the glass manufacturer that raw material processing and melting maintain consistency for quality production of a particular glass composition. Various methods of determining glass oxides, homogeneity, and color have been established for manufacturing, control, laboratory, and research and development purposes.

A. Glass Analysis

Routine glass composition checks are performed in the glass production facilities by daily density determinations. The density of a glass is a function of its chemical composition, and since density measurements are very accurate, a slight composition change can be easily determined. Density measurements also indicate differences due to annealing; therefore, glass samples for density measurement should follow identical annealing procedures.

The sink–float density comparison is the most rapid and accurate technique for determining densities and density differences of glass. Density measurements of glass samples are made during production, at 8-h intervals or as necessary, depending on individual situations. Density measurements should be repeated when density differences exceed control limits.

A complete analysis of routine production glass is performed on a weekly, bimonthly, or monthly basis for control purposes. More frequent analysis is required for research projects, controlled composition variations, or production abnormalities.

Glass oxide constituents can be analyzed by x-ray fluorescence, atomic absorption, or wet chemical methods. Complete glass composition analysis can be obtained most rapidly by x-ray fluorescence techniques. Wet chemical analytical methods require considerably more time and attention.

Following sample preparation, the results of complete glass composition analysis by x ray fluorescence can be obtained in approximately two hours. X-ray fluorescence equipment may include a multisample holder for the purpose of analyzing several unknown compositions simultaneously. Approved sample preparation procedures must be strictly followed to ensure a homogeneous sample representative of the unknown composition. The glass must be weighed, melted into a "button," ground, and polished.

The physical properties of a glass are determined for use in control tools, developmental studies, research projects, and to provide the glass manufacturer with values pertinent to controlled modifications, consistency, and quality of glass production. Thermal shock, chemical durability, viscosity, and optical properties (color, purity) are frequently employed as routine quality-control procedures.

V. Coatings

After the glass-container-forming process, wherein the desired stable shape and size has been attained, the container is then transferred through a series of processes that strengthen the glass for vendor use and, eventually, the consuming public.

A. Hot-End Coating

Immediately after the container is formed, it is passed through a tunnel or hood coater apparatus containing stannic chloride (anhydrous tin tetrachloride) vapor. Other materials of approximate quality and purpose are also available and their application may vary. The vapor is applied by means of four spray nozzles, two on each side of the hood, positioned at an angle to provide complete coverage on the container body, with minimum coating to the container opening or the finish surface. This process takes place while the glass is still hot. Depending upon the size and weight of the container, a temperature range of 800–900°F is necessary to complete the chemical reaction between the glass surface and coating material.

Hot-end coating materials are not necessarily applied to all types of glass containers; however, when applied, a very small quantity of coating material on the glass provides an invisible and protective surface which reduces its susceptibility to damage throughout the remaining processes and in trade use.

The quantity of coating material and its location on the glass container is subject to quality-control specifications and procedures. The coating thickness is determined in coating thickness units (CTUs) or, perhaps, other descriptive values using one, or several, coating thickness detectors. The detectors employ a reflective light principle with a resultant thickness dimension display.

Exiting the hot-end coating phase, the containers are readied for stress removal by alignment equipment immediately prior to entering an annealing chamber.

1. Stacker

The containers are immediately conveyed to a unit which transfers them into the annealing chamber (lehr). The transfer unit (stacker) may be of several makes and descriptions, however, its purpose is always to introduce the ware in an efficient and predetermined order into the lehr. The stacker transfer assembly must be sturdy, dependable, readily adaptable to various container sizes and spacings, and able to function without damage to the product. The successful transfer of the ware into the annealing chamber begins the very critical process of preparation for eventual use.

2. Annealing

Strain develops in the glass as a result of the forming processes and the rapid decrease in glass temperature. This stress must be relieved in the glass before it can be handled safely. Stress is removed by controlled glass temperature increases to slightly above the stress relief temperature requirement, then gradually cooling the glass through the annealing regions.

The annealing lehr (Fig. 6) is designed to reheat the glass container to the strain and stress removal temperature of 950–1050°F for a typical glass container composition, to hold that temperature for the time necessary, and then to slowly cool the glass through its critical temperature ranges.

The annealing lehr is usually tunnel shaped and of various widths and heights, with a metallic woven belt traveling on rollers or slide straps, heavily insulated heat transfer sections, controlled cooling regions, and a platform section where the glass is readied for further processing. The

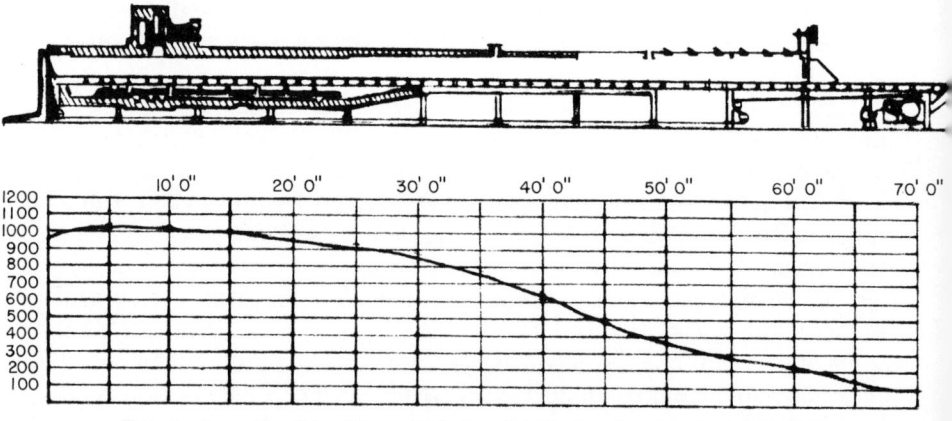

FIG. 6. Annealing lehr. [From Holscher (1953). Reprinted with permission.]

lehr heating sections are controlled by various energy input systems commanded by temperature sensors within the lehr assembly. These sensors signal a controller, programmed to maintain established temperatures by relaying commands to the energy systems. The cooling sections contain blowers, louvers, and vents which are similarly controlled.

The length of time for which the glass is subjected to the necessary temperature ranges is controlled by the spacing of the glass ware on the belt, length and width of the temperature zones, and speed at which the belt travels. The glass must be heated to temperatures of 950–1050°F for long enough to remove the forming stresses. It then enters the annealing region, where the cooling process begins and the temperature of the glass is below its strain point (approximately 995°F). Stresses in the glass are not likely to be introduced at temperatures below the strain point; therefore rapid cooling of the glass begins and continues throughout its travel to the platform or packing station.

As the glass exits the lehr tunnel, its temperature has decreased to 750–900°F. From the platform section, samples of the glass are checked for strain with a polariscope, according to quality-control practices. Polarized light, when it passes through the glass, will remain undisturbed unless strain conditions disrupt its path. The quantity of interference (strain) can be observed through the analyzer. If the strain is above acceptable annealing values, adjustments in either time or temperature can be made to the lehr.

B. Cold-End Coating

Glass container surfaces can be extensively damaged by container–container contact or by contact with abrasive handling materials and/or equipment. Damage to the surface could result in failure of the container; therefore, it is desirable to protect the surface throughout the processing of the container.

The strength of the glass surface is protected by the application of hot-end and cold-end coating materials (glass lubricants). The combination of hot-end and cold-end lubricants provides a semipermanent coating on the ware that allows the customer the ability to operate his packaging lines at speeds of up to 1200 bottles a minute without damage to the glass container. The glass surface lubricants are supplied to the glass at a temperature not so high as to fracture the glass, nor too low to complete the curing of the coating. The temperature range of 150–180°F is sufficient for the application of cold-end coating materials available today. Soluble lubricants are usually sprayed on the glass at lower temperatures and permanent-type lubricants are applied at higher temperatures. Either lubricant can be applied by utilizing atomization spray nozzles traveling transversely over the glass container. The quantity of coating applied can be varied by adjusting the air, liquid concentration, or speed at which the nozzle travels.

Caution must be exercised when selecting a glass lubricant because many label adhesives are not compatible with all coating materials; therefore, coating specifications must be coordinated between the producer and customer.

After completion of the glass lubricating process, the glass container has reached the end of the lehr platform, where it is in its permanent state, at ambient temperature, and can be safely handled by service personnel. The glass container can now be inspected and packaged, or transferred onto a single-line conveyor system for automatic inspection processes and packaging.

1. Inspection

The inspection procedures begin with glass samples being taken to a central laboratory where they are thoroughly inspected per manufacturer and customer specifications. Central laboratory functions involve such tests as capacity, weight, dimensions, shape, appearance, stress, internal pressure, thermal shock, friction, and impact. Should the container fail to meet specifications, a corrective order is taken to production management for immediate attention and the ware is culled.

Inspection procedures also include sample retention for density, composition analysis, and other complex analytical determinations. Descriptions covering these tests and frequencies are given in Section IV.A.

After transfer from the lehr platform onto the single line, the glass container is directed through a series of inspection devices which automatically reject individual defects. These devices inspect for such defects as checks and splits, variations in glass thickness, birdswings, blisters, dips in sealing surface, and variations in shape. Birdswings, splits and checks, and blisters, when present, will reflect light that has been directed on a specific area of the container to a photoelectric sensor. If a defect is sensed in a container, the sensor unit will command rejection of that item. The condition of the container sealing surface is determined through momentary pressurization of the container; if a pressure leak exceeding a predetermined value exists, the container will be rejected.

In addition to this autoinspection, as the container is conveyed on the single-line, it passes a manual inspection station where an inspector will observe the container for visual defects, i.e., chips, broken areas, bottle shape, and general appearance.

2. Packaging

The glass container, having passed this rigid examination, is conveyed to the packaging area and cartoned or palletized per customer requirements. Packaging of the container can be completed by manual or auto-

Automatic packaging is accomplished as the containers exit the single-line conveyor onto the packing accumulator, where they are indexed into the packer equipment and (1) inserted into the carton or (2) bulk palletized. Various models of automatic packaging equipment, capable of many functions, are available to the glass producer.

VI. Pollution Control and Its Effect on Industry

Over the past decade, government, under pressure from environmental groups, has taken an increased interest in waste streams generated by industry. These waste streams include air, water, hazardous waste, toxic waste, and solid waste. Although the Environmental Protection Agency (EPA) at one time was substantially involved with noise pollution, the federal government has virtually eliminated EPA involvement in this area, based on the feeling that the problem (1) should be handled by local government and (2) was already being dealt with by other agencies. It is important to note that the noise regulations developed by the EPA have not been rescinded. Therefore, there is a real possibility that the EPA could again become involved. Although the noise problem is not addressed here, it should not be precluded from consideration.

It is important to note that environmental regulations are promulgated not only by federal legislation, but by state and local legislation as well. The regulations are arranged in a hierarchy. Federal law mandates that state and local regulations be more stringent than or as strict as federal law. The rule of thumb is that the most stringent regulation applies. There is little or no local and state consistency in these regulations, however, and all applicable regulations must be closely monitored to ensure that a facility is in compliance.

A. Air Pollution

Clean air was the first environmental problem to be addressed by government. The original Clean Air Act of 1967 was amended by Congress in 1969. The amendments, which were passed into law in 1970, so extensively revised the 1967 Act that the 1970 version is commonly known as the Clean Air Act.

Undoubtedly the greatest impact on the glass container industry comes from controls that are required for furnace emissions. These controls can range from process and furnaces designs to scrubbers, electrostatic precipitators, and bag houses. There are now two sets of standards in existence—one for facilities which were in existence in 1970 and one for new sources (New Source Performance Standards—NSPS). The latter are more stringent than the former.

The major pollution emission from glass furnaces consists of particulate matter containing sodium sulfate (Na_2SO_4), nitrogen oxides (NO_x), and sulfur oxides (SO_x).

Pollution from glass furnaces stems from the combustion of fuel and the vaporization of raw materials. The pollutants formed by fuel combustion are SO_x, NO_x, and CO; SO_x and NO_x will be considered here. There are two possible sources of NO_x: nitrogen in the combustion air and nitrogen contained chemically in the fuel. Some SO_x emissions are caused by oxidation of sulfur compounds in the fossil fuels. The remainder are formed by vaporization of the glass melt.

Particulate formation, the major type of air pollutant released in glass manufacturing, is formed when Na_2O and SO_2 are combined in the vaporization state over the glass, then condensed in the stack.

Emission limits vary from state to state, but the general equation for calculating the limit is $E - np^x$, where E denotes allowable emissions (lbs/h), p the process rate (tons/h), n an integer, and x a factor less than 1.

Although the cost of installing pollution control equipment runs into millions of dollars, there have been significant benefits as a result of the air regulations. One significant method of cutting pollutant emissions is to raise cullet levels in the batch. Because of this increase in cullet, lower operating temperatures can be maintained, resulting in significant fuel savings and raw material cost reduction. Another benefit, is the use of better firing techniques and increased overall furnace efficiency. Raw material changes have also been a boon to fuel efficiency. Although results vary, fuel conservation has been in the range of 10–20%.

Electrical boost has also served to reduce pollution related to normal firing practices, allowing the lowering of operating temperatures of the furnace; however, melting costs are increased.

It is obvious that the utilization of furnace and process changes can be of benefit to both the community and the company. Because of tough environmental laws and the rising costs of fuel, glass manufacturers have been forced to economize and improve their operations. Otherwise, more pollution control equipment could be required. This would involve high capital expenditure and operating costs, without certainty of economic payback.

B. Water Pollution

The Federal Water Pollution Control Act was passed into law in 1972, and the result has been that Americans now pay considerably more for water pollution than is realistic. The limits now set for the discharger are more stringent than is required for the protection of the waterways. This is due to the informal adage adopted by EPA of "zero discharge by 1985."

As with air pollution laws, there are hierarchies of government agencies and regulations, again, with little consistency among the states.

There are only two options available to the glass industry regarding the discharge of their process and to cooling waters. The options are NPDES (National Pollutant Discharge Elimination System) discharges, and discharges to sanitary sewers. In general, noncontact cooling water is of little concern to government agencies. Normal monitoring requirements include temperature, pH, and flow. Process waters (cullet quench) are a different matter. The main pollutants are suspended solids (mainly glass fines), oil, and grease.

The glass industry has dealt with this problem in several ways. First, the glass manufacturers have, where feasible, discharged process waters into sanitary systems. This does not totally eliminate the need for pollution controls, and depending on the facility, can be of great cost.

Because of sanitary and municipal water costs, glass plants have begun recirculating process waters for cullet quench. This does create operating problems, which can usually be resolved economically. Substantial cost savings can be attained by reducing sewer effluent and water demand.

Where sanitary systems are not available, plants must rely on process ponds, plate pack separators, oil skimmers, and other methods to remove pollutants prior to discharging the water from the plant property. Water "cleaning" can be costly, but there are opportunities for significant cost reductions.

C. Hazardous and Toxic Wastes

In comparison to air and water pollution, the glass industry has little "ongoing" impact on hazardous waste (RGRA) and toxic wastes (TSCA), nevertheless, both do effect the glass industry. Continuous policing of plants for toxic wastes and filing proper documentation with the various governmental agencies is required.

This problem is being handled in two ways. The most obvious is the elimination of toxic materials by design, or substitution of nontoxic material wherever possible. Second, when the disposal of hazardous waste is required, disposal of spent material to legitimate recyclers (which is permissible) because the material has the characteristics of toxicity rather than being toxic in and of itself.

D. SOLID WASTES

The glass industry has always been a leader in recycling its solid wastes. Waste paper and wood are sold to recyclers; the same is true of waste oils. Reject glass is returned to the furnace for use as a raw material. The principles and utilization of recycling have benefited both the glass industry and society.

E. CONCLUSION

The overall effects of pollution and its attendant regulations have had and will continue to have a major effect on the glass industry. Millions of dollars have been spent on pollution projects, and the paperwork involving government forms, questionnaires, and regulations is astronomical. Large and small glass companies have been hard pressed to keep up with, and meet, the myriad of regulations.

One of the greatest problems facing the larger multistate companies is the total lack of consistency, not only on the state and local levels, but among the EPA regions themselves. (A condition acceptable in Region I may not be acceptable in Region V.)

A second point is that the EPA seldom if ever takes into account the cost effectiveness of the laws it promulgates.

The key to the glass industry's ability to survive the economic hardships placed on it by government appears to be in recycling and in taking maximum benefit of all opportunities for conservation.

References

Burch, O. G. (1964). Glass! What is it—how did it evolve—where is it now—what is its future? *Glastech. Ber.* **37**(2), 39–44.

Day, R. K. (1953). "Glass Research Methods." Industrial Publications, Chicago.

Holscher, H. H. (1953). Feeding and forming. *In* "Handbook of Glass Manufacture" (F. V. Tooley, ed.), Sect. X, p. 362. Ogden Publ.

Holscher, H. H. (1965). Hollow and specialty glass: Background and challenge. *Glass Ind.* **46,** 9–26.

Kephart, W. W., and DeNapoli, F. J. (1981). Glass container industry specifications for raw materials in the 1980's. *In* "Proceedings of Minerals and Chemistry of Glass and Ceramics: The Next Decade" (P. W. Harbin, ed.), p. 71. Industrial Minerals.

Morey, G. W. (1938). "The Properties of Glass," Chapter I. Van Nostrand-Reinhold, Princeton, New Jersey.

Scholes, S. R. (1952). "Modern Glass Practices," Industrial Publications, Chicago.

Spain, R. W. (1958). "Better Glass Making." Industrial Publications, Chicago.

CHAPTER 4

Tubing and Rod Manufacture

R. W. Wilson

CORNING GLASS WORKS
CORNING, NEW YORK

I. Danner Process	138
II. Updraw Process	140
III. Vello and Downdraw Processes	141
IV. Tubing/Rod Drawing Operations	144

In the simplest terms, the drawing of tubing and rod involves stretching a stream of molten glass under controlled forces while it is cooling so that as it approaches room temperature and becomes rigid the desired dimensions of diameter and wall thickness will be achieved. Unlike other forming processes such as blowing, pressing, centrifugal casting, etc., no molds for containing or shaping the stream are used. Besides containing and shaping, molds also provide these other operations with the major source of cooling the molten glass by heat conduction to the relatively cold mold. Lacking molds, the drawing operation depends largely on radiation, and to a lesser extent air convection, as its only source of cooling. As a result, much higher initial viscosities in the forming operation are required here than in the other processes. These viscosities can range from 10,000 to 300,000 P depending on several factors, principally the composition of the glass being drawn, the wall thickness desired, and the mass flow rate of the glass.

There are four major tubing/rod drawing processes in use commercially today. These processes are quite similar in the drawing operation itself but differ radically in the way in which this initial forming viscosity is obtained. The general principles of each of these processes and how the initial drawing viscosity is achieved will be discussed first and then the common features of the drawing operation itself later.

I. Danner Process

The Danner process is probably the oldest continuous tubing/rod process still in use today. Here molten glass is conveyed in a relatively short forehearth from the front end of the glass melting unit to a bowl to the bottom of which is attached a short spout which forms a ribbon-shaped glass stream. Immediately below the tip of the spout is the rotating Danner mandrel, a tubular refractory piece tapered at the lower end inclined downward at a slight and adjustable angle of about 20° to the horizontal, onto which the stream flows. It is rotated and supported axially over its entire length by a hollow, high-temperature-resistant metal shaft which is held at its upper end by a motor-driven rotating mechanism.

The spout and mandrel are enclosed in a heated temperature-controlled refractory muffle chamber opened only slightly at the back end to let the mandrel shaft pass to the rotating mechanism and at the front end for the tubing or rod to issue.

The rate of flow of glass to the Danner process is directly controlled thermally by maintaining the viscosity of the glass flowing through the spout. This can differ for different glass compositions, flow rates, and other factors, but is usually in the range of 1000 P. Major changes in flow rate are sometimes obtained by regulating the vertical position of a needle in the entrance of the spout to act as a valve.

The ribbonlike stream flowing from the spout on contacting the rotating mandrel partly overlaps the layer of glass picked up by the mandrel on the previous revolution. In the period of time it takes the mandrel to make this revolution, the layer flows down the inclined mandrel a slight amount owing to the action of gravity. After repeated revolutions, an undulating surface contour in the form of a spiral is clearly discernible on the glass surface for some distance from the original contact point. This unevenness, under the influence of surface tension and gravity forces, gradually decreases with increasing distance, so that, if the initial spout stream viscosity is low enough and the rate of cooling of the glass on the mandrel is not too great, the surface of the glass near the lower end of the mandrel will appear quite smooth.

The speed of rotation of the mandrel is determined by a number of other factors. To maintain a uniform thickness of the layer of glass around the mandrel at a given distance from the contact point, it must rotate fast enough to prevent undue sloughing of the glass to the underside of the mandrel and yet not so fast that instabilities result in the lapping of the spout stream or tubing/rod flow from the lower end of the mandrel. To give some idea of the magnitude of this speed, Danner processes have frequently operated in the neighborhood of 10 rpm. The consistency of this rotational speed is one of the most critical factors determining the performance of the Danner process.

As the glass flows down the inclined mandrel, it cools mostly by radiation to the surrounding refractory muffle, which is maintained at a somewhat lower temperature to control the rate of cooling of the glass. As the glass cools and its viscosity increases, the thickness of the layer of glass gradually increases so that as it approaches the lower end of the mandrel and the initial tube/rod drawing viscosity of roughly 50,000 P is achieved, it is calculated to be some three or four times thicker than it was after first contacting the mandrel at the 1000-P spout viscosity.

Having reached the initial drawing viscosity, our description of the Danner process is complete except to note that (1) the very end of the mandrel is usually tapered to varying degrees to a smaller diameter at the tip depending on the range of diameters and wall thicknesses of tubing to be drawn or whether rod is to be made and that (2) the metal shaft supporting the mandrel is hollow to allow the introduction of blowing air (or vacuum) to provide a pressure differential between the inside and outside of the tubing of a few inches or less of water column to allow for additional control of tubing diameter and wall thickness.

The Danner process has one distinct advantage over all the other methods of drawing tubing/rod. The circular symmetry inherent in the rotation of the mandrel provides a way of cooling the glass quite uniformly over the 50:1 or so viscosity range necessary for drawing. As a result, if good geometrical and mechanical integrity of the mandrel, shaft, and rotating mechanism are maintained, excellent circular uniformity of wall thickness of the tubing can be achieved.

While this method of cooling glass may be the most efficient it also presents some disadvantages.

(1) The maximum production rate of tubing from the process is limited by the radiating area of the mandrel. As these rates are increased, the engineering problems associated with supporting and rotating larger and larger mandrels (as large as 2 ft in diameter and 8 ft long) with precision become increasingly difficult.

(2) Volatilization of low-melting-point oxides from the leading edge of the lapping stream and the free surface of the glass on the mandrel causes cords and a silica-rich layer on the outside surface of the finished tubing/rod which have deleterious effects in certain later applications, especially in the case of hard borosilicate glasses.

(3) The slight angle of inclination of the mandrel reduces the component of the force of gravity, causing the glass to flow down the mandrel to about a third of that for a vertical flow process. This produces a greater

cross-sectional area of glass at the tip of the mandrel, which will require a greater degree of stretching in the subsequent drawing operation, resulting in greater variability of the final dimensions of the tubing/rod, especially in the case of small diameters.

Because of these disadvantages and others, the Danner process has been replaced in many applications by the Vello process, which will be described later.

II. Updraw Process

This process, used principally for drawing large-diameter tubing, is mostly of historical interest. It has in recent years been largely replaced by the downdraw process to be described later.

In the updraw process, glass at the low viscosity of a few thousand poises, is conveyed in a relatively short, temperature-controlled forehearth from the front end of the melting unit to a large circular shallow bowl. In the center of the bowl protruding from the bottom is a cylindrically shaped hollow, refractory piece tapered at its upper end called the cone. Cones of different top diameter are used for different ranges of sizes of tubing to be made. The extent of the protrusion of the cone into the glass is adjustable so that the top land of the cone may be, depending on operating conditions, somewhat above to somewhat below the level of glass in the bowl. Mounted from above and extending a few inches or less below the glass level in the bowl is a centrally positioned large refractory tube called a sleeve. In this configuration, a region of glass between the inside of the sleeve and the outside of the cone is created which can be cooled to the desired high tube drawing viscosities independently of the glass between outside the sleeve and inside the bowl that is controlled at roughly the same low viscosity as the forehearth. To provide this cooling, inside the area of the sleeve is mounted from above an annularly shaped water cooler which can be positioned horizontally and vertically with respect to the sleeve and the tubing (which passes through inside the annulus) for minor adjustments in tubing dimensions. Another adjustment in tubing dimensions is provided by a blowing air tube mounted inside the cone which can both cool the inside surface of the cone and tubing, and create a differential in pressure between the inside and outside surfaces of the tubing.

The tubing/rod is drawn from the land or tip of the cone through the inside of the water cooler annulus and out of the region of the sleeve by a motor-driven mechanism mounted some 10–20 ft directly above which contacts the by then cooled, hardened tubing/rod by frictional contact with the outside surface. The speed of the drawing mechanism is closely controlled for a particular size of tubing/rod, and can be changed to make different sizes. The tubing/rod is cut manually by thermal crackoff immediately after it passes through the drawing mechanism.

The fact that the direction of motion of the glass in the updraw process is diametrically opposite that of the force of gravity acting on it gives this process a unique advantage over other tubing/rod drawing processes—slight vertical adjustments in the position of the cone tip relative to the glass level outside the sleeve can produce significant changes in the diameter to wall ratio of the tubing being drawn. By this means, a very wide range of tubing sizes can be produced on the same cone, thereby minimizing the number of equipment changes needed to produce the wide range of tubing diameters and wall thicknesses frequently required in specialty glass applications.

This advantage, however, can turn into a serious disadvantage—slight variations in the glass level in the bowl from tank glass level fluctuations and other causes can adversely affect the diameter and wall thickness of the final product.

In opposing gravity, other disadvantages of the process arise:

(1) Production rates are restricted to about half that of the downdraw process.

(2) Accidental breakage of the draw from cutoff mishaps, stones, etc., results in contamination of the cone and sleeve from the fallen cold glass, sometimes requiring a complete change of equipment.

(3) Unlike other drawing processes, there is no direct control of the mass flow of the process. It is a result of all the other variables in the process.

III. Vello and Downdraw Processes

The ways in which the all-important initial forming or drawing viscosity is achieved in the Vello and downdraw processes can be described together since the same methods are used in both processes. In fact, one of the major advantages of these processes is that the complete range of tubing diameters, from less than 1 mm to over 10 in., can be and frequently is made using the same forehearth or glass delivery system—only a change in equipment in the subsequent drawing operation is required. The design of the delivery systems in use differs widely, ranging from (1) platinum-lined tubes for corrosive glasses in which refractory contamination can adversely affect devitrification, to (2) enclosed electrically fired forehearths for glasses which are susceptible to surface volatilization, to

(3) free-surface, gas-fired, wind-cooled forehearths, some of which are the very longest widest, and deepest in use for high-rate production of the common soda–lime and lead glasses.

The purpose of the delivery system is to conduct the glass from the front end of the melting unit, cool it under carefully controlled conditions and deliver it to a so-called bowl at essentially the initial drawing viscosity. At the bottom of the bowl through which the glass will flow is an orifice ring, usually round (except in the case of lens-type thermometer tubing), which can be changed to different sizes for different production rates. If tubing is to be drawn, an annular opening is obtained by mounting approximately concentric with the ring a bell (so called from its original shape). This bell is attached to the end of a hollow, high-temperature-resistant metal shaft which extends upward through the depth of glass in the bowl and is mounted at its upper end in the bell–shaft adjustment mechanism. This device provides a means for positioning the bell horizontally with respect to the orifice ring. Vertical adjustment is also provided so that one can make minor changes in glass flow rates at the desired viscosity without having to change orifice rings.

Bells originally resembled bells, but today, bells of a wide variety of shapes are used, ranging from pencil-tipped for thermometer tubing and small-bore capillary tubing in the Vello process to up to 12 in. in diameter (at the bottom) for large-diameter tubing in the downdraw process.

As in the case of the Danner process, blowing air pressure (or vacuum) of a few inches of water column is introduced at the top end of the hollow bell shaft to control the diameter and wall thickness of the drawn tubing over a fairly wide range of sizes without the need for changing bell sizes.

If only rod is to be drawn, the bell and bell shaft are not necessary and in many cases rod is drawn directly from a somewhat smaller orifice ring.

Mounted directly below the orifice ring and bell and surrounding the issuing tubing/rod for a short distance is an enclosure called a muffle, which is open at the bottom to let the glass stream pass and can be heated or cooled for fine control of the initial drawing viscosity of the tubing/rod.

The rate of flow of glass to the Vello and downdraw processes is directly controlled thermally by maintaining the viscosity of the glass flowing through the annulus formed by the orifice ring and bell shaft. Minor changes in flow rate can sometimes be achieved by vertical adjustment of the bell relative to the orifice ring. Major flow rate changes require changes in orifice ring sizes.

The principles employed in drawing tubing/rod by the Vello and downdraw processes make possible several advantages over the other methods.

4. TUBING AND ROD MANUFACTURE

(1) As was mentioned earlier, the complete range of tubing diameters can be made from a single delivery system by changing equipment for one process to the other—small tubing by Vello, large tubing by downdraw.

(2) A wide range of production rates is obtainable. One of the limits to production rates is the cooling capacity of the delivery system. It must be able to cool the glass from roughly 1000 P at the melting unit to 50,000 P (or so) initial drawing viscosity. It is not a great engineering accomplishment to design a large forehearth with sufficient radiating area from the free surface of the glass in the forehearth channel to handle 6000–7000 lb/h of lime glass for fluorescent tubing or a small one to handle the dimensional precision of diode bead and case tubing at 100 lb/h or less. Actually, rates of 5–10 lb/h are sometimes used in making samples of tubing from experimental glass compositions.

(3) Unlike the Danner process, the glass flow to the forming parts (orifice ring and bell) of the Vello and downdraw processes is in the same direction (vertically downward) as the force of gravity. As a result a smaller initial annular cross section is formed, less stretching is required, and closer dimensional control of the diameter of the final drawn tubing/rod is obtained.

(4) With the use of platinum-lined delivery systems glass compositions which are corrosive to refractories and easily devitrifiable can be produced by these processes, though at low production rates.

(5) Cross-sectional shapes other than round can be made. This is pointed up by the manufacture of lens-type thermometer tubing, most of which used in the United States, now by the Vello process. This feature together with others makes the drawing of thermometer tubing perhaps the most interesting of all tubing operations.

The magnifying cylindrical lens shape is formed by an orifice resembling, but certainly not exactly like, an enlargement of the final tubing cross section. The bore is not round, but elliptically shaped by a tiny uniquely shaped bell tip. The white, yellow, or red stripes frequently found in clinical or other lens mercury-filled thermometers are produced by introducing remelted colored opal glass into the stream of clear glass in the Vello bowl before it flows through the orifice. Common spirit-filled thermometer tubing is made in the same way except a round orifice is used. It is of major importance that no twisting action, typical of other tube drawing, occur in the tubing after it leaves the orifice.

Perhaps the most serious problem occurring in the operation of the Vello and downdraw processes is the control of cross-sectional wall thickness uniformity of the drawn tubing. This is a particularly important property, especially in the case of tubing from which small pharmaceutical containers are to be made. It also can affect the straightness of the tubing. Lacking the circular symmetry inherent in the rotating mandrel of the Danner process, the Vello and downdraw processes depend to a great extent on the uniformity of the viscosity of the glass in the bowl for cross-sectional wall thickness distribution. Deviations in viscosity uniformity in the bowl can be compensated for to a great extent by horizontal adjustments in the position of the bell relative to the orifice ring, but subtle variations in wall thickness distribution require changes in the way the glass is cooled in the forehearth or delivery system before it reaches the bowl. This is one of the reasons why the delivery systems have means for both overall cooling of the glass from the melting unit to the bowl and heating devices such as electrically fired electrodes, electrically heated wire elements, or gas-fired burners to control the different rates of cooling at different locations along the length of the glass delivery system. These rates of cooling are determined largely empirically by measuring and regulating the temperatures indicated by thermocouples or optical pyrometers strategically located at various points in the delivery system.

IV. Tubing/Rod Drawing Operations

Having reached the initial drawing viscosity by the four processes briefly described above, the subsequent drawing operation in each, with the exception of the updraw process described above, is quite similar. In the Danner and Vello processes, the tubing/rod travels for the most part in a horizontal direction and can be essentially identical as far as the drawing operation itself is concerned and will be described later.

In the downdraw process the tubing/rod, after leaving the bell, travels directly downward under the action of gravity and cools mostly by radiation to the much cooler surroundings. After traveling some 10–20 ft, depending mainly on the mass flow rate and the diameter of the tubing/rod, the glass has hardened sufficiently that it can be gripped and its speed of travel controlled by frictional contact with wheels turned by an accurately regulated motor-driven mechanism. Immediately below the pulling wheels the tubing/rod is severed manually, as in the case of the updraw process, by thermal crackoff. To confine the direction of travel and maintain straightness of the tubing/rod, a system of loose-fitting adjustable guide bars is located at regular intervals in the length of draw between the muffle and pulling wheels. The production rate of tubing/rod from the downdraw process is limited by the vertical length of the draw and for other reasons, and so is used only on those sizes of tubing/rod which are too large to be made on the Vello process. Tubing diameters from about 3 to over 10 in. of a hard borosilicate glass at rates of up to 12 lb/min, and diameters of less than ½ in. of other compositions at rates of a few pounds per hour can be made by the downdraw process.

In the Danner or Vello process, the glass flow after leaving the tip of the mandrel or bell describes a curve resembling a catenary while it is falling, under the combined action of gravity and stretching tension, to the runway some distance below on which it is thereafter supported. This distance or so-called catenary drop can vary tremendously depending on operating conditions, particularly the initial drawing viscosity and diameter and wall thickness of the tubing/rod being drawn. Drops of from less than a foot for small-diameter aluminosilicate glass rod at 10,000 P to over 35 ft for most lead glass tubing at over 200,000 P are used. It is sometimes necessary to adjust the length of catenary drop to cover the wide range of sizes sometimes required from one draw.

The runway, which is usually horizontal, but in some cases inclined in a straight line either upward or downward, consists for the most part of freewheeling grooved rollers (mounted in as straight a line as possible) on which the glass tubing/rod rests. In cases of high flow rates and low initial drawing viscosities, the glass tubing/rod may be supported by an upward-moving air stream during the first section of the runway.

The glass is drawn along the runway by a pulling machine located at a point where the glass has sufficiently cooled to be rigid enough to be nondeformable. The pulling machine consists of accurately driven pairs of wheels or a pair of tractor belts which provide the drawing traction by frictional contact with the rigid tubing/rod. In order to produce a wide range of sizes of tubing accurately on a draw the speed of the pulling machine must be variable over a wide range of 10 : 1 or more yet controllable for a particular size of tubing/rod to accuracies of a few tenths of a percent.

After passing through the pulling machine the tubing/rod is cut to rough length by a cutter or chopper synchronized with the pulling machine. For speeds of tubing/rod of 20 ft/sec or less the cutting operation is usually performed by a combination of abrasion and thermal checking of the glass. At higher speeds of up to 50 ft/sec on small diameters a chopper which essentially crushes the glass is used.

Following this, the tubing/rod falls onto a takeout conveyor belt traveling at right angles away from the draw. If accurate lengths of tubing are required, a small amount is trimmed off each end by thermal crackoff and the ends are subsequently fire polished by gas burners farther down the conveyor. If accuracy of length is not necessary, the raw ends of the tubing are sometimes fire polished to minimize breakage.

This is usually the final operation in drawing tubing/rod—unlike most other glass-forming operations annealing is not needed. However, in a few select cases such as larger sizes of tubing which are to be cut later automatically into short pieces, very heavy wall tubing, or large-diameter rod, annealing is carried out to reduce the residual stresses in the glass.

It is important that the glass tubing/rod be sufficiently cooled (well below the strain point of the glass) on the runway, not only that it not be deformed by contact with the pulling machine wheels or tractor belts, but that the straightness be maintained, breakage minimized, and asymmetrical stresses not be set up in the glass when it is being carried away by the takeout conveyor. However, it is also important that consideration be given to the rate at which the glass is cooled at various sections of the runway to avoid excessive residual stresses in the finished tubing, particularly if it is to be cut into short pieces in subsequent operations. For these reasons, most runways are equipped with means for providing adjustable convective air cooling at some sections to enhance the normal cooling by radiation and at other sections means for muffling the effects of radiation. To be able to accomplish this degree of cooling control for the wide range of production rates and diameters experienced in Danner and Vello tubing/rod manufacturing, the lengths of runways vary greatly—from about 20 ft for slow rates of large-diameter tubing/rod to some 400 ft for high-speed draws.

The way the dimensions, diameter, and wall thickness of the tubing are obtained can be most easily explained by considering the mass flow equation

$$W = \pi \rho w(D - w)V,$$

where W is the mass flow rate of glass, ρ the density of the glass, w the wall thickness, D the outside diameter, and V the velocity of the tubing at the pulling machine.

Temperature effects are usually less than a few tenths of one percent and have been neglected. After measuring W and knowing ρ and the normal specifications for w and D, one can determine the velocity V the tubing must have at the pulling machine. Adjusting the speed of the pulling machine to give this value and ascertaining that no slipping in the frictional contact of the pulling wheels or tractor belts and tubing is taking place, one obtains the desired value of the tubing cross section $\pi w(D - w)$. To find the desired value of each term w and D in this product, it is necessary to adjust the amount of blowing air introduced at the top of the bell or mandrel shaft. This air flows through the hollow shaft then through the inside of the tubing over the entire length of the draw and exits to the atmosphere when the tubing is cut off. By this means, a pressure differential above atmospheric pressure is established inside the tubing which

4. TUBING AND ROD MANUFACTURE

affects the way in which the glass stretches after leaving the bell or mandrel tip and determines the diameter to wall thickness ratio D/w of the tubing at the cutoff point. Increasing the blowing air pressure increases the diameter to wall thickness ratio by increasing the diameter and decreasing the wall thickness. Then with the desired tubing cross-sectional area already established by the pulling machine speed the desired values of diameter and wall thickness are obtained by adjusting the blowing air pressure until the diameter averaged over a short period of time equals the desired value. This is usually carried out automatically by continuous measurement of the diameter of the tubing at some point on the runway before it reaches the pulling machine. Negative values of blowing air pressure or vacuum are rarely used except in the case of rod, thermometer tubing, and capillary tubing—to obtain smaller diameter to wall thickness ratios smaller bells or mandrel tips are required.

The diameter averaging process mentioned is necessary to take into account the fluctuations in diameter caused by random variations in the many process parameters affecting it such as the mass flow rate, initial drawing viscosity, cooling rate on the catenary, runway friction, blowing air pressure, pulling speed, etc. However, under very carefully controlled conditions a standard deviation of one-third to one-half of one percent diameter variation can be achieved.

CHAPTER 5

Glass Spheres

Charles Hendricks

LAWRENCE LIVERMORE LABORATORY
LIVERMORE, CALIFORNIA

I. Introduction	149
II. Rayleigh Technique	150
III. Commercial Processes for Producing Glass Spheres	152
IV. Production of High-Precision Spheres by the Drop-Generator Process	154
A. The Drop Generator	156
B. High-Temperature Vertical Tube Furnace	157
C. Hollow-Shell Production from Liquid Drops	158
D. Glass Formation and Fining	160
E. Glass Composition and Surface Control	163
F. Measurements	164
G. Filling Glass Spheres with a Gas	165
H. Mass Production of Glass Spheres	165
I. Range of Glass-Sphere Sizes	166
J. Spheres Produced from Specialized Materials	166
V. Conclusion	167
References	167

I. Introduction

Both hollow and solid glass spheres are widely used in industrial and consumer products. Some of the better-known uses of small solid spheres are in "beaded" movie screens and reflective paint and plastics. Less well-known are the uses of small glass spheres as plastic fillers and to impart "smoothness" in prepared mixers for such drinks as Tom Collinses and daiquiris. Hollow glass spheres are used as fillers in plastics for such widely diverse products as boat hulls and microwave antenna lenses. For many applications, acceptable limits on the physical parameters of the spheres—surface finish, wall uniformity, size distribution, and composition—are quite broad and do not unduly affect the overall cost of the product.

Recently, however, glass spheres have been used in the production of high-precision fuel capsules (targets) in the field of inertial-confinement fusion (ICF) (Nuckolls et al., 1972; Hendricks, 1979). The targets contain a mixture of the hydrogen isotopes deuterium and tritium which, if sufficiently compressed and heated, will fuse together to form helium and neutrons and simultaneously release a large quantity of energy (Emmett et al., 1974). The compression and heating can be done by laser beams or by intense beams of ions or electrons.

As the national ICF program developed in the early 1970s, it became apparent that a critical path element was the target that contained the deuterium–tritium (DT) fuel. Design specifications required glass spheres with heretofore impossibly smooth surfaces and uniform walls. (Beck and O'Brien, 1968). The peak-to-valley variation in the surface was to be 100 Å or less, wall thicknesses were to be uniform to better than 1% of the thickness, sphere diameters were to range from 50 μm to more than 500 μm, and wall thicknesses were to range from 0.5 to over 30 μm. Furthermore, the DT-filled spheres should not leak over periods of time ranging from several months to years! Clearly, the requirements for ICF targets meant that then-existing production techniques for glass spheres had to be improved several-fold.

This chapter on glass spheres will describe the Rayleigh technique for producing uniform liquid drops; early and current commercial processes for producing glass spheres; and the drop-generator process, which is currently the only process capable of efficiently making glass spheres that meet ICF target requirements.

II. Rayleigh Technique

A technique for generating uniform liquid spheres was first thoroughly analyzed in detail by Lord Rayleigh (1879) and is usually ascribed to him as the "Rayleigh technique," even though the generation of liquid spheres was studied by many scientists more than a century ago. Briefly, as described by Rayleigh, a circularly cylindrical jet is formed by forcing a liquid through a nozzle or orifice. If care is taken in the making of the orifice, a smooth jet can be produced that will maintain its integrity for a long distance away from the orifice before it breaks up into almost random-sized drops. If a capillary wave of suitable wavelength is launched on this jet, the unbroken length of the jet will shorten markedly, and, instead of random-sized drops being produced, the jet will disintegrate into a stream of uniform drops. Normally, with no externally applied wave, a jet breaks up in response to a set of random stimulations or "noise" resulting from either internal fluctuations or mechanical vibra-

5. GLASS SPHERES

tions from the surrounding environment. A disturbance on a jet will grow in amplitude under the action of surface tension until the amplitude is large enough to cut the jet into drops.

Because it is important to the overall process of producing uniform glass spheres, Rayleigh's analysis of the breakup of a jet into uniform drops is briefly presented here.

Rayleigh assumed a jet could be represented by the expression

$$\lambda = a + C(t)\cos mz,$$

where a is the undisturbed jet radius; $C(t)$ is the time-dependent jet envelope; and m is the wave number $2\pi/\lambda$, where λ is the wavelength of a disturbance on the jet. The parameters associated with the fluid in the jet are density ρ, surface tension T, and viscosity η, the last of which, in our case, is assumed to be negligible. For a jet traveling with a constant velocity v along the Z axis, the equation for the behavior of the jet with radial perturbations of the surface small compared to a is

$$\ddot{C}(t) - \frac{T}{\rho a^3} \frac{(1 - m^2 a^2) ma I_1(ma)}{I_0(ma)} C(t) = 0,$$

where $I_0(ma)$ and $I_1(ma)$ are the modified Bessel functions of order 0 and 1. Small displacements of the surface grow at a rate $e^{\mu t}$ depending on the wavelength λ, with the growth rate μ expressed by the relationship

$$\mu^2 = \frac{T}{\rho a^3} \frac{(1 - m^2 a^2) ma I_1(ma)}{I_0(ma)},$$

and C is proportional to $e^{\mu t}$. The larger the value of μ, the more rapidly the jet breaks up into drops. When the applied wavelength is $\lambda = 1.43 \times 2\pi a$, the growth rate factor μ is largest. Liquid jets can be stimulated to generate uniform drops with a wide range of applied wavelength. The disturbance can be applied to the jet by several means, the most common of which is by vibrating the orifice through which the liquid flows. This may be accomplished by a piezoelectric transducer, by a magnetic driver such as an earphone or loudspeaker, or by some other vibrating device. The pressure of the upstream liquid may be modulated or an oscillating electric field may be applied to the jet externally or by electrodes internal to the liquid stream.

Rayleigh was known to have applied a disturbance to a jet by means of a plucked tuning fork either held in air some distance from the jet or held in contact with the table on which the apparatus was supported. The rate of growth for small disturbances of a jet is very high for most low-viscosity liquids, and a surprisingly small coupling between a driver and a jet may be sufficient to produce uniform drops. For positive control in the presence of noise disturbances coupled to the jet, higher-amplitude driving of the desirable wavelength may be necessary to assure uniformity of the generated drops.

As the drops form at the tip of the liquid jet, it is possible to deliberately place a predetermined electrical charge on each drop (Schneider *et al.*, 1965; Schneider and Lindblad, 1965). This is done by enclosing the jet and part of the stream of drops inside a conducting cylinder so as to form a cylindrical capacitor consisting of the unbroken jet and the surrounding conductor. If the liquid of which the jet is formed is a sufficiently good conductor, the charge per unit length on the jet can be considered to be a direct linear function of the electric potential between the conducting cylinder and the jet, and the charge carried away by each drop will primarily reflect the instantaneous potential at the time the drop separates from the jet. Because downstream drops are capacitively coupled to the jet, and particularly to a drop just separating from the jet, the charge on the downstream drops will also affect the charge on the separating drop. When there is a large potential difference between the conducting cylinder and the jet, the additional effects of charged downstream drops may be very small. However, if the applied potential is suddenly reduced to zero, with the expectation of zero charge on the next drop, it is found that that drop actually acquires a small charge opposite to that of the previous drops.

Consider a stream of drops being produced at a rate of f drops per second from a jet of initial radius a. The spacing between drops will be v_d/f, where v_d is the drop velocity. If the drops are formed in a gaseous medium, aerodynamic effects act to reduce the velocity of the drops and can lead to collisions between drops. To avoid collisions and coalescence of drops, it is expedient to charge, deflect, and capture as many as fifteen out of sixteen drops in the stream, allowing only every sixteenth drop, for example, to follow a complete trajectory through the system. Thus each surviving drop will be independent and unaffected by previous or following drops.

III. Commercial Processes for Producing Glass Spheres

Glass spheres can form quite spontaneously under certain conditions. For example, a large fraction of the fly ash that issues from coal-fired power plants consists of hollow, almost spherical, spheres of a glassy material. The walls of these spontaneously generated spheres are nonuniform, there are many irregularities and inclusions (bubbles and unfused particulates) in the walls, and the surfaces are generally very rough. Nevertheless, the presence of the hollow glass spheres in the furnace output gases is an indication that even very uncontrolled processes can give rise to such spheres.

A production process for making hollow spheres was patented in the late 1930s. Basically, the process relies on forming a slurry of silicates, borates, and other glass-forming chemicals. The slurry is then dried and pulverized and may be sieved to yield a more-or-less uniform dry powder. The power is introduced into a high-temperature environment, such as a gas flame or furnace, where the particles melt. Simultaneously, gases are formed in the molten material from the water of hydration and by decomposition of some of the particulate material, and these gases expand to form hollow molten spheres of glass. As the particles leave the high-temperature region, they cool rapidly and can be collected as hollow glass spheres.

Unfortunately, this slurry process has many uncontrolled variables that lead to a wide distribution of sizes, wall thicknesses, and surface configurations in the final product. Experience with the process indicates that parameter control is very difficult, that uniform size and wall thickness are almost impossible to obtain, and that surface quality is not high.

A commonly used commercial process for producing glass spheres—the flame process—also requires several steps in which parameters are relatively poorly controlled. An aqueous solution of silicates, borates, and a substance that is thermally decomposable to gaseous products (e.g., urea) is spray dried in a large, heated, well-ventilated chamber to produce large quantities of well-mixed particulate material. Particle size distribution is very broad, although the use of good spray nozzle techniques tends to produce a limited droplet size and, therefore, somewhat limited final dry particle size. Coalescence of drops before they are completely dry can lead to some very large particles, while some very small drops produced in the spray produce comparably small dry particles.

To produce glass spheres by the flame process, quantities of the dry particles are introduced into a gas–air flame by means of a powder injector at the flame nozzle. As the particles are carried into the high-temperature region of the flame, the particles begin to melt and form a molten glass surface. The thermally decomposable material produces gas bubbles, and, as the particles move through the flame regions, expansion and coalescence of the gas bubbles in the molten glass forms a thin shell. The bubbles continue to expand even as the particle moves into cooler regions of the flame, and some thinning of the walls will occur; if the expansion is complete before the viscosity becomes too high, surface tension forces will tend to make the soft shell into a sphere.

It is clearly fortuitous that a large percentage of the spheres produced by the flame process are spherical (or very nearly so) and have reasonably uniform walls. In the commercially available glass spheres we have examined, the sizes range up to about 200 μm in diameter, and wall thicknesses are between 0.5 and 2.0 μm. Because of surface and wall imperfections and size, and because of strict wall thickness requirements, we were able to find a few spheres that could be used as ICF targets only by sorting through 10^8–10^{11} commercial spheres by sieving and optical techniques.

Considering the very large number of commercially produced spheres that had to be sorted through to find a few suitable for use as ICF targets, a more efficient process for producing ICF-quality target spheres was clearly required. The drop-generator process proved to be just such a process.

IV. Production of High-Precision Spheres by the Drop-Generator Process

The glass spheres used in fusion targets must meet very critical standards, which, as late as 1974, could not be met by any commercially produced material. To bring a useful perspective to the necessity for a better means of producing glass spheres, the required shell parameters for ICF targets are discussed.

For an ICF target sphere, exact dimensions of the diameter, wall thickness, and surface uniformity are critical aspects of any experiment. For example, the diameter may be specified to $\pm 5\%$, but must be measured to 0.5%; i.e., the exact size may not be absolutely critical, but, for experimental results to be interpreted correctly, the size of the shell must be very accurately known.

Similarly, the wall thickness may be specified to a few percent, but, for a thickness of a few micrometers, must be uniform within and measured to a few hundred angstroms. Surface finish on the glass spheres is also a critical parameter for fusion targets, and a peak-to-valley roughness of more than a few hundred angstroms cannot be tolerated in most experiments.

A typical ICF target is a hollow glass sphere filled with a DT mixture at pressures ranging from 10 to more than 100 atm. Walls are typically 1–35 μm thick and may be coated with various polymeric or metallic layers up to more than 100 μm thick. Table I lists typical values for diameters and wall thicknesses of target spheres and for the quality specifications that must be met (Hendricks *et al.*, 1975).

In searching for a process that could be used to efficiently make large quantities of high-precision glass spheres, experimenters decided in 1974 to try the liquid-drop generation and control techniques developed by

TABLE I

Design Specifications, Tolerances, and Measurement Accuracy Required for Glass-Sphere ICF Targets

	Specification and tolerance	Measurement accuracy
Diameter	50–2000 μm ± 5%	0.5 μm or ±1% of radius
Sphericity	1.0 μm or ±1% of radius	1.0 μm or ±1% of radius
Concentricity of inner and outer surfaces	±0.05 μm	10 nm or 1% of wall thickness
Wall thickness	0.5–30 μm ± 5% of wall thickness	±0.05 μm
Surface finish	<20-nm peak-to-valley	10 nm
Gas fill density	2–210 mg/cm^3	±5%

Hendricks and his colleagues in the Charged Particle Research Laboratory at the University of Illinois (Schneider and Hendricks, 1964; Babil, 1971; Hendricks and Babil, 1972). This liquid-drop technique had already been used to produce very uniform hollow spheres of several materials, including frozen hydrogen, copper, water and ice, and various epoxy formulations. To avoid the problem of producing hollow uniform spheres from viscous molten glass, an aqueous solution of glass-forming chemicals was used as the starting point for the experiments. Initially, a solution of sodium silicate, boric acid, and urea was tried (Hendricks and Dressler, 1976).

The work done earlier by Hendricks and Babil (1972; Babil, 1971) had shown that, when a series of drops of various solutions were dried in a hot vertical column, thin hollow spheres of the solute could be formed. Several earlier papers (Ranz and Marshall, 1952; Charlesworth and Marshall, 1960) treated the theory of the formation of a skin on the surface of an evaporating drop and the diffusion of vapor through the skin from the enclosed solution. Hendricks and Babil (1972; Babil, 1971) found that, in some cases, evaporating drops of solution produced shriveled spheres that looked much like miniature raisins, particularly in the case of polymeric materials such as polystyrene and others dissolved in organic solvents. In other cases, dried drops produced solid spheres with no voids.

It seemed reasonable that if a stream of very uniform liquid drops could be produced of a solution of glass-forming chemicals, and if the solvent could be evaporated to leave a set of dried particles whose masses were quite uniform, the particles could be heated to form uniform glass spheres. Initially, urea or another thermally decomposable gas former was included to aid in the blowing of a bubble in the glass sphere as it melted.

The liquid-drop process, as it has been developed to date, has the following aspects:

(1) A solution (aqueous or other solvents) of glass-forming compounds, such as sodium silicate and various other necessary chemicals, is mixed and made stable.

(2) Using the techniques described first by Rayleigh and more recently by others (Schneider and Hendricks, 1964), a drop generator is used to form and control uniform spherical drops of the solution.

(3) The solvent is evaporated from the drop in a vertical furnace to produce an intermediate particle, either a porous dry particle or a fragile hollow shell of the solute.

(4) The intermediate particle is passed vertically through a high-temperature vertical furnace to fuse the materials into a glass.

(5) Water vapor or other gases evolved during the glass-forming phase act to blow the molten particle into a hollow glass shell.

(6) Temperature profiles, drop sizes, chemical compositions, and other system parameters may be chosen to produce high-quality glass spheres with specified sizes and wall thicknesses.

The liquid-drop process has proven to be so efficient that its yield of usable spheres, according to our highly stringent criteria, can be as high as 99%.

A. The Drop Generator

The basic techniques of generating streams of uniform liquid drops were described above (see Section II). A solution of glass-forming and shell-blowing chemicals is formed by the generator into a liquid jet, which breaks into a stream of uniform drops that are spaced only a few drop diameters apart. This close drop spacing occurs whether the jet is disintegrating randomly or is driven by a deliberately introduced perturbing wave. Drops so closely spaced would be susceptible to collisions and coalescence in the furnaces because of aerodynamic drag and turbulence, and wide variations in the size distribution of the resultant glass spheres would result. Consequently, provisions are made to circumvent collision problems by removing all but every mth drop, where m can be any integer.

The drop-producing system consists of several primary components. A reservoir and fluid-feed system deliver the glass-forming solution to the actual drop generator at a rate controlled by the pressure applied to the liquid in the reservoir. The drop generator itself consists of both an orifice, through which the liquid issues in the form of a cylindrical jet, and a piezoelectric driver, which is arranged to launch a capillary wave on the jet either by modulating the pressure in the liquid-flow channel to the orifice or by vibrating the orifice itself to modulate the flow velocity in the jet. At typical operating frequencies, 1000–2000 drops/sec will be passed into the furnace; consequently, it is necessary to maintain careful control of the frequency and amplitude of the signal used to stimulate the jet to ensure that drop mass will be constant and that satellite drops will not be produced.

Downstream from the orifice and surrounding the jet at its breakup point is a cylindrical ring used to selectively induce an electrical charge on the drops as they are formed; this charge is used to implement the process by which some of the drops are removed from the stream to avoid coalescence. Immediately following the charge ring is a pair of deflector electrodes and a catcher for collecting the deflected drops. An electrode on one side of the drop stream is held at ground potential, and an electrode on the other side is at a high negative potential. Drops on which a negative charge is induced by a positive voltage of the charge ring are deflected toward the grounded electrode, where the drops enter a trough or catcher from which they are removed by a vacuum line. During normal operation, only every sixteenth or thirty-second drop is left uncharged and undeflected to travel into the furnace; the other drops are deflected into the catcher and removed.

B. High-Temperature Vertical Tube Furnace

The glass spheres are produced in a vertical tube furnace (Fig. 1) that has an inside diameter of 7.5–10 cm. Several materials have been tested as furnace tubes, with the most successful being stainless steel, fused quartz, or mullite for temperatures up to 450°C and fused quartz or mullite for temperatures up to 1500°C.

The vertical furnace consists of two major sections: the upper section and the lower section. The upper section is about 3 m long with a cool region (20–30 cm) at the top where drops are injected, and a warm region at the bottom (2.5–3.0 m) in which the temperature is maintained at 200–400°C. The upper section is heated by flat heating tapes that are wound around the tube and covered by aluminum foil.

The lower section of the furnace is about 2 m long with the upper 1 m surrounded by a large electrically heated tube furnace. Temperatures in this region can be as high as 1300–1500°C. The lowermost part of the furnace is another cool region through which the spheres pass to a collector that can be easily removed from the remaining structure. There is also

Furnace regions (from top to bottom):

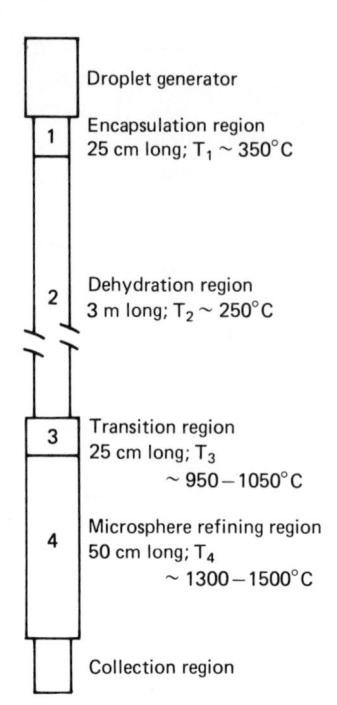

- Droplet generator
- 1. Encapsulation region — 25 cm long; $T_1 \sim 350°C$
- 2. Dehydration region — 3 m long; $T_2 \sim 250°C$
- 3. Transition region — 25 cm long; $T_3 \sim 950–1050°C$
- 4. Microsphere refining region — 50 cm long; $T_4 \sim 1300–1500°C$
- Collection region

FIG. 1. Schematic diagram of the vertical drop furnace.

an outlet in this lower cool region through which furnace gases may be pumped to provide a vertical downflow in the furnace.

Thermocouple sensors at several locations along the furnace provide temperature information used to control the power supplies to the various heating elements along the furnace tube.

C. Hollow-Shell Production from Liquid Drops

The generation of hollow spheres of various materials by evaporation of the solvent from liquid drops has been described by Hendricks and Babil (1972; Babil, 1971). In their experiments, solutions were formed into drops that then passed through a heated column, and the resulting particles were collected and examined microscopically. Inorganic materials generally could be formed into thin, smooth, fragile spheres that were essentially spherical, provided that column temperatures and transit times of drops and particles through the heated column were suitable. Polymeric organic materials could also be formed into spheres, but the spheres had a tendency to collapse into raisinlike shriveled spheres upon drying.

5. GLASS SPHERES

It was also established that drops of various salt solutions gave rise to nonhollow, porous particles in the drying column under some conditions. Using as a starting point the work of Hendricks and Babil, experimenters found that glass spheres can be produced by first drying the drops to a thin, dry, fragile shell in the warm (200–400°C) part of the vertical furnace and then melting the particle to form a glass shell in the high-temperature (1300–1500°C) section of the column.

As an example of the formation of a glass sphere by the liquid-drop process, consider a 200-μm-diam drop of aqueous solution injected into the top of the furnace column. A rapid vaporization of water occurs from the surface of the drop, and, when the outer surface loses water more rapidly than solute can diffuse inward in the drop, the concentration of solute at the surface will rise. When the surface layer becomes saturated with solute, further evaporation results in deposition of the solute at the surface as a more-or-less solid or gel-like layer. Subsequent evaporation is no longer from a liquid, but is through the crust, or membrane, that has formed on the drop. This second phase of the drying process is, of course, controlled by the properties of the crust, the temperature in the column, the humidity in the column atmosphere, and the solvent characteristics (usually water, but may be other solvents as well).

The evaporation of water through a layer on the surface of the drop has been discussed extensively by Ranz and Marshall (1952), Charlesworth and Marshall (1960), and Hendricks and Babil (1972). Temperatures in the upper portion of the furnace column must be well controlled and balanced between being too high and too low. The temperature must be high enough to ensure rapid drying so that the column length may be kept reasonably short and to ensure that the solute vapor pressure will be high enough to keep the shell from collapsing, and the temperature must be low enough to ensure that the drop will not boil and destroy itself by shattering into small drops or fragments and to ensure that the solute vapor pressure will be low enough to avoid rupturing the shell. Particles that are insufficiently dried in this upper section may explode when they enter the high-temperature lower section.

The temperatures for the various sections of the furnace have been found empirically, with guidance to the appropriate ranges being provided by previous work. In the upper zone of the drying section, a temperature of 300–400°C will provide rapid solvent evaporation to form a shell on the drop surface. Once the skin has formed, a temperature of 200–300°C is sufficient to ensure rapid drying of the particle without rupturing the skin by an excessively high vapor pressure within the drop. The rate of production of solvent vapor must be almost exactly matched by the rate of solute diffusion through the skin on the drop; i.e., the internal pressure must be only a small fraction of an atmosphere above the ambient pressure in the column.

As the solvent leaves the drop, more of the solute is deposited in the shell, and the total area of the shell can actually increase during the evaporation process. Shells as large as 2 mm in diameter and only 1000 Å thick have been observed to form from drops that were initially only 200 μm in diameter.

D. Glass Formation and Fining

The processes that occur for hollow, almost spherical, shell precursors and for nonhollow particles as they enter the high-temperature regions of the furnace and form glass spheres are different and will be discussed separately. Let us first consider the hollow-shell situation.

As the spheres come into the high-temperature section, they are rigid, very dry, and somewhat porous. As its temperature rises, the shell begins to melt, and the reactions between the flux oxides and the glass formers begin to form a glassy material. Gases are evolved as a result of decomposition of some of the chemicals in the shell and as the water of hydration of some materials is driven off. The shell enters the higher-temperature region with a relatively low temperature (e.g., 600 K), with an internal pressure of 1 atm, and with a diameter as large as 2 mm. If the shell were gastight, an increase in temperature to, say, 1600 K, even with no internal evolution of gas, would increase the internal pressure to over 2 atm—a pressure certain to destroy the shell. We must, therefore, conclude that the shell is porous at least during the early stages of glass formation.

As the materials in the shell melt, the shell shrinks, and gases initially in the shell and evolved gases permeate through the wall until the diameter decreases to the point where the shell is thick enough to become impermeable to these gases. The shell will continue to shrink until the internal pressure is equal to the external pressure (1 atm) plus the overpressure resulting from surface tension forces of the molten glass. This state is reached rather quickly—probably in less than 0.5 sec.

If the glass-forming process occurs over too long a period (i.e., if the furnace temperature is too low), the spheres will either be misshapen and will have bubbles in the walls or, at best, will have wall thickness defects. If temperatures are too high in the initial glass-forming region, the hollow spheres may collapse to form solid spheres. This collapse could be the result of excessive internal pressures and high rates of gas evolution before the shell has had a chance to become nonpermeable so that, by the time the sphere could contain the gases and maintain sufficient internal pressure to support the wall, the gases are gone. The collapse might also be caused by a part of the wall blowing out, thus allowing the entire shell to collapse, perhaps into an irregular shape that then continues into the remainder of the furnace and is made spherical by surface tension forces.

The glass-forming processes are complex and not totally defined, but seem to be as described. There is no doubt of the occurrence of some permeation of gases through the glass as it is formed and probably even some "blowing" to a larger diameter as the decomposition gases are evolved during the melting process.

Let us consider now the case of nonhollow dry particles entering the furnace to form spheres. The processes leading to formation of a hollow sphere of glass from either a porous, nonhollow intermediate particle (like a miniature snowball) or a dry, irregular particle of mixed glass-forming chemicals are somewhat different from those for the hollow-shell intermediate particle. Again, all the processes are not totally understood, but, as in the methods described earlier for producing commercial glass spheres, the outer surface of the nonhollow particle probably melts first and forms a layer of impermeable molten glassy material around the particle. As the inner temperature increases, or, at least, as the inner material begins to melt and form a glass, gases will be evolved by decomposition, by water of hydration being driven off, or both. These gases will form microscopic bubbles which will expand with concomitant thinning of adjacent walls. Continued bubble expansion leads to perforation of the bubble walls, and, eventually (in perhaps 0.5–1.0 sec), the many small bubbles will form one single bubble that will expand until the internal pressure equals the pressure due to surface tension forces plus the pressure in the hot column.

Both the liquid-drop and dry-particle processes are capable of providing high-quality glass spheres, but, in both cases, the initial hollow glass spheres cannot possibly be as perfect as we see the final product to be. It must be concluded that further refining of the spheres takes place downstream in the remaining high-temperature regions of the furnace.

In the lower-temperature regions of the furnace, vaporization of the solvent in the drop is a gradual process in which all of the solvent does not evaporate instantaneously. It is very unlikely that the evaporation will take place in such a way as to produce a bubble in the exact center of the liquid drop and maintain the bubble at that location during the entire dehydration process. Nevertheless, a large fraction (99% in some batches) of the glass spheres produced have walls in which the inner and outer surfaces are concentric to within better than 1 or 2% of the wall thickness, and the inner and outer surfaces are immeasurably close to being perfectly spherical. Surface tension will produce spherical inner and outer surfaces, but, in a static system of a fluid shell bordered by two nonconcentric spherical surfaces (e.g., a bubble inside a liquid drop), there is no force that will tend to center the two surfaces—i.e., to force the wall to be a uniform thickness. There must, therefore, be dynamic, varying processes occurring in the high-temperature region that lead to forces that tend to center the spherical gas bubble inside the liquid shell, whose outer surface is also spherical.

In the high-temperature region of the furnace, after the spheres are formed, gas bubbles in the walls will either diffuse to the outer surface and disappear or expand until they break through one surface or the other. This diffusion and expansion can introduce local irregularities into the wall thickness that will smooth out if the viscosity is low enough for a sufficiently long time; however, the disappearance of nonconcentricities of inner and outer surfaces would still remain unexplained.

Wang, Elleman, and others at the California Institute of Technology Jet Propulsion Laboratory have shown that several modes of oscillation of one fluid sphere inside another, or of a bubble inside a liquid sphere, will lead to centering of one fluid in another or of a bubble in a fluid shell (Saffren *et al.*, 1981; Lee *et al.*, 1981). This process is relatively rapid and requires only a few oscillations for the centering to take place. Therefore, to apply the results of the JPL group to the case of molten glass spheres whose inner and outer spherical surfaces are initially nonconcentric, but become so as the spheres pass through the high-temperature furnace, we need only find a serendipitous effect to introduce a fluctuation into the spheres. The requirement is for a fluid flow to occur in the molten wall of the glass shell for centering to occur. Let us recall that glass blowers produce uniform walls by alternately heating, blowing, and shrinking a piece of glassware. An expansion–contraction process in the furnace possibly can provide the same effect. As the hollow spheres travel through the furnace, they are in an environment in which the ambient pressure is relatively constant at very nearly 1 atm; however, they are also in a nonisothermal environment as they travel through the furnace. As the spheres fall through the high-temperature region, local temperatures may vary by many tens of degrees longitudinally and perhaps more than that radially. Thus, the temperature in the shell may rise and fall several times after it is first formed and fused into glass. Since the external pressure is constant, the shell must expand and contract in response to the temperature variations, which causes increases and decreases in the internal pressure of the shell. This variation can account for the somewhat surprising uniformity of the glass spheres produced in *all* the several methods of production. The centering forces are weak and, therefore, can only remove small asymmetries from the spheres. However, if the earlier steps are carefully controlled to produce good-quality intermediate particles, the final product can be of exceptionally high quality.

E. Glass Composition and Surface Control

Table II gives the percentage, by weight, of the primary composition used in the drop-generator process. This composition produces a relatively soft glass that is subject to weathering and surface degradation unless special care is used. However, the initial composition of the liquid drops can be varied to form a wide variety of glasses.

The difference between the solution values and the final glass compositions, as shown in Table II, is interpreted as the result of evaporation of some of the sodium and potassium in the high-temperature section of the furnace. This particular composition was chosen because it produces a relatively low-viscosity glass at the temperatures available in the furnace column. While there are glass formulations with lower viscosities in the temperature range of interest (e.g., a pure SiO_2/Na_2O system), such a glass is very susceptible to surface weathering from water vapor in the atmosphere. This surface deterioration is probably due to the formation of NaOH at the glass–air interface with subsequent etching of the underlying glass layers by the NaOH. It should be noted that the surface of the most commonly used glass, which contains boron, will degrade in the presence of water vapor with the production of what are apparently sodium tetraborate crystals on the surface. However, by washing the surfaces of the glass spheres with an experimentally developed solution, surface degradation can be retarded. The wash procedure involves several cycles of washes with a solution of $0.5N$ HNO_3 + $0.1N$ NH_4F followed by washes, in sequence, with hot distilled water, acetone, and, finally ethanol.

Many different glass formulations can be used to form hollow spheres using the liquid-drop technique. However, for the particular system used at Lawrence Livermore National Laboratory, it has been found

TABLE II

Chemical Composition of Solution and Glass Shells in the Liquid-Drop Process

	Solution (wt. %)	Glass shells (wt. %)
SiO_2 (as sodium silicate)	66.3	70.6
Na_2O	22.7	21.9
B_2O_3 (as boric acid)	2.9	2.0
K_2O (as potassium hydroxide)	8.0	5.4
Li_2O (as lithium hydroxide)	0.1	0.1

empirically that high-quality spheres can be produced with the composition listed here. Altering the SiO_2/Na_2O ratio apparently raises the melting point of the glass, which leads to difficulties in formation of the shell. Changing the amounts of boron, potassium, and lithium also decreases the quality and yield of glass spheres. Increasing the boron content leads to the same problems with viscosity as does shifting the silicon-to-sodium ratio; reducing the lithium, potassium, and boron decreases the durability of the glass; and increasing the lithium and potassium content causes problems in subsequent processes that involve heating the sphere, such as in the DT filling procedure.

While the concentricity of the spheres is dependent on the entire high-temperature process, it is clear that the region in which the spheres first melt and form a glass is very critical to the quality of the end product. If the dehydrated spheres in this region are highly spherical and have walls with nearly concentric inner and outer surfaces, the glass spheres will be more likely to have highly uniform walls.

F. Measurements

Because the glass spheres are transparent, the most effective way to examine and measure the spheres in air is to use an optical interference microscope. This microscope provides an interference pattern of the shell that can be quantitatively and qualitatively analyzed for surface concentricity, wall thickness, and wall uniformity. A perfect glass shell (uniform walls, perfect sphere) will exhibit interference fringes that are perfectly circular and exactly centered on the geometric center of the shell. The number of interference fringes is a measure of the thickness of the wall.

To set the best possible operating temperatures in the various regions of the vertical furnace, we performed a series of tests in which glass spheres were produced and samples were examined for various defects. Although this empirical technique was a lengthy process, it enabled us to achieve the very high yields required by the ICF program. A disadvantage of the empirical technique was the lack of an accurate model by which parameters could be set for producing spheres with different wall thicknesses and diameters. For each new size, the empirical process had to be repeated. Recently (1982), however, a model has been developed that can be used to predict temperatures, solution concentration, drop size, column gas flow rates, etc., to produce spheres with particular sizes and wall thicknesses. Some adjustment is required to "tweak" the system to achieve high yields, but the parameter space can be chosen quite well before experiments begin.

Throughout the experiments leading to production of high-quality spheres, it has been advantageous to provide a downdraft gas flow in the furnace to prevent the very light, buoyant spheres from being caught in convective currents and either carried up the furnace or wafted to the walls, where fusion to the surfaces would take place. The downdraft, which is provided by an external pump attached to a side vent at the bottom of the column in the collector region, also serves to sweep out water vapor and evaporated alkali vapors from the furnace. These vapors, if not removed, could condense on the surfaces of the glass spheres, leading to rapid deterioration of these surfaces.

G. Filling Glass Spheres with a Gas

As the dried particles are formed in the upper, warm part of the furnace, their porosity provides a means by which various gases can be introduced into the final glass spheres. The composition of the gases inside the hollow spheres before they enter the melting region of the furnace is nearly the same as in the surrounding furnace atmosphere. This condition has been used to introduce various diagnostic gases—in particular, argon and bromine—into the glass spheres. Up to about 0.5 atm of argon and lesser quantities of xenon and bromine have been introduced by this method. By pressurizing the entire furnace, it should be possible to increase the pressure of a particular gas or gases in a completed shell. This is a very useful technique because gases other than the hydrogens, helium, and neon do not readily permeate through the glass spheres, even at elevated temperatures.

Permeation of normal hydrogen, deuterium, tritium, helium, and neon through the glass spheres at temperatures of 300–450°C has been used to fill the spheres with these gases, all of which move quite readily through the glass walls at these temperatures in a heated, pressurized chamber. After equilibrium has been reached between the high-pressure chamber atmosphere and the gas inside the glass spheres, the temperature is reduced while the pressure is maintained (in some cases as high as 100 atm). When the spheres are cold (room temperature), the chamber gases are pumped out and the gas-filled spheres are removed. Because the permeation rates are exponential functions of the temperature, the gases remain in the glass spheres for long periods (months to years) without significant outward permeation. Even longer gas retention times can be achieved by storing the filled spheres at reduced temperatures (e.g., -30°C) in a freezer.

H. Mass Production of Glass Spheres

Before a process for producing spheres of a particular composition and size can be transferred from a developmental stage to a production stage, a rigorous test procedure must be conducted on the end-product spheres. Spheres that are produced in a new process are first examined under a low-power stereo microscope. If they appear acceptable, they are dry sieved to obtain a distribution of spheres in the desired size range. The sieved spheres are then subjected to a high-pressure process designed to destroy any shells having severe wall defects. In this process, the shells are placed in a Petrie dish containing Freon TE, and the dish is inserted into an apparatus that operates at up to 2000 psi with dry nitrogen. The spheres are then washed with the experimental solution described earlier to stabilize the surface and remove any detritus from the crush and sieving procedures. At least 50% of the spheres must survive the test procedure at this point for the batch to be deemed even marginally acceptable.

After the wash, the spheres are carefully examined by both optical and scanning electron microscopes to establish the quality of the surfaces. The glass spheres are then exposed to air for 1 week, subjected to a temperature of 450°C at ambient pressure for 24 h (which indicates the probable survival rate during DT fill procedures) and allowed to cool. They are then reexamined with the optical and scanning electron microscopes and, to be acceptable, must exhibit little or no breakage or surface deterioration. Finally, the spheres are examined carefully under an interference microscope to determine concentricity and wall thickness and uniformity.

A developmental process is deemed ready for production only if it can be used to produce, on a routine basis, a final yield of from 50 to 90% high-quality spheres that pass all of the above tests. Even after a process is in production, several of the above tests are performed on a regular basis as a preselection and quality-control procedure.

I. Range of Glass-Sphere Sizes

The liquid-drop technique has been used to produce glass spheres ranging in diameter from a few tens of micrometers to a few hundred micrometers. Dry-particle methods have yielded good-quality spheres more than 1 mm in diameter. Using the liquid-drop process, wall thicknesses can be chosen from under 1 μm to more than 30 μm.

J. Spheres Produced from Specialized Materials

A wide variety of materials may be used with the liquid-drop technique for producing spheres. Hendricks and Babil have formed spheres of polymeric materials and of inorganic salts by using this technique. Metallic spheres have also been produced by the same method.

A difficulty that can arise with some glass formulations is a propensity for the solutions to gel in the reservoir before the drops are made. At least two methods can be used to circumvent the gelation problem. One method is to use organic compounds that do not gel to provide the glass-forming constituents. Such compounds include siloxanes, for introducing silicon, and chelating agents, by means of which many other ions can be introduced. A wide variety of compounds can be chosen that will not lead to gelling of the solutions.

A second method is to keep the constituents separate until immediately before the drop is formed. Two or more reservoirs can be employed, with the solutions being mixed as they pass into the drop-formation system. Most solutions of interest do not gel instantly, and the constituent solutions can be mixed in small circulation zones a few milliseconds before the drops are formed.

V. Conclusion

While several processes are available for the production of hollow spherical glass shells, the liquid-drop process described here presently provides batches of shells with the smallest standard deviations in size, wall thickness, and wall uniformity. Because the spheres can be produced at rates of several thousand per second by the liquid-drop process, and because up to 99% of the spheres in a batch meet the very stringent requirements for use as ICF targets, this process has been very useful in the ICF program.

Improvements are still needed in the liquid-drop process. Specifically, better means are required for selecting system parameters and predicting the resultant output, and better efficiency is needed in changing from the production of one sphere size to another. Nevertheless, the liquid-drop process is a major advance in the long-term goal of developing energy-producing inertial-confinement fusion reactors.

References

Babil, S. (1971). Controlled generation of uniform solid aerosol particles with radii in the 0.5 to 20.0 micron range. Ph.D. Thesis, Department of Electrical Engineering, University of Illinois, Urbana.
Beck, W. R., and O'Brien, D. L. (1968). U.S. patent 3,365,315.
Charlesworth, D. H., and Marshall, W. R. (1960). *AIChEJ.* **6,** 9.
Emmett, J. L., Nuckolls, J. H., and Wood, L. L. (1974). *Sci. Am.* **230,** 24.
Hendricks, C. D. (1979). *J. Nucl. Mater.* **85,** 79.
Hendricks, C. D., and Babil, S. (1972). *J. Phys. E* **5,** 905.
Hendricks, C. D., and Dressler, J. L. (1976). Production for glass balloons for laser targets. *Lawrence Livermore Lab. [Rep.] UCRL* **UCRL-78481.**
Hendricks, C. D., Behymer, R. D., Brown, J. A., Heston, G. W., McCann, E. R., and Weinstein, B. W. (1975). Fabrication and characterization of laser fusion targets. *Lawrence Livermore Lab. [Rep.] UCRL* **UCRL-76679.**

Lee, M. C., Feng, I., Elleman, D. D., Wang, T. G., and Young, A. T. (1981). *Proc. Int. Colloq. Drops Bubbles, 2nd, 1981,* p. 107.

Nuckolls, J. H., Wood, L. L., Thiessen, A. R., and Zimmerman, G. B. (1972). *Nature (London)* **239,** 139.

Rayleigh, Lord (J. W. Strutt) (1879). *Proc. R. Soc. London* **29,** 71.

Ranz, W. E., and Marshall, W. R. (1952). *Chem. Eng. Prog.* **48,** 141.

Saffren, M., Elleman, D. D., and Rhim, W. K. (1981). *Proc. Int. Colloq. Drops Bubbles, 2nd, 1981,* p. 7.

Schneider, J. M., and Hendricks, C. D. (1964). *Rev. Sci. Instrum.* **35,** 1349.

Schneider, J. M., and Lindblad, N. R. (1965). *J. Sci. Instrum.* **42,** 635.

Schneider, J. M., Lindblad, N. R., and Hendricks, C. D. (1965). *J. Colloid Sci.* **20,** 610.

CHAPTER 6

Solder Glass Processing

N. N. SinghDeo*
EXEL MICROELECTRONICS, INC.
SAN JOSE, CALIFORNIA

R. K. Shukla
INTEL CORPORATION
LIVERMORE, CALIFORNIA

I. Introduction	170
II. Solder Glass Requirements	171
III. Solder Glass Evolution	172
A. Background	172
B. Glass Fillers	172
C. Vitreous and Devitrifying Glasses	174
D. Solder Glass Compositions	179
IV. Cerdip Process	180
A. Raw Glass Manufacturing	182
B. Glass Glazing	182
C. Cerdip Package Assembly	183
D. Stress Testing	186
V. Glass Properties and Processing Effects	186
A. Thermal Expansion	187
B. Glass Flow and Devitrification	187
C. Chemical Durability	191
D. Outgassing	194
VI. Moisture Outgassing	194
A. General Considerations	194
B. Moisture Measurement Methods	195
C. Moisture Sorption Mechanisms	197
D. Dry Solder Glass Processing	199
VII. Soft Error	200
A. General Considerations	200
B. Soft Error Mechanism	201
C. Radioactive Impurities in Solder Glasses	202
D. α Activity Measurements	202
E. Low-α Solder Glasses	203

* Present address: Indy Electronics, Inc., Manteca, California.

VIII. Future Perspectives 204
   A. Strength 204
   B. Lower Seal Temperature 204
   C. Moisture Level in Sealed Cavity 205
   D. Soft Error 205
   References 205

I. Introduction

The term *solder glass* is derived from the fact that these glasses are one of the lowest-melting glasses available in the industry and are used for joining pieces of glass, metal, or ceramics. The ceramic-to-ceramic joint is utilized primarily in the semiconductor and electronics industries, where sealed ceramic packages protect relatively fragile and environmentally sensitive silicon integrated circuits (ICs). Solder glasses are also utilized in the television industry in joining the face plate of a television to its cone. The television tube is hermetically sealed with the aid of solder glasses for the purpose of attaining and maintaining high vacuum conditions which are required for the proper operation of the electron gun. Solder glasses are primarily used for creating hermetic seals for systems in which metallic solders cannot be used (owing to their low electrical resistance and to their lack of adherence to substances such as glasses and ceramics). Organic adhesives, on the other hand, have high permeability to moisture (Traeger, 1976), making them unsuitable for hermetic sealing.

Emphasis in this chapter will be on the processing of solder glasses after the manufacture of the raw glass. Glass chemistry will be covered to the extent that it affects the user. Processing of solder glasses in the semiconductor industry, in the form of a sealant for "cerdip packages," will be the central theme in the chapter since very stringent demands are made of them when they are used in this form.

The chapter is outlined as follows: First, the basic requirements for a glass to be used as a solder glass will be described, followed by a description of the evolution of solder glass systems to the point at which they exist today. Next, a detailed description of the role of solder glasses in cerdip manufacturing will be given, followed by a thorough discussion of the properties of solder glasses as they affect and are affected by different solder glass processing variables. Two aspects of solder glass technology which have received great attention in the last decade, namely, the outgassing of occluded and dissolved gases from solder glasses (moisture, in specific) and "soft error" contribution will be dealt with in greater detail in the later sections. The chapter will conclude with suggestions for further research and development in the field of solder glasses.

II. Solder Glass Requirements

The term solder glass describes a specific use of glass to create a hermetic seal rather than a specific composition or property. Solder glasses, to be effective as sealants, must meet the following general requirements:

(1) *Low seal temperature* The glasses should be such that they flow and form adherant seals with materials they are intended to seal at temperatures well below their melting points. The flow temperature should be below the range which would damage the system they are intended to protect by forming a hermetic seal, e.g., the IC chip inside a cerdip package. This restriction greatly limits the selection of glasses, as will be seen later.

(2) *Thermal Expansion matching and mechanical durability* The sealed ware is often required to undergo thermal and mechanical shocks and still retain its hermetic integrity. The extent of these treatments depend on the intended use of the sealed ware and can be fairly stringent, e.g., for electronic packages in military use. Good thermal expansion matching (as indicated by closeness in thermal expansion coefficients) of the sealant to the other components of the package is considered mandatory in order to reduce or eliminate residual thermal stresses.

(3) *Chemical durability* The solder glass seal is subjected to harsh chemical attacks during the process of manufacturing the sealed ware. This is particularly true of semiconductor IC packages, which go through an electroplating process after hermetic sealing. The solder glass seal is required to withstand such chemical attacks without losing its hermetic integrity and its high electrical resistance.

(4) *Outgassing control* The glass during processing should not produce any reaction products which are deemed harmful to the system being sealed. For example, outgassing of moisture is of particular concern in IC packages (Section VI).

In addition to the general requirements stated above, solder glasses may have other special requirements imposed on them depending on their specific applications, e.g., soft error prevention in dynamic random access memories (DRAM) in a cerdip package (Section VII). Owing to such requirements, solder glass formulations and their processing have become a unique and technologically important area for research and development.

III. Solder Glass Evolution

A. BACKGROUND

Even though glasses, in general, have been one of the oldest known materials to mankind because of their rather unique properties, the use of glasses for the purpose of hermetic sealing is only of relative recent vintage, inspired mainly by the growth of the electronics and semiconductor industries (Alma and Prakke, 1946). Covalently bonded glass-to-metal seals have been utilized in the vacuum tube industry. For the bond to be effective the processing temperature must be in the vicinity of 1400°C. The process also requires controlled cooling from the bonding temperature to room temperature to reduce residual thermal stresses. These process steps are quite expensive compared to a process in which the entire operation could be performed at a significantly lower temperature. This started the evolution of solder glasses. Researchers at Owens Illinois and Corning Glass Works undertook the development of glass systems which would have good flow properties at temperatures below 500°C and would have sufficiently low thermal expansion coefficient to create a thermally matched seal to the face plate of a television tube. This resulted in selection of the lead–zinc–borate compositions as solder glasses. Lead glasses, in general, are known to have high electrical resistivity coupled with relatively low softening and working temperatures (Table I). Hence the $PbO-ZnO-B_2O_3$ family found ready application in the television industry (Martin and Zimar, 1961).

B. GLASS FILLERS

In the late 1950s, a group of researchers at the Fairchild Semiconductor Company thought that the solder glasses developed by the television industry could also be used to join pieces of pressed and sintered alumina

TABLE I

APPROXIMATE SOFTENING AND WORKING POINTS FOR TYPICAL GLASSES

Glass system	Softening point (°C)	Working point (°C)
Borosilicates (Pyrex)[a]	820	1220
Soda–lime–silica[a]	700	1000
High-lead glass (lead silicate)[a]	380	500
Solder glass (lead–zinc–borate)[b]	320–450	450–570

[a] Source: Holloway (1973).
[b] Source: Frieser (1975).

6. SOLDER GLASS PROCESSING

to obtain hermetically sealed casing for environmentally sensitive silicon integrated circuits. The electrical connections of the chip to the outside would utilize metallic "leadframe" embedded through the glass seal, much like the electrical connections through glasses in vacuum tubes, as shown in a cross-sectional view of a cerdip package in Fig. 1. One problem from such an application standpoint was clearly evident. The thermal coefficients of expansion of these glasses were greatly different from those of the materials to which they were intended to bond. The thermal coefficient of expansion for lead–zinc–borate glasses is in the range of 70–120 $\times$ 10^{-7}/°C, compared to the lower expansion coefficient of alumina-based (pressed and sintered) ceramic substrates $\sim$65 $\times$ 10^{-7}/°C.* The metallic leadframe embedded through the glass seal is either Kovar (Fe–Co–Ni alloy) or Alloy 42 (Fe, 42% Ni) with thermal coefficient of expansion in the range of 40–47 $\times$ 10^{-7}/°C. Development of solder glasses for this application proved to be a great challenge since, in general, a change in glass composition to reduce the thermal expansion coefficient usually results in an increase in the working temperature of the glass. The solution found was to artificially lower the thermal expansion coefficient of the glass without sacrificing its low processing temperature capability (high PbO content) by uniformly dispersing an inert secondary phase in the glassy matrix. The additives, known as *glass fillers,* have very low expansion coefficients and are usually refractory materials. Table II lists some common fillers and their thermal expansion coefficients. The solder glasses for such applications, therefore, are "composites" in nature, consisting of a glass base with inert and low-thermal-expansion additives (fillers). The quantity or volume fraction of the additives needed would depend on their respective thermal expansion coefficients. If the thermal

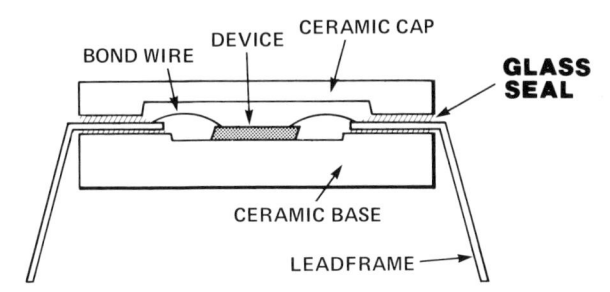

FIG. 1. Cross-sectional view of a cerdip integrated-circuit package showing the glass sealant.

* The ceramic substrate is typically 92–96% alumina with glassy binders and coloring agents (Graham and Tallan, 1971).

TABLE II

Coefficient of Thermal Expansion[a] for Various Glass Fillers

Filler	CTE × 10⁷/°C	Reference
β-eucryptite (Li$_2$O · Al$_2$O$_3$ · 2SiO$_2$)	−65	Rabinovich (1979)
Boron carbide (B$_4$C)	43	Rabinovich (1979)
Silicon carbide (α-SiC)	40	Rabinovich (1979)
Zircon (ZrSiO$_4$)	45	Graham and Tallan (1971)
Cordierite (2MgO · 2Al$_2$O$_3$ · 5SiO$_2$)	23	Graham and Tallan (1971)
Tin oxide (SnO$_2$)	40	Samsonov (1962)

[a] All the values are mean CTE for polycrystalline materials in the temperature range of 25–400°C, except β-eucryptite (25–325°C).

expansion coefficient of an additive is very low (or negative, as in the case of β-eucryptite), comparatively lower quantities of the additive are needed to match the thermal expansion coefficient of "composite" solder glass to that of the substrate (alumina). When solder glasses are utilized as package sealants in the semiconductor industry, they typically contain 20 vol. % of additives. The CTE of these glass-filler composites can be predicted by using Kingery's model. In applying this model (Kingery *et al.*, 1976), it should be noted that the effectiveness of a filler in controlling the CTE of the composite depends on both the CTE and the Young's modulus of the filler, as confirmed by Rabinovich (1979) for various possible fillers.

Figure 2 shows SEM micrographs of three different solder glasses after cooling from seal temperatures. Figure 2a shows a two-filler glass system (KC-405), whereas Fig. 2b corresponds to glass with a single filler (XS1175-M1). Figures 2c and 2d correspond to a devitrifying glass (7583) before and after devitrification (Section III.C), which is also a single-filler system. Table III lists some of the code names and manufacturers of commercial solder glasses.

C. Vitreous and Devitrifying Glasses

The addition of fillers to the base glass provided the CTE parity needed for matched seals in IC packages in the 1960s. Later on, increased silicon IC complexity required the seal temperatures to be brought down even further (<450°C), which has led to the evolution of two distinct families, the vitreous and the devitrifying glasses. (Prior to that the glasses were mostly devitrifying.)

Devitrifying glasses are analogous to thermosetting polymers; i.e., they solidify after softening, when held at high temperatures for some length of time. Vitreous glasses, on the other hand, continue to remain fluid at comparable high temperatures and therefore can be reheated and "worked" a few times without any appreciable loss in their flow characteristics. Figure 3 shows a schematic representation of the viscosity versus time for two different types of solder glasses.

Devitrification occurs by nucleation and growth of thermodynamically stable crystalline phases in the glass at temperatures sufficiently high that ionic mobility is adequate to support it. This results in the formation of a glass–ceramic body consisting of fine crystallites embedded in a glassy matrix. The presence of crystallites that are closely packed and of random orientation makes the resultant glass–ceramic (devitrified glass) stronger from a crack propogation viewpoint than the original glass (Hasselman and Fulrath, 1965). Devitrification has been shown to increase the glass seal strength by as much as 150% (Forbes, 1967).

Lamson and Ramsey (1978) have shown that the modulus of rupture (MOR) values of devitrified solder glasses also improved with reduction in the crystallite size of the devitrified crystals (reduction in crystallite grain size from 34.4 to 15.2 μm increased MOR from 6.02 to 14.28 kpsi).

The devitrification process of $PbO-ZnO-B_2O_3$ glasses has been studied extensively by various researchers (Forbes, 1967; Ramsey, 1971, 1972; Andrus and Powell, 1972), and leading to one of the key conclusions that higher ZnO content (>%10) promotes devitrification. The exact mechanism of nucleation and growth of crystallites in $PbO-ZnO-B_2O_3$ glasses is not well understood, although it is believed to be occurring first at the surface (Rabinovich, 1979; Pavlushkin and Kalmanovskaya, 1976).

The crystal structures of the different phases occurring in the $PbO-ZnO-B_2O_3$ glasses have been discussed by Petzoldt (1966). Ramsey (1971) has shown that the phases found in a typical devitrified glass (CV97 glass manufactured by Owens Illinois) are $2PbO-ZnO-2B_2O_3$ (2:1:2 phase) and $PbO-2ZnO-B_2O_3$ (1:2:1 phase). The same study (Ramsey, 1971) suggests that the 1:2:1 phase is undesirable as far as mechanical strength of the seal is concerned.

The choice between vitreous and devitrifying glasses depends on two basic factors. If seal strength and retention of hermetic seal integrity are of prime importance, then a devitrified type of seal glass will be preferred. If lower seal temperature is of greater significance, then a vitreous glass will be the right choice. Everything else remaining the same, i.e., thermal expansion matching, chemical stability, etc., a devitrified glass is mechanically stronger than a vitreous glass. Overall, the choice of a vitreous glass over a devitrified variety does involve some sacrifice in the mechanical strength of the glass seal. The trend in the semiconductor industry is

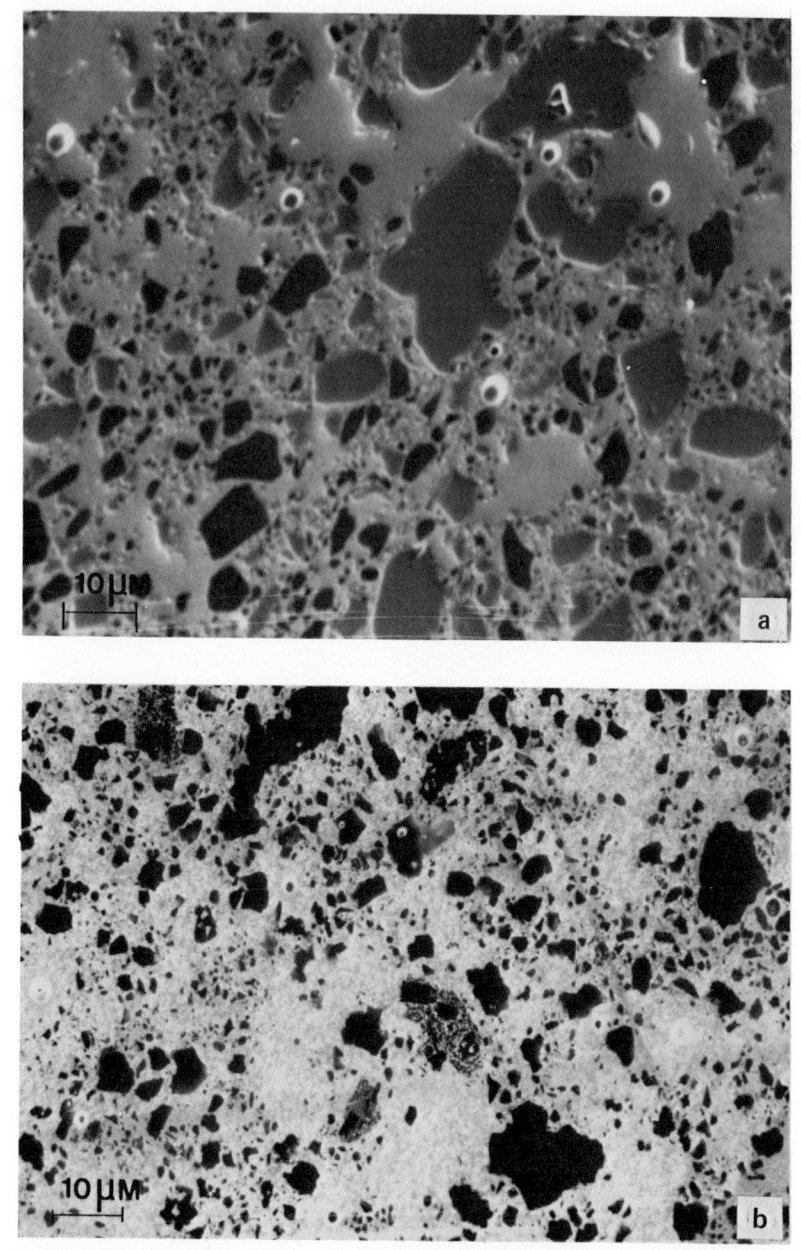

FIG. 2. SEM micrographs (700×) of polished cross sections of various solder glasses: (a) LS2001 (KC405) vitreous glass with a two-phase filler system; (b) XS1175-M1 vitreous glass with a single filler; (c) 7583 glass (devitrifying), single filler, prior to devitrification; (d) 7583 glass after devitrification.

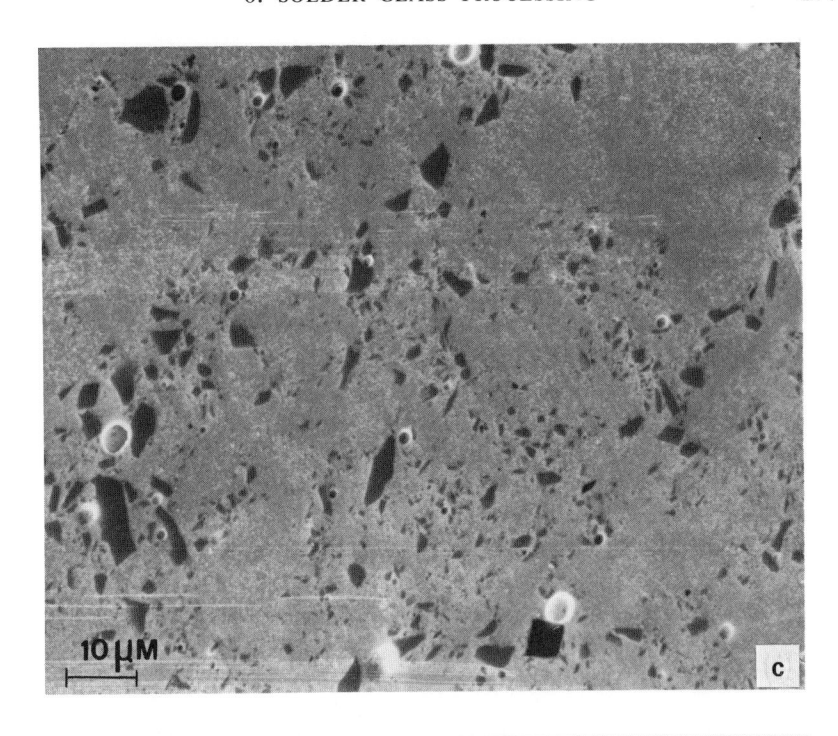

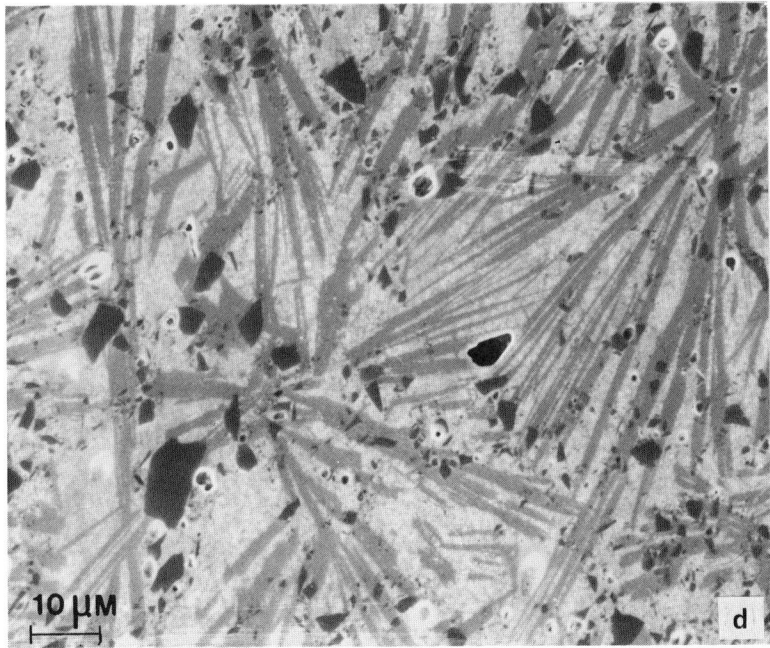

FIG. 2 (*continued*).

TABLE III

Some Common Commercial Solder Glasses Used in Semiconductor Packaging

Glass code(s)[a]	Manufacturer[b]	Devitrifying?	Filler(s)
7583	CGW	Yes	$ZrSiO_4$
CV111	OI	Yes	$ZrSiO_4$
XS1175-M1	OI	No	β-eucryptite
SG-200	OI	No	β-eucryptite, $PbTiO_3$
LS0113 (KC1M, NCG560)	NEG	No	$PbTiO_3$, ZrO_2
LS2001 (KC405)	NEG	No	Willemite, SnO_2
TG191BF (KC402)	CGW	No	Cordierite
LS0803 (KC400, NCG-564)	NEG	No	Not available
LS0110 (KC1, NCG-556)	NEG	No	Not available

[a] Manufacturer's codes. Alternative codes (in parentheses) are used by glazers.
[b] CGW: Corning Glass Works (U.S.A.), OI: Owens Illinois Corporation (U.S.A.), NEG: Nippon Electric Glass Company (Japan).

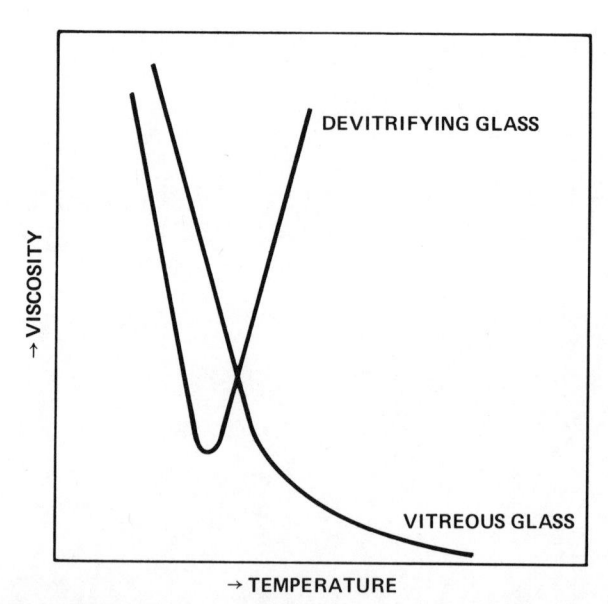

Fig. 3. Schematic drawing of solder glass viscosity versus time at seal temperatures for vitreous and devitrifying glasses.

D. SOLDER GLASS COMPOSITIONS

Figure 4 shows a ternary phase diagram for the $PbO-ZnO-B_2O_3$ family, with the phase fields for vitreous and devitrifying glass compositions at temperatures in the vicinity of 500°C indicated. Solder glasses usually contain 70–80 wt. % PbO, which provides good flow at lower temperatures. Zinc oxides ranging anywhere from 10 to 15%, is added to promote devitrification in devitrifying type of glasses. Silicon dioxide is a glass former, and vitreous glasses contain a relatively larger quantity of it compared to the devitrifying type of glasses. Addition of SiO_2 to the glass network also improves the chemical resistance of these glasses. Some glass manufacturers also add metal (Fe, Co) oxides in small amounts to "stain" the glass a darker color. The effect of various additives on the devitrification and chemical resistance properties of solder glasses has been critically reviewed by Frieser (1975) and Shirouchi (1969).

Table IV lists the chemical composition of various commercial solder glasses as measured by emission spectrometry. The analysis is only semiquantitative and does not detect nonmetallic elements (P, As, halides, etc.). Note that the presence of fillers (20–30 vol. %) in the glass has

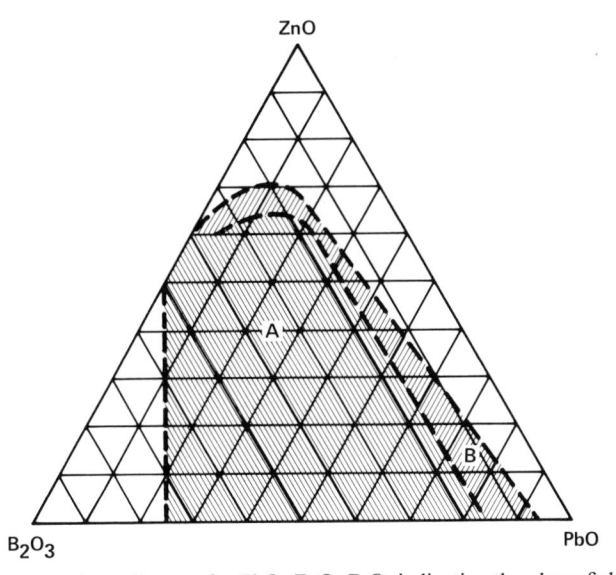

FIG. 4. Ternary phase diagram for $PbO-ZnO-B_2O_3$ indicating the phase fields for vitreous and devitrifying solder glasses. (A, clear glass region; B, devitrifying glass region.)

TABLE IV

Emission Spectrographic Analysis of Common Solder Glasses (Including Fillers)

Element[a]	CV111	7583	KC1M	TG191BF	XS1175-M1
B	10	12.5	12.5	12.5	10
Mg	—	—	0.02	2.5	0.02
Al	0.025	0.3	0.5	3.0	5.0
Si	8.5	7.5	4.0	12.0	15.0
Ti	0.04	0.025	8.0	—	0.5
Cr	—	—	0.25	0.03	—
Fe	0.01	0.01	0.75	0.05	0.08
Co	—	—	0.15	0.05	—
Zn	8.5	10.0	—	3.5	2.5
Li	—	—	—	—	2.0
Zr	12.5	8.5	7.0	—	0.02
Hf	0.25	0.2	—	0.5	—
Sn	—	—	—	—	—
Ba	2.0	2.0	0.01	—	0.5
Pb	Balance	Balance	Balance	Balance	Balance

[a] All elements are reported as wt. % of the corresponding oxide.

lowered the concentration of PbO compared to the expected values from the $PbO-ZnO-B_2O_3$ phase diagram.

Table V lists some important properties of various commercial glasses. It may be observed that the seal temperature and the dwell time at the seal temperature are generally lower for the vitreous glasses than for the devitrifying glasses.

IV. Cerdip Process

Figure 5 shows different types of solder-glass-sealed packages which are used for encasing integrated circuits. Of these, the "cerdip" package is the most common and shall be utilized as an example to demonstrate typical processing of solder glass sealants. From the glass viewpoint, cerdip processing can be broadly classified into three major steps–raw glass manufacturing, cerdip parts manufacturing (glazing), and assembly of these with the IC chip (sealing, electroplating, etc.) to create a finished, hermetic package.

6. SOLDER GLASS PROCESSING

TABLE V
Properties of Solder Glasses[a]

Glass code	Seal temperature (°C)	Softening point (°C)	Density (g/cm³)	CTE (30–250°C) 10^7 °C^1
CV111[b]	485	380	5.92	68.8
7583[b]	485	370	6.0	84.0
XS1175-M1	430	385	4.7	60.5
KC1M	450	400	6.8	64.0
KC402	415	350	5.2	68.0
KC405	415	390	6.16	67.5
KC400	430	350	7.2	65.9
KC1	450	397	4.8	52.5
SG-200	400	345	5.61	63.0

[a] Data supplied by glass manufacturers.
[b] Devitrifying compositions (all the others are vitreous).

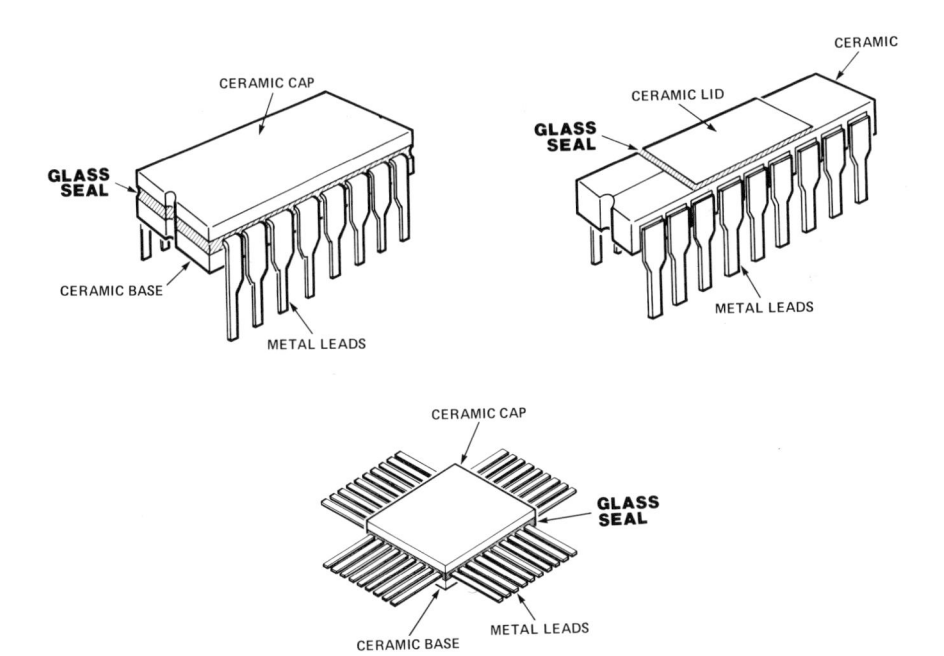

FIG. 5. Illustrations of variously configured electronic packages showing solder glass hermetic seals.

A. Raw Glass Manufacturing

Figure 6 shows the process flow chart for raw glass manufacturing. The glass batch is melted. Since lead is a highly toxic material and the components of the Pb–Zn borate glasses have high vapor pressure, appropriate precautionary steps are undertaken to guard against hazards to personnel. The glass after the melting and the subsequent solidification operations, is milled in a ball mill to attain a certain range in particle size. The milled and sized glass is blended with modifiers, which also have a particle size range. The particle size distribution of glass powder and filler is strictly controlled because it affects further processing (typically, 100% through 100 mesh and ~90% through 400 mesh). The finished product is a composite glass powder which is sent to the glazer for the next processing step. Powdered glass is stored in dry atmosphere, since these glasses readily adsorb moisture.

B. Glass Glazing

The glass powder (with fillers) is used by a glazer to produce components of a cerdip package, namely, bases and lids (known as piece parts).

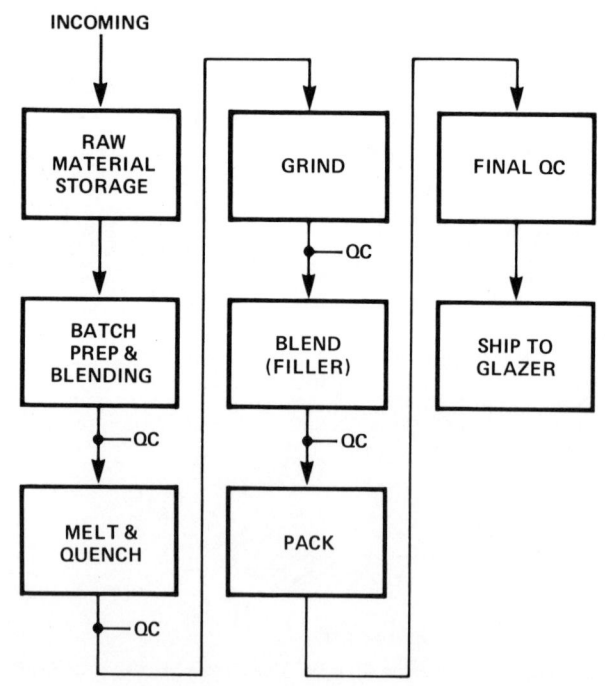

FIG. 6. Flow chart showing the key steps involved in the manufacture of raw solder glass powder.

This is accomplished by screen printing a layer of glass slurry (comprising solder glass, organic resins, and solvents to produce a paste of controlled rheology) on the alumina ceramic, followed by drying and glazing steps (Fig. 7). To control the glass layer dimensions to the required specifications, screen printing and drying operations are repeated several times (two or three), thus building the ultimate thickness in several small steps. This is followed by glazing. During the drying process, most of the organic solvents are driven off. This is a low-temperature operation, usually taking place below the glass softening point.

During glazing the glass is heated above its softening point, which causes the glass powder to sinter and lightly bond to the ceramic substrate. The cerdip caps and bases piece parts produced in this manner are then sent to an integrated-circuit house for further assembly.

C. Cerdip Package Assembly

Figure 8 shows schematically the various key steps involved in the assembly of a cerdip package; these include the following:

(1) *Lead attachment* In this operation the lead frame (Fe, 42% Ni alloy) is embedded into the glass layer on top of the base. This operation requires the glass to be fairly fluid and typically is carried out at high temperatures. For example, the glass could see temperatures as high as 600°C for 1–3 sec in high-speed operations or could see temperatures around 450°C for close to a minute in low-speed operations. The glass property of significance at this stage is its flow. The glass should be able to flow readily at this stage. It is not important at this stage that the solidified

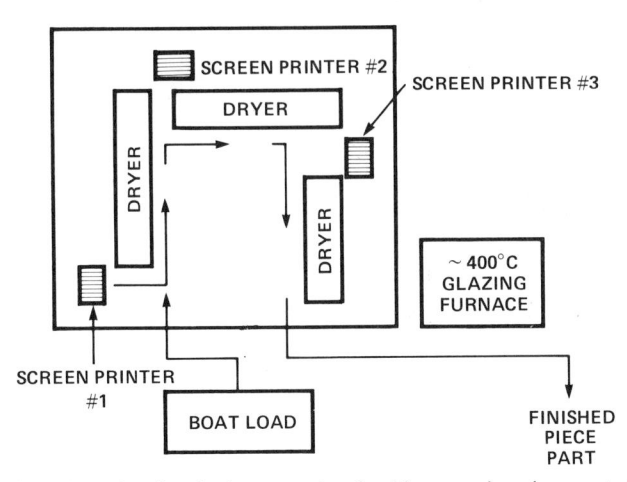

Fig. 7. Flow chart showing the key steps involved in preparing piece parts for hermetic packaging.

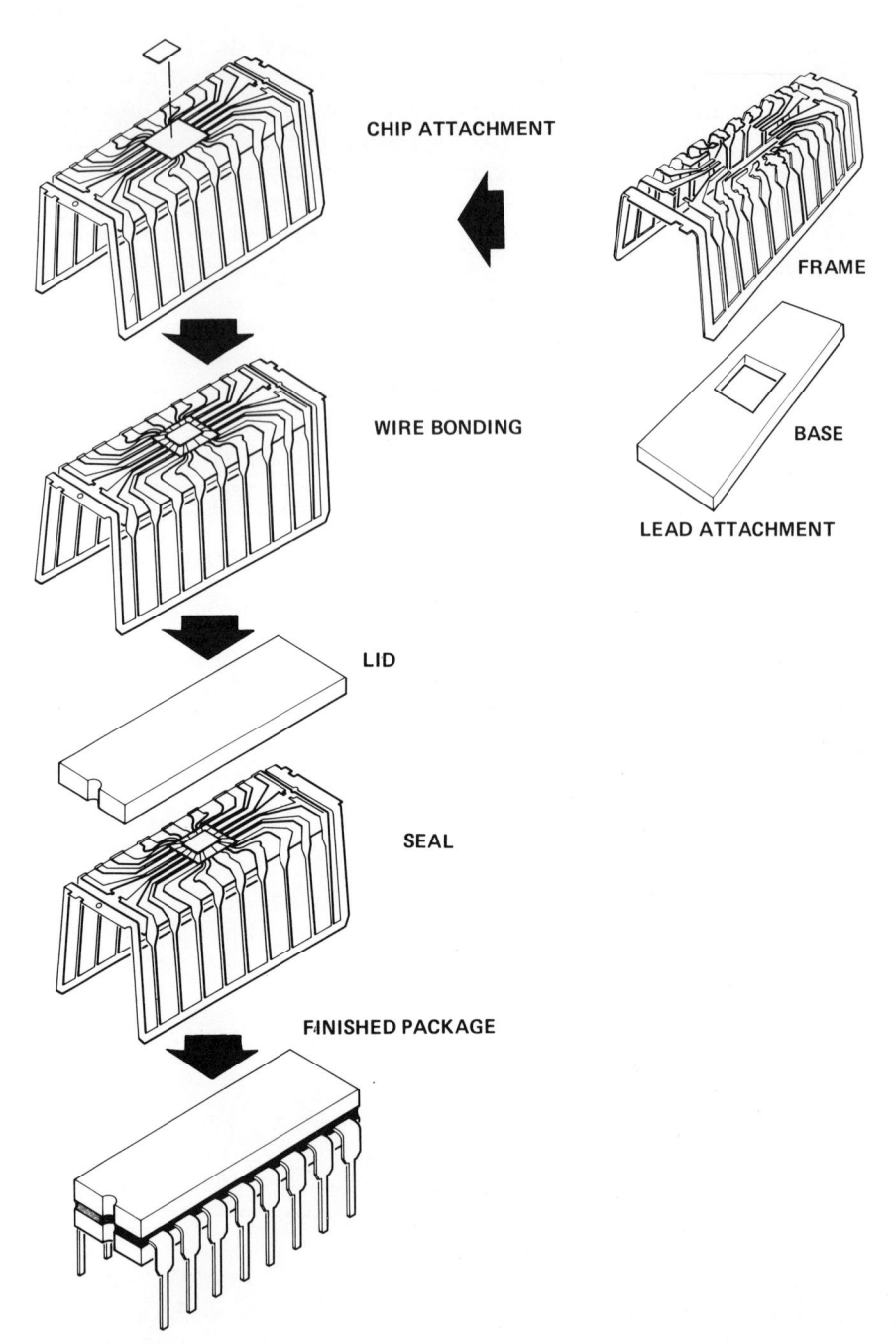

FIG. 8. Schematic representation of key steps in cerdip assembly which involve solder glass.

glass be stress and crack free, since it can relieve the stress and heal the crack subsequently on remelting during final seal. For devitrifying glasses, it is especially important to control the time and the temperatures seen by the ceramic base at lead attachment, since it reduces the glass "life" (i.e., the "glass flow" time prior to complete devitrification during seal), thus increasing glass viscosity at seal.

(2) *Chip attachment* The silicon chip (IC) is attached to a cavity in the center of the base by means of a solder material. The most common solder used for this purpose is Au–Si alloy of eutectic composition, which melts at 363°C. However, to attain faster throughput, this operation is mostly carried out at temperatures over 400°C. If the temperature at die attach is greatly in excess of 400°C and the time is also relatively long, the attendant loss of glass flow at this operation can cause incomplete seals at the seal operation.

(3) *Wire bonding* This operation involves attachment of fine wires (~30-μm diam) ultrasonically from the chip surface (bond pads) to the lead fingers on the leadframe. This operation is done at room temperature and hence is benign toward the glass. The only requirement imposed on the glass is that it should continue to hold the metallic lead fingers firmly in such a manner that the ultrasonic vibration does not dislodge the leads from the glass.

(4) *Sealing* In this step a ceramic lid is attached to the assembled package, which is described in the previous three sections, to create a hermetic seal. The base unit with the attached chip and the bonded wires is put face down on the lid. A holding fixture known as a "seal boat" keeps the base and lid aligned properly. The seal boat (with several of these cerdip components) goes through a belt furnace, in which heating is usually done by a convective system. The glass layers on the lid and base melt during their passage through the furnace and produce a hermetic seal. The control of sealing parameters (temperature profile, atmosphere, etc.) is of paramount importance in controlling the behavior of the solder glass.

(5) *Lead finish* Usually, the cerdip parts are electroplated after the seal operation. The most common coat applied on the leadframe is tin, although in some cases an Au coating is also applied. Sometimes the tin-plated parts are tin reflowed by passing the parts through a conveyer furnace containing an inert atmosphere. Wave soldering of the tin-plated parts is also done by passing them through a molten Pb–Sn solder wave. Finally, after the lead coating operation, the tie bar that connects the leads at the bottom is sheared off so that individual leads become electrically isolated on the outside of the package. From the glass viewpoint, the lead finishing operation imposes the following demands. The electroplating operation subjects solder glass seal to harsh mineral acids, while wave

solder operation can cause severe thermal shock, jeopardizing the hermetic seal integrity.

D. Stress Testing

The assembled hermetic package is subjected to different kinds of tests to simulate, in an accelerated manner, various environments that it would be expected to withstand during its life span. An understanding of such tests is necessary to properly design the hermetic seal for success. These tests also impose measurable performance requirements on the solder glass seal. The key tests involved are the following:

(1) *Insulation resistance* This test is intended to determine the behavior of the glass sealant under conditions of high humidity and fluctuating temperature. The temperature range of cycling is from 100 to 0°C. The humidity is maintained at 95% RH. The end point is an electrical test, where a bias of up to 300 V is applied between the adjacent leads of the package and the leakage current is determined. The glass would be considered to have successfully completed the test if it stays highly resistive (current below 10^{-7}A) at the conclusion of the test.

(2) *Temperature cycling* This test simulates the sudden temperature changes to which a solder glass could be exposed. The temperature range of cycling in the most stringent case is -65 to $+150°C$. The medium in which the temperature extremes are obtained could be either air or a fluorocarbon liquid.

(3) *Torque testing* In this test, a vise grips the bottom part of the package and the cap is torqued by a calibrated torque wrench. The maximum torque is usually recorded. Intermediate torque readings indicating the onset of failure have also been recorded in some instances. The onset of fracture is detected by sensitive audio detectors. The maximum and intermediate torque readings have been used as rough guidelines for the mechanical strength of the package. These readings have been used to compare different glass types, processing parameters, and over-all designs.

In addition to the above-mentioned standard tests, similar other test procedures are carried out to test the seal integrity by utilizing either destructive or nondestructive test methodologies.

V. Glass Properties and Processing Effects

Discussions in the area of assembly and the stress tests that are performed on the assembled cerdip package emphasize several areas that were suggested in the introduction to this chapter, i.e., thermal compatibility of the different components of the package (especially the glass, which is the weakest of the components in tension), flow, or viscosity of the glass, chemical durability of the glass, and electrical resistance of the glass. The key glass properties and how they need to be controlled in order to meet such stringent criteria in an effective manner are described in the following subsections.

A. THERMAL EXPANSION

As has been noted earlier, different kinds of modifiers are used to bring about better thermal matching of the glass with the other components of the cerdip package, alumina in particular. Hogan (1971) has indicted that for the glass seals to be good, the thermal expansion coefficients of the different components of the seal, i.e., glass and ceramic, should not vary by more than 100–500 ppm. Even though the solder glass fillers are considered to be inert, some dissolution of fillers into the glass at high temperatures is possible. Figure 9 shows schematically the different thermal treatments solder glass experiences in assembly of a cerdip. Obviously, prolonged treatments at these temperatures can alter the thermal expansion coefficient of the glass as a result of chemical reaction with the fillers. Since the fillers are usually crystalline materials, any rounding of edges and corners of filler particles in the glass, after heat treatment, would indicate dissolution. The glass manufacturers recommend guidelines for glass seal time–temperature profiles; usually these guidelines do not produce appreciable additive dissolution, but a check of the thermal expansion of the glass after a simulated seal cycle should be made to verify their claim.

B. GLASS FLOW AND DEVITRIFICATION

For a proper seal to be created, the glass must be quite fluid during sealing. In the case of a vitreous glass, its viscocity continues to decrease

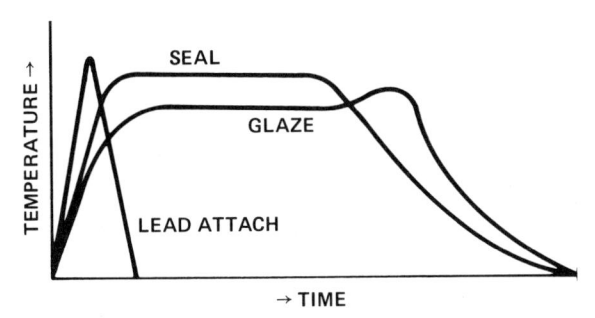

FIG. 9. Schematic of thermal treatments undergone by solder glass during cerdip manufacturing.

with increase in temperature, as stated earlier, but an upper limit on seal temperature is needed to limit filler dissolution, thus limiting the ultimate fluidity that can be obtained. In the case of devitrifying glasses, the glass viscosity changes drastically at seal temperatures during devitrification, which will be dealt with in the following sections.

Thermal analysis of solder glasses (DTA/DSC) is of great importance in selecting seal temperatures and has been treated by several authors (Ramsey, 1971, 1972; Andrus and Powell, 1972; Forbes, 1967). Figures 10 and 11 show DSC curves obtained for solder glasses XS1175-M1 (vitreous), SG-200 (vitreous), and 7583 (devitrifying). The exotherm in Fig. 11 corresponds to the maximum devitrification rate (and hence is not present in the DSC of vitreous glasses in Fig. 10) and can be taken as the point at which rapid increase in viscosity begins. Since nucleation/growth is a time–temperature-dependent phenomenon, the heating rate affects the temperature at which maximum devitrification takes place. The higher the heating rate, the higher the maximum devitrification temperature (Ramsey, 1971). The effect of heating rate on the devit peak (maximum devitrification temperature) is quite significant, as shown in Fig. 12. The difference between 9 and 24°C/min moves the devit peak for a CV97 type of

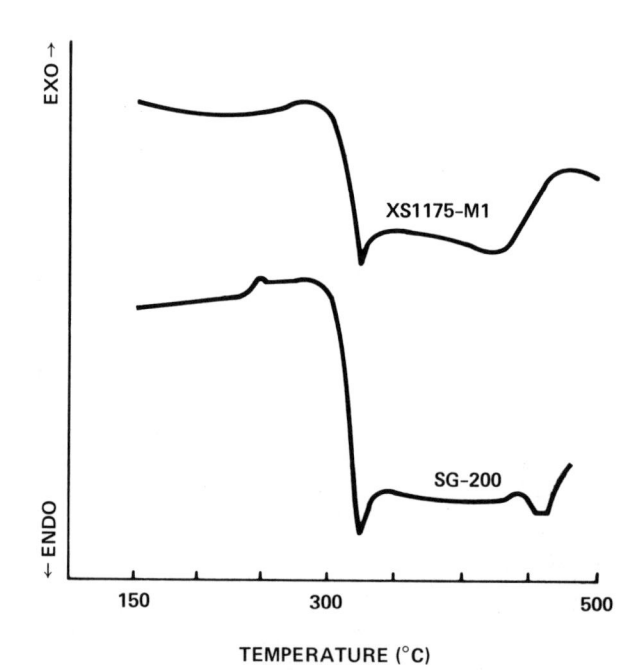

FIG. 10. DSC curves for two vitreous glasses, XS1175-M1 and SG-200. Heating rate is 20°C/min, ambient air.

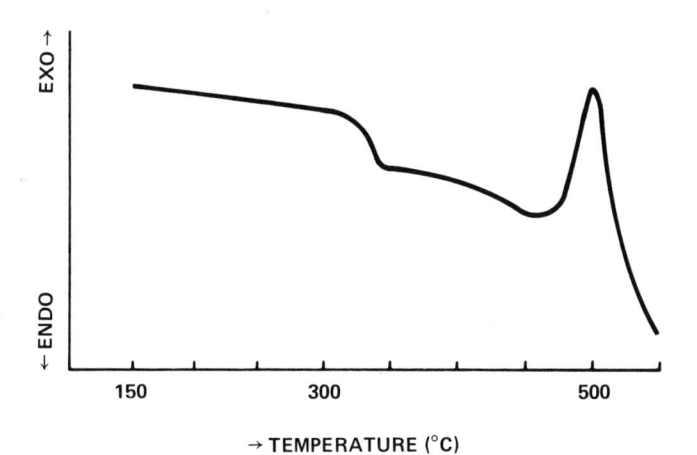

FIG. 11. DSC curve for a devitrifying glass (7583) showing the exotherm (devitrification peak). Heating rate is 10°C/min, ambient air.

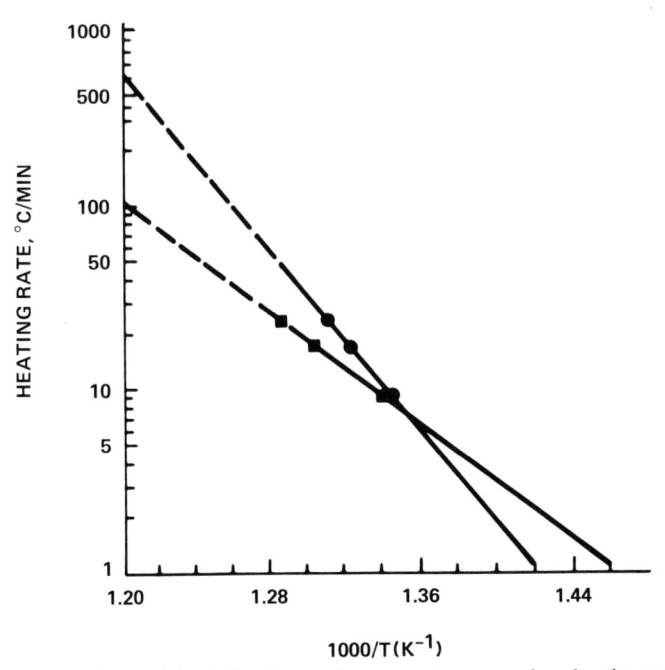

FIG. 12. Dependence of devitrification peak temperature on various heating rates for the solder glasses: ●, CV97 and ■, CV98. [After Ramsey (1971).]

devitrifying glass from 460 to 490°C. Increase in the devit peak temperature extends the temperature range for good fluidity. Hence heating rates of up to ~120°C/min are used in sealing devitrified glasses to obtain good glass flow during heating to the seal temperature. The devitrification process is nucleation and growth related. Hence the rate of devitrification is temperature dependent. The higher the temperature is, the higher is the rate of devitrification. Figure 13 shows the sealing range recommended by Kyocera International (1983) on their data sheet for CV111 glass from Owens Illinois. It can be observed from the diagram that as the seal temperature is increased, the hold time is correspondingly reduced. The intent, of course, is to obtain total devitrification to attain maximum strength in devitrifying types of glasses.

Another variable that influences glass flow is the contact angle of the glass fluid with the ceramic (alumina) substrate: the authors have found that the contact angles of solder glasses vary slightly. The authors' experiments have yielded contact angles of ~30° in glasses such as XS1175-M1 on alumina substrate in air. Other data from the literature suggest that the solder-glass contact angle varies, but when the glass is readily flowing, the contact angle is in the 30–40° range (Broukal, 1962; Espe, 1951; Pask and

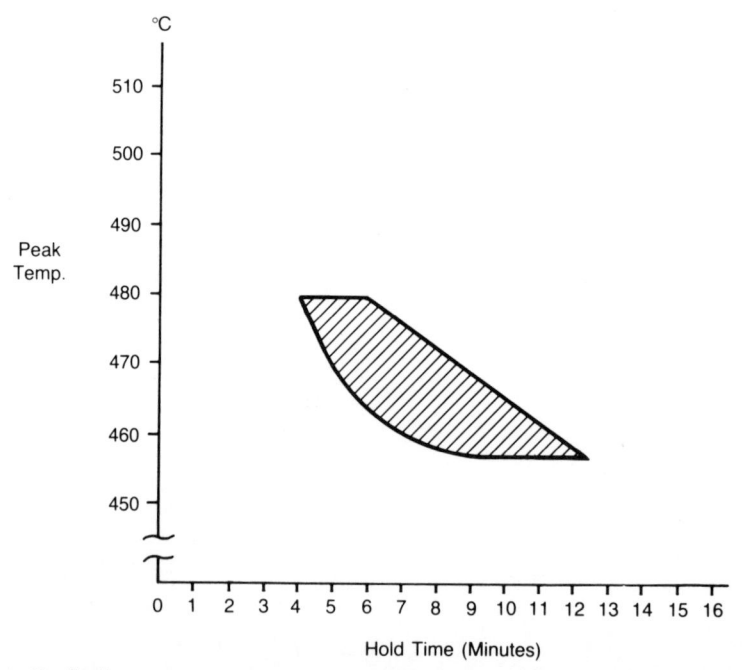

FIG. 13. Sealing-parameter range (time and temperature) recommended for a devitrifying glass, CV111, by Kyocera International (1983).

Fulrath, 1962). These contact-angle values imply imperfect wetting between the solder glass and the ceramic substrate. The implications are twofold. First, care needs to be exercised during the piece-part manufacturing stage so that no impurities (which may further increase the contact angle) are present on the surface of the ceramic. Second, a relatively higher seal temperature is recommended during glazing, since data (Broukal, 1962) show that the contact angle gets smaller at higher temperatures. Hence a high glazing temperature (without impairing other properties) is recommended.

The relatively high contact-angle values require that an adequate amount of glass is available, in both cap and base, to form seals that are free from internal voids. On the other hand, since the solder glass is the weakest member of the package sandwich, its thickness should be maintained as low as possible to increase the joint strength. The glass thickness used in a given piece part is chosen as a compromise between the above two divergent factors.

If glass thickness is not sufficient to maintain the equilibrium contact angles, it can break up into higher "harmonics," causing islands of glass to appear (Fig. 14), as well as internal voids, as stated earlier. The proper design of leadframe which spreads the glass around it can minimize this phenomenon (Fig. 15), thus allowing lower glass thicknesses to be used on the piece parts.

C. Chemical Durability

Lead glasses have low chemical resistance to mineral acids in general. The behavior of solder glasses in different chemical environments is shown in Table VI. It should be apparent that among the acid treatments sulfuric acid is the least harmful, while HNO_3 is the most. This is generally true of all lead-based glasses since $PbSO_4$ is insoluble in water while $Pb(NO_3)_2$ is not.

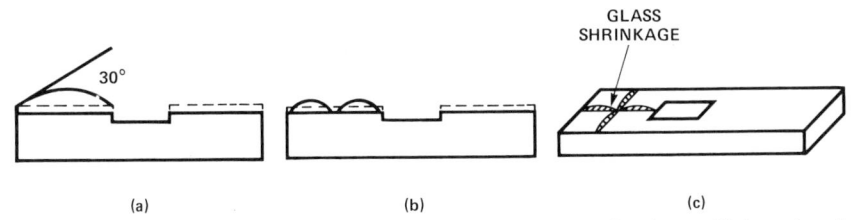

FIG. 14. Effect of contact angle: (a) equilibrium contact angle when sufficient glass is present, (b) second harmonic (glass separation) when insufficient glass is available to maintain equilibrium contact angle, and (c) glass depressions caused by insufficient glass on the substrate.

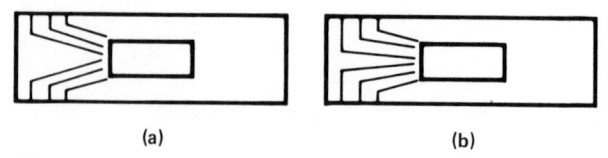

FIG. 15. Different leadframe designs: (a) large areas of glass not covered by leadframe (poor design) and (b) good design of leadframe, with glass clustering prevented.

The process of electroplating cerdip packages has been designed with this in mind; e.g., tin plating of H_2SO_4-rich baths and gold plating in approximately neutral baths (Lea Ronal Aural 92) are also done successfully. Tin–lead electroplating has not been successful because the plating bath contains fluoboric acid which rapidly attacks the glass during the plating process, thus reducing the seal strength. Attempts to plate nickel in sulfamate or Watts baths or to plate copper in cyanide baths have not been successful, again owing to attack on the glass by the chemicals in the plating solution, which either attack the seal glass or render it conductive at the surface. Exhaustive attempts with several different plating systems have not been tried because the glass is quite sensitive to chemical treatments and the electroplated tin has served the purpose quite well.

Plated metal on glass surfaces has been observed occasionally in tin-plated parts. This can cause electrical shorting between adjacent leads, thus defeating the purpose of having glass as an electrical insulator. This phenomenon is caused by the presence of elemental lead on the glass surface, which causes tin to electroplate over it. This is possible if the sealing furnace atmosphere is deficient in oxygen, since lead oxides are easily reduced. Surface reduction of solder glasses can also be caused by improper electroplating conditions, which are described as follows.

(1) If the current density is quite high, excessive hydrogen evolution at the cathode (metal) leads to PbO reduction in the glass area adjacent to the metal leads.

(2) If the electroplating bath agitation is improper or the load is too large, the gaseous hydrogen evolved at the cathode does not have an opportunity to escape out of the electrolyte, thus causing reduction of the PbO.

Besides the reduction of PbO to metallic lead, surface crystallites which are conductive in vitreous glasses may be present. The surface crystals, in turn, participate in creating metallic films on glasses during the electroplating process. The conductive crystals are usually associated with ZnO contamination at seal.

TABLE VI
Chemical (Acid) Durability of Solder Glasses[a]

Treatment	XS1175-M1	CV111	LS0113	TG191BF	SG-200	KC405
50% H_2SO_4, 1 h, at 95°C	8.65	6.45	2.32	5.89	0.88	—
50% H_2SO_4 + 25% HNO_3, 5 min, at 25°C	2.37	0.88	2.12	2.11	1.50	—
5% HNO_3, 5 min, at 25°C; then 50% H_2SO_4, 15 min, at 95°C	80.9	74.4	96.4	56.7	62.7	—
1N HCl, 5 min, at 25°C	1.3	3.4	0.7	0.9	—	—
18N H_2SO_4, 5 min, at 50°C	1.1	1.4	0.2	0.4	—	—
10% HNO_3, 10 min, at 20°C	—	—	210	—	—	138
10% HCl, 10 min, at 20°C	1.0	6.4	3.0	—	—	1.9
10% H_2SO_4, 10 min, at 20°C	0.14	0.08	0.1	—	—	0.6

[a] Measured as normalized weight loss (mg/cm^2) after washing off the reacted layers; data supplied by glass manufacturers.

D. Outgassing

As mentioned before, the glass powder is blended with various organic solvents and resins to prepare a paste for screen printing. The commonly used solvents are pine oil, α-terpeniol, etc.; the common resins or binders are poly-α-methyl styrene or cellulose acetate. Release of these organics prior to the point at which glass forms part of the seal is quite important because the residual organics increase the porosity of the sealed glass, thus reducing its mechanical strength. Ideally, all the organics should be removed during the glazing stage of piece-part manufacture. In order to accomplish this, manufacturers of piece parts often resort to using only solvents, but this has an adverse effect on the rheology of the glass paste as far as screen printing is concerned and more prints are needed to produce the same glass thickness. Hence current industry processes use glass pastes with resins. Some safeguard against excessive occlusion of organics can be achieved by subjecting the parts to a further postglazing treatments. The extent of postglazing that can be tolerated again will be limited by modifier dissolution in both vitreous and devitrifying glasses and by loss of flow characteristics in devitrifying glasses.

The base glass is allowed to flow in the lead attachment operation and some residual organic outgassing occurs during this operation, but the time is usually too short to accomplish a great deal of reduction of the organic content. If the time is extended by a decrease in temperature (to retain glass life), glass fluidity is not sufficient to allow escape of gases to any great degree. If both temperature and time are large, the glass usually runs over the side of the piece part, causing it to stick to the "seal boats" at sealing. The seal profile used for cerdip sealing should be designed with opportunity being given for outgassing prior to sealing. This implies a nonlinear heating rate to the required seal temperature.

Control of outgassing is possible to a certain extent by appropriate selection of heating schedules (time–temperature profiles) during the sealing operation. Besides the outgassing of organic species from solder glasses, evolution of moisture during sealing can also occur. Because of the recent interest in moisture outgassing properties of solder glasses, the next section is devoted to a detailed discussion of this subject.

VI. Moisture Outgassing

A. General Considerations

Moisture outgassing from solder glasses during sealing of a cerdip is of great practical importance to the electronics industry from the viewpoint of product reliability. The presence of high moisture levels inside a sealed cavity of an integrated circuit is a well-documented cause of device failure, primarily due to electrochemical corrosion (Ebel, 1982; Kolesar, 1974, 1976; Eisenberg et al., 1966; Koelmans, 1976). In the extreme case, moisture levels as low as 1500 ppmv have been shown to induce device failure in certain applications due to a noncorrosive mechanism (Shukla and SinghDeo, 1982).

Solder glasses are known to be a primary source of water vapor inside a sealed cerdip even when the sealing atmosphere is kept dry. Oxide glasses invariably contain water bound in their structure as hydroxyl groups. In addition, moisture can be present on the surfaces of glass in both chemisorbed and physically adsorbed forms.

Removal of sorbed gases from glasses has been important to high-vacuum technology from the days when Edison invented the light bulb. The phenomenon of outgassing from glasses was extensively studied by pioneers such as Langmuir (1918) and Todd (1960), and the early work on moisture desorption from glasses was critically reviewed by Holland (1964). However, solder glasses are more complex from the outgassing viewpoint because of the presence of fillers, binders, and residual organics from the glazing processes. Therefore there has been renewed interest recently in the moisture outgassing behavior of commercial solder glasses.

B. Moisture Measurement Methods

Cavity moisture level limits and their measurement are dictated by the fact that the moisture levels determine the dew point of the package cavity. Below the dew point, water can condense on the device surface, attacking the passivation layers. The passivation layers of devices contain P_2O_5 as an ingredient, which can produce phosphoric acid in the presence of water, leading to aluminum metal line corrosion.

Because of the reliability hazards associated with high moisture levels in the package cavity of an IC, most users have imposed an upper limit on the free moisture content in the cerdip cavity to 5000 ppmv. With improvements in technology, this limit will eventually be lowered to 500 ppmv in the future. Measurement of moisture levels as low as 5000 ppmv in cavity volumes as small as 0.05 cm^3 is not a trival problem. Simple calculations will show that such amounts of water correspond to only a few monolayers of moisture adsorbed on the internal surfaces of the package. This has led to some highly specialized methods of moisture detection and measurement, e.g., mass spectrometry (Thomas, 1982; Kiely et al., 1981; Pernicka and Raby, 1980), moisture-sensor integrated circuits (Lowry et al., 1979; Jachowicz and Senturia, 1981; Kovac et al.,

1977; Davidson and Senturia, 1982), and infrared adsorption (Bossard and Mucha, 1981).

Measurement of the water sorption kinetics of solder glasses has been carried out primarily by two methods:

(1) *Microbalance–mass spectrometry* This method involves simultaneous measurement of mass change and residual gas analysis in a controlled, high-vacuum environment. The design of measurement equipment has been described in detail by Vasofsky et al. (1979). The method offers the advantage of determining not only the rate of outgassing, but also the composition of gases evolved. Table VII shows typical outgassing data from XS1175-M1 glass at various temperatures in vacuum (Shukla et al., 1980). As the temperature goes up, H_2O and CO_2 are the principal constituents outgassed, with H_2O being dominant at high temperatures. This is also consistent with findings of Vasofsky (1979).

(2) *Moisture evolution analysis (MEA)* Moisture sorption properties of materials can be effectively studied by using this technique (Shukla et al., 1980). The method measures water desorption rates of a sample subjected to an arbitrary time–temperature profile at atmospheric pressure. The apparatus schematic is shown in Fig. 16. Moisture is detected in this method by passing a carrier gas (N_2 or O_2) over the heated sample and then through an electrolytic P_2O_5 cell. The exact amount of water absorbed by P_2O_5 is determined coulometrically. The MEA technique is

TABLE VII

MASS SPECTROMETRIC ANALYSIS OF OUTGASSED PRODUCTS OF SOLDER GLASS XS1175-M1[a]

	Outgas temperature[b] (°C)			
Species	25	110	220	430
Hydrogen	0	0	0	0
Methane	0.5	40	87	1116
Water	401	4137	5678	23,127
Nitrogen	0	0	325	5365
Oxygen	0	4	110	191
Hydrocarbons as butane	20	200	381	633
Carbon dioxide	176	4020	6237	8098
Benzene	0	0	0	2505
Phenol	73	142	142	1450

[a] Glass sample obtained from a glazed part, composition of outgassed products in parts per million by volume.

[b] Dwell time at each temperature before analysis was 10 min.

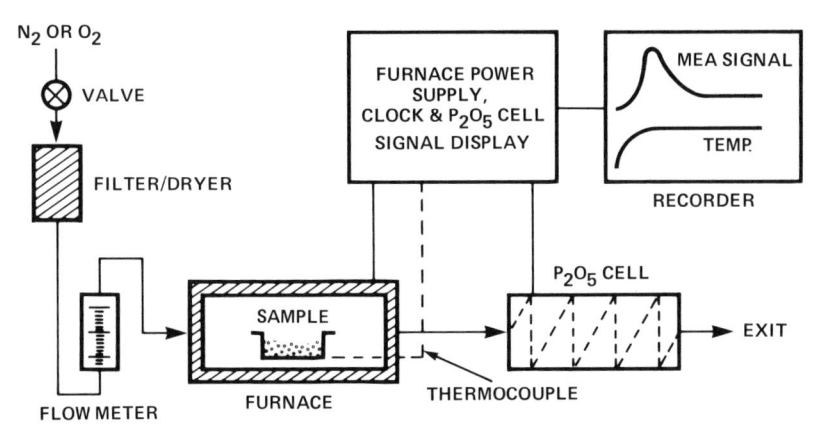

FIG. 16. Schematic of moisture evolution analyzer (MEA) apparatus.

simple and fast and offers the advantage of controlling the carrier gas over the sample during outgassing (inert, oxidizing, etc.). This allows one to determine the contribution of residual organics to the overall moisture evolution by the solder glass. The drawback of MEA is that it does not give the composition of the evolved gases since it is only sensitive to moisture.

C. MOISTURE SORPTION MECHANISMS

Figure 17 shows the moisture evolution profile, obtained by the MEA method (Shukla *et al.*, 1980), of XS1175-M1 glass subjected to a three-step time–temperature profile. It is evident from the figure that moisture evolution from successively higher steady-state temperatures reaches a maximum before approaching zero asymptotically. This implies that the water vapor is liberated from at least three regimes.

(1) *Low-temperature region* At temperatures below the softening point of the glass, moisture is primarily desorbed from the glass surface. It can be shown from the known activation energy data of surface desorption of moisture from glasses (Holland, 1964) that this contribution is due to both physical and chemical desorption and has been found to be dependent on glass surface area.

(2) *Medium temperature region* At temperatures near the softening point of a glass (second plateau in Fig. 17), the desorption of moisture is partly due to oxidation of residual organics (binders and vehicles) used at the screen printing operation, as well as to bulk desorption of absorbed water. By measuring the difference in evolved moisture content from this regime in N_2 versus O_2 ambient, one can distinguish bulk absorbed mois-

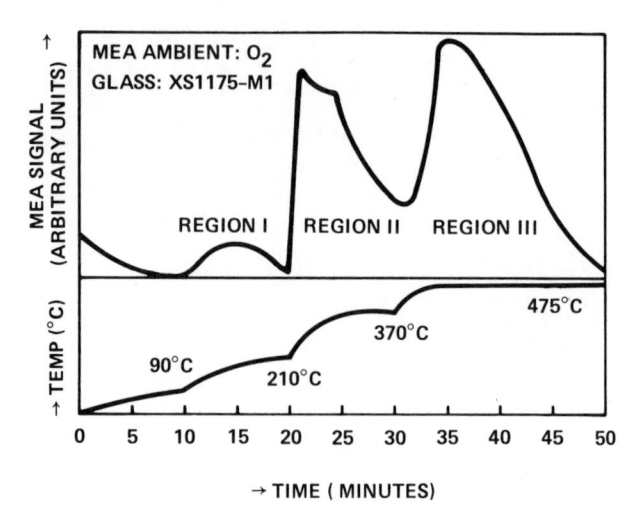

FIG. 17. MEA data for XS1175-M1 glass indicating different mechanisms of moisture evolution (regions I, II, and III).

ture versus moisture generated due to organics oxidation. Obviously, these results would be highly dependent on the glazing vendors' processes due to the different quantity of organics used, as was shown by Shukla et al. (1980).

(3) *High-temperature region* At temperatures above or near the seal temperature (last plateau in Fig. 17), solder glass viscosity is rather low and hence the diffusion of water molecules through glass would be rapid. The low viscosity allows hydroxyl (—OH) bonds in the oxide network to be broken readily, causing moisture to be released. In this regard, the glass system can be considered to be in pseudoequilibrium with moisture at a given temperature, with a fixed concentration of OH ions in the molten glass network. This is further confirmed by the observation that the higher the steady-state temperature, the higher net moisture is released.

The understanding of moisture evolution kinetics and its temperature and time dependence is a key to producing a "dry process" for cerdips. In addition to the moisture desorption mechanism proposed above, various other mechanisms have also been proposed to account for the high moisture levels associated with certain glasses (Vasofsky and Lowry, 1980; Lowry et al., 1978). Thus the glazing of XS1175-M1 glass in vacuum leads to a porous structure, causing excessive entrapment (and subsequent evolution during heating) of moisture (Vasofsky, 1979). An interesting difference has been found between vitreous and devitrifying solder glasses. Cerdips with devitrifying glass seals have higher cavity moisture levels in general. It has been proposed that during recrystallization of devitrifying glasses, "bonds" are broken and reformed, thus releasing large amounts of entrapped moisture (Lowry *et al.*, 1978). However, studies of the MEA characteristics of XS1175-M1 versus CV111 glasses have failed to confirm this hypothesis. It is quite likely that the primary source of higher moisture levels of devitrifying glasses is their higher seal temperatures, which would release higher moisture levels (chemical pseudoequilibrium hypothesis).

D. Dry Solder Glass Processing

The discussion of moisture evolution mechanisms described above should act as a guide in defining "dry cerdip" processing requirements in the following three categories.

(1) *Base glass and fillers* The bulk moisture in the glass occurs as a result of water dissolution in the glass melt itself. To prevent high moisture evolution rates at seal temperatures, the raw glass must be as dry as possible. It is extremely important to control the humidity, and the use of water-releasing (hydrates, hydroxides, etc.) minerals in the glass melt must be avoided. In fact, it is recommended that the glazer make a check of the moisture levels of his raw glasses a routine part of initial raw material inspections.

In addition to the moisture content of the raw glasses, moisture evolution from fillers must be studied in detail to reduce the adsorbed moisture in the raw glass. For example, use of dry alumino-silicate fillers (high activation energy of moisture desorption) in vitreous glasses has been attributed to the dryness of vitreous over devitrifying glasses, which are reported to have lower activation energy for moisture desorption (Lowry *et al.*, 1978).

(2) *Glazing and screen printing* The role of binders and vehicles in creating moisture in a sealed cavity of cerdip has already been discussed. The glazing atmosphere must be oxidizing and dry to eliminate the extent of residual organics. The choice of binders and vehicles must be such that they are clean burning and leave no hydrocarbon residues. With vitreous glasses, high glazing temperatures (close to seal temperature) can be used with greater ease because they have no effect on glass life, leading to dryer products than can be obtained for devitrifying glasses. The glazing process should be done in such a way that no excessive voiding or pores are produced on the glass surface, since such internal surfaces are traps moisture. The ease of achieving these conditions with vitreous glasses has made them popular in dry solder glass applications.

(3) *Sealing* Ultimate moisture-level control lies in the hands of the

user of glazed parts, since the seal profile can be controlled to minimize high-temperature moisture evolution behavior. In this regard, high-temperature vacuum prebaking before sealing has been suggested to dry out the solder glasses (Vasofsky and Lowry, 1980), but the kinetics of moisture absorption of dry glasses produced by baking needs to be taken into account when storage conditions are not kept dry.

More advances on all three fronts—glass manufacturing, screen printing and glazing, and cerdip sealing—need to be made to achieve lower moisture levels, projected to be less than 500 ppmv in the next few years.

VII. Soft Error

A. General Considerations

The use of solder glasses in microelectronic packaging is affected by a rather unique failure mode of integrated circuits. The failure mode, known as *soft error,* occurs when an α particle emitted from a radioactive impurity in the packaging material passes through the memory area of a dynamic RAM on CCD circuits. The energy transfer from the α particle to the bulk silicon results in generation of electron–hole pairs. Under certain conditions, the extra electron–hole pairs generated by the passage of ionizing radiation through silicon can create a change in critical charge in a memory storage capacitor, causing it to flip from a "1" to a "0" state. This produces a random single-bit error in the device under operation and is known as soft error. (In contrast, a *hard error* is usually nonrandom in location and related to permanent defects incorporated into the device structure.)

The phenomenon of soft errors in dynamic random access memories (dynamic RAMs) was first described in the pioneering work by May and Woods (1978, 1979). They found that this phenomenon was prominent in devices housed in cerdip-type packages only. Devices housed in plastic or side-brazed laminated ceramic packages (not utilizing the solder glass seals) were found to be relatively unaffected. Further analysis showed that the source of the radioactivity that produced soft error in devices housed in cerdip packages was primarily the solder glass. Since solder glasses are derived from oxides of heavy metals and other naturally occurring minerals, they can contain trace levels of radioactive impurities which can contribute significantly to the soft error rates of devices. This part of the chapter is devoted to a detailed discussion of the soft error phenomenon since it has strongly affected the solder glass technology for cerdips.

B. Soft Error Mechanism

The interaction of radioactive rays with matter during their passage through a medium is a well-studied subject. However, most of the literature has dealt with the radiation damage effects at high doses of radiation, which can cause atomic displacement, defects in lattice structure, and ionization damage. The subject of soft error deals with radiation doses so far considered benign, and even though the effect is electrically detectable in devices, the damage is not permanent. Moreover, soft error is caused by only one of the radioactive rays (α rays).

Alpha particles are doubly-charged helium nuclei which are emitted as a consequence of radioactive decay of an unstable nuclei. They are highly energetic, ranging in kinetic energy up to ~10 MeV for natural α-decay processes. Owing to their large mass and double positive charge, α particles interact heavily with matter as they pass through a medium, losing all of their kinetic energy within a relatively short range. The α-particle range in silicon as a function of incident energy is well known (Fano, 1964) and is comparable to current device dimensions (<10 μm) for the kinetic energy range of interest.

Figure 18 shows a schematic of the basic process of α-particle-induced soft error. Thus, as α particles travel through top layers of silicon, charge carriers are generated in the vicinity of the deletion layer. For N-channel technology, the electrons diffuse toward the edge of the depletion region and upon reaching it are swept into storage regions. If the number of electrons collected by this method exceeds the critical charge level, a soft error results. The dynamics of charge collection from α-particle tracks in integrated circuits has been described by Hsieh et al. (1981).

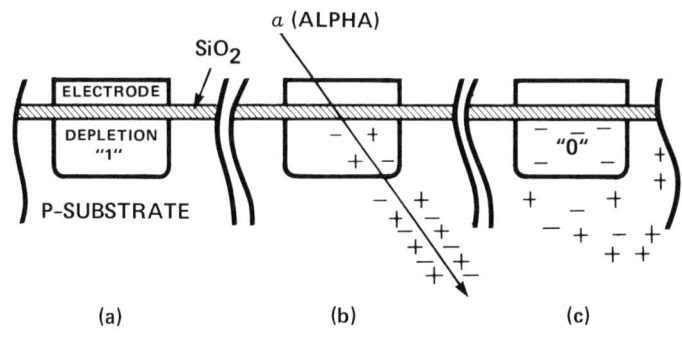

FIG. 18. Schematic showing soft error induced by passage of an α particle through a MOS capacitor: (a) capacitor in depletion mode ("1" state), (b) charge generation due to α-particle passage through the circuit, and (c) drift of electrons into the depletion area, causing inversion to a "0" state.

C. Radioactive Impurities in Solder Glasses

A comparison of the α spectra (intensity of α rays as a function of energy) obtained from solder glasses with a number of uranium and thorium minerals showed them to be essentially identical (Meieran et al., 1979), implying that uranium and thorium are the radioactive parents responsible for α emission from the solder glasses.

Meieran et al. (1979) compared the α activity of different types of solder glasses and showed that those glasses which have high levels of zirconia (a filler) have also a high level of α activity. Goddard's radiometric series shows that U and Th can be present as substitutional impurities in zirconium compounds derived from natural minerals because of chemical similarities among these elements. In addition, solder glasses involve the use of natural minerals of heavy elements and are therefore likely to have trace levels of radioactive impurities from geological sources, even when zirconium is not present.

Meieran et al. (1979) verified the presence of U and Th in solder glasses (using mass spectrometry) in the amount of 1–100 ppm of uranium by weight. Alpha activity generated from solder glasses containing zirconia roughly corresponds to activity levels expected from 1–100 ppm of uranium and thorium.

It must be pointed out that owing to the extremely low levels of radioactive impurities in solder glasses, special chemical analysis techniques have to be used. These methods are emission spectroscopy, mass spectroscopy, neutron activation analysis, electron microprobe, among others, and have been discussed elsewhere (Volborth, 1969; Wainerdi and Uken, 1971; Rankama, 1954).

D. α Activity Measurements

Measurement of the α activity of glasses can be made in terms of two quantities, α-disintegration rates per unit weight of material (e.g., picocuries per gram), and α flux (measured in terms of α counts per square centimeter of sample per hour). In the field of solder glasses the use of the α-flux quantity is more common than the disintegration rates of the parent nuclei. Since all α particles measured in α counting arise within a small depth from sample surface (emissions from deeper regions are absorbed internally), the technique of α counting is strictly a surface measurement and the volume of material is not important.

Because of the extremely low levels of radioactivity in solder glasses, special care has to be taken to successfully measure the α-emission rates. Various techniques for α measurements have been discussed in elsewhere (Meieran et al., 1979) and will not be elaborated upon here.

E. Low-α Solder Glasses

Measurement of α fluxes from various non-zircon-containing solder glasses in the past has shown activity in the range of 1–10-α/cm² h levels. In the presence of Zr as a filler element, values as high as 50 α/cm² h have been measured. For various 16K DRAMs, it has been reported that 1 α/cm² h corresponds roughly to a failure rate of 1%/1000 h of device operation due to soft error. This implies that solder glass α activities must be reduced by orders of magnitude to make failure rates comparable to hard errors (<0.02%/1000 h).

With ever-tightening requirements for low α activity of solder glasses, tremendous efforts have been made in the industry to lower α activity by careful control of raw materials. However, the high cost of purifying natural minerals to remove radioactive impurities means that alternative sources of materials must be investigated. For example, use of synthetic fillers offers a solution. Use of entirely new families of "cold" fillers (titanates, tin oxide, etc.) is being pursued in the industry as a means of achieving low α activity. This should offer a challenging area for new solder glass developments.

However, in developing low-α-activity solder glasses, the following points have to be considered.

(1) Even if highly purified/synthetic materials ingredients are used with very low α activity, contamination during processing can increase the ultimate α activity of the solder glass.

(2) As the α activity of solder glasses is decreased, its measurement with accuracy becomes a practical limit in itself. With the proportional gas counters, α levels as low as 0.001 α/cm² h can be measured successfully. It must be pointed out that the α activity of solder glasses is very low by common standards; i.e., the background in a measurement system for α counting is comparable to the α activity of low-α-activity solder glasses. Due to high levels of background (cosmic rays, noise, etc.), it is doubtful that levels below 0.001 α/cm² h will be measured with reasonable accuracy under commercial conditions. Solder glasses with α activity as low as ~0.1 α/cm² h are already available in the industry; it is expected that the lower practical limit on low-α solder glasses is about 0.01 α/cm² h.

(3) Since the discovery of the α-induced soft error mechanism, device designers have independently tried to make the devices less prone to this failure mode by design changes, with some success (McPartland et al., 1980). One approach has been to incorporate error detection and correction (EDAC) (Levine and Meyers, 1980) at a system level of integration. In addition, the device trend has been toward lower gate-oxide thickness (SiO_2 thickness at MOS gate), which tends to increase charge

storage capability (increased capacitance). Hence decreasing gate-oxide thickness should decrease the soft error rate for the same α flux. It is conceivable that in the future device design will place less stringent demands on the α activity of packaging materials.

(4) With the very low levels of α activity associated with modern solder glasses, the occurrence of soft error due to cosmic-ray background becomes important and would place a natural lower limit on the α activity of concern. Masters (1980) reports the cosmic-ray background to be ~ 0.003 α/cm^2 h as measured by the ZnO scintillation method. This coupled with device design improvements would imply that efforts to reduce the α activity of solder glasses to levels below ~ 0.005 α/cm^2 h are not practical.

(5) Since the α rays can be effectively absorbed by a few micrometers of a solid, an alternative solution to the problem is to apply a thin protective coating to the device surface. Various polymetric as well as inorganic coatings have been tried (White *et al.*, 1981). Since the seal temperature of cerdips is rather high ($>400°C$), this places a serious limitation on the use of polymeric coats (e.g., polyimides), which would need comparable thermal stability.

VIII. Future Perspectives

A. Strength

The cerdip solder glass systems are presently at a very interesting crossroads. A shiftover has started from the devitrified variety to the vitreous variety in the semiconductor packaging area. This was necessitated by the fact that present semiconductor device complexity requires lower temperature of seal or, if the temperature is retained constant, lower dwell times at that temperature. The changeover to the vitreous glass system has introduced a mechanically weaker glass system relative to the previous devitrified workhorse, e.g., CV111. The users of this package may experience some loss of hermeticity in the very demanding packages using vitreous types of glasses. Thus the issue of glass strength must be addressed, and stronger glass systems must be developed.

B. Lower Seal Temperature

Further lowering of seal temperature seems to be the direction now, not only for the sake of meeting the demands of the semiconductor device engineers, but also for cost-reduction purposes where lower-cost alternatives for the materials used in the cerdip piece parts are feasible at lower seal temperatures, e.g., using polymeric chip attachment materials instead of gold. A note of caution may be added at this juncture. If seal-temperature lowering brings about further loss in strength, this is not a solution to the problem.

C. Moisture Level in Sealed Cavity

Reduction of moisture content in the sealed cavity of the cerdip was a problem five or ten years ago, but that problem has become fairly well understood and the industry is making rapid progress toward attaining the goal of 500 ppmv of moisture. The packages are not there yet but appear feasible in the near future.

D. Soft Error

From the soft error standpoint, the sealed glasses are at a quite low level of radioactivity. Commercial glass systems are being quoted at α-activity levels of 0.1 α/cm^2 h. It is quite likely that the previous trends may push these values even lower, but the impetus seems to have slowed since the semiconductor device designers are making the devices less prone to soft error and error correction methods are being widely applied.

References

Alma, G., and Prakke, P. (1946). *Philips Tech. Rev.*, **8**(10), 289–295.
Andrus, K. B., and Powell, H. E. (1972). Paper presented at *Annu. Meet. Am. Ceram. Soc., Ceram.-Met. Syst. Div., 1972, Washington, D.C.*
Bossard, P. R., and Mucha, J. A. (1981). *Annu. Proc. Reliab. Phys. [Symp.]* **19**, 60.
Broukal, J. (1962). *Silikatechnik* **13**, 428–432.
Budd, S. M., Exelby, V. H., and Kirwan, J. J. (1962). *Glass Technol.* **3**, 124.
Davidson, T. M., and Senturia, S. D. (1982). *Annu. Proc. Reliab. Phys. [Symp.]* **20**, 249.
Ebel, G. H. (1982). *NBS Spec. Publ. (U.S.)* **400-72**.
Eisenberg, P. H., Brandwie, G. V., and Mayer, R. A. (1966). *N.Y. Conf. Electron. Reliab. 7th, 1966*, p. 37.
Espe, W. (1951). *Feinwerktechnik* **55**, 303–306.
Fano, U. (1964). *N. A. S. = N. R. C., Publ.* **1133** (Nucl. Sci. Ser.—Rep. 39).
Forbes, D. W. A. (1967). *Glass Technol.* **8**(2), 32–42.
Frieser, R. G. (1975). *Electrocomponent Sci. Technol.* **2**, 163–199.
Graham, H. C., and Tallan, N. M. (1971). *In* "Physics of Electronic Ceramics" (L. L. Hench and D. B. Dove, eds.), Part A. Dekker, New York.
Hasselman, D. P. H., and Fulrath, R. M. (1965). *Lawrence Livermore Lab. [Rep.] UCRL* **UCRL-11775**.
Hogan, R. E. (1971). *Chem. Tech.* **1**, 41–43.
Holland, L. (1964). *In* "The Properties of Glass Surfaces," Chapter 4. Wiley, New York.
Holloway, D. G. (1973). "The Physical Properties of Glass." Wykeham Publications, London.

Hsieh, C. M., Murley, P. C., and O'Brien, R. R. (1981). *Annu. Proc. Reliab. Phys. [Symp.]* **19**, 38.
Jacowicz, R. J., and Senturia, S. D. (1981). *Sens. Actuators* **2**, 171–186.
Kiely, J., Flinn, P., and Sun, B. (1981). *Annu. Proc. Reliab. Phys. [Symp.]* **19**, 67.
Kingery, W. D., Bowen, H. K., and Uhlmann, D. R. (1976). "Introduction to Ceramics," 2nd ed., p. 603. Wiley, New York.
Koelmans, H. J. (1976). *J. Electrochem. Soc.* **123**, 168–171.
Kolesar, S. C. (1974). *Annu. Proc. Reliab. Phys. [Symp.]* **12**, 85.
Kolesar, S. C. (1976). *J. Electrochem. Soc.* **123**, 155–167.
Kovac, M. G., Chleck, D., and Goodman, P. (1977). *Annu. Proc. Reliab. Phys. [Symp.]* **15**, 85.
Kyocera International (1983). "Product Information." San Diego, California.
Lamson, M. A., and Ramsey, T. H., Jr. (1978). *Electron. Packag. Prod.* January, pp. 84–96.
Langmuir, I. (1918). *J. Am. Chem. Soc.* **40**, 1361.
Levine, L., and Meyers, W. (1980). *Computer* October, pp. 43–50.
Lowry, R. K., Van Leeuwen, C. J., Kennimer, B. J., and Miller, L. A. (1978). *Annu. Proc. Reliab. Phys. [Symp.]* **16**, 207.
Lowry, R. K., Miller, L. A., Johas, A. W., and Bird, J. M. (1979). *Annu. Proc. Reliab. Phys. [Symp.]* **17**, 97.
McPartland, R. J., Nelson, J. T., and Huber, W. R. (1980). *Annu. Proc. Reliab. Phys. [Symp.]* **18**, 261.
Martin, F. W., and Zimar, T. (1961). *Symp. Am. Sci. Glassblowers Soc., 6th, 1961,* pp. 32–41.
Masters, B. J. (1980). *Annu. Proc. Reliab. Phys. [Symp.]* **18**, 269.
May, T. C., and Woods, M. H. (1978). *Annu. Proc. Reliab. Phys. [Symp.]* **16**, 33.
May, T. C., and Woods, M. H. (1979). *IEEE Trans. Electron Devices* **ED-26**(1).
Meieran, E. S., Engle, P., and May, T. (1979). *Annu. Proc. Reliab. Phys. [Symp.]* **17**, 13.
Pask, J. A., and Fulrath, R. M. (1962). *J. Am. Ceram. Soc.* **45**, 592–596.
Pavlushkin, N. M., and Kalmanovskaya, M. A. (1976). *Izv. Akad. Nauk SSSR, Neorg. Mater.* **12**, 2042–2046.
Pernicka, J. C., and Raby, B. A. (1980). *NBS Spec. Publ. (U.S.)* **400-72**, 308.
Petzoldt, J. (1966). *Glastech. Ber.* **39**, 130–136.
Rabinovich, E. M. (1979). *Bul. Am. Ceram. Soc.* **58**(6), 595–598, 605.
Ramsey, T. H. (1971). *Bul. Am. Ceram. Soc.* **50**(8), 671–675.
Ramsey, T. H. (1972). *Solid State Technol.* **15**(1), 29–33, 43.
Rankama, K. (1954). "Isotope Geology." McGraw-Hill, New York.
Samsonov, G. V. (1962). "New Refractory Materials for Smelting and Casting Refractory Metals" (in Russian). Izd. Akad. Nauk, SSSR.
Shukla, R. K., and SinghDeo, N. N. (1982). *Annu. Proc. Reliab. Phys. [Symp.]* **20**, 122–127.
Shukla, R. K., SinghDeo, N. N., Sharma, N. K., and Blish, R. (1980). *NBS Spec. Rep. (U.S.)* **400-72**, 213–219.
Shirouchi, Y. (1969). *Fujitsu Sci. Tech. J.* **5**, 123–165.
Thomas, R. W. (1982). *NBS Spec. Publ. (U.S.)* **400-72**, 126.
Todd, J. (1960). *J. Appl. Phys.* **31**, 51.
Traeger, R. K. (1976). *Proc. Electron. Components Conf.* **25**, 361.
Vasofsky, R. W. (1979). *Annu. Proc. Reliab. Phys. [Symp.]* **17**, 91–96.
Vasofsky, R. W., and Lowry, R. K. (1980). *Annu. Proc. Reliab. Phys. [Symp.]* **18**, 1–9.
Vasofsky, R. W., Czanderna, A. W., and Thomas, R. W. (1979). *J. Vac. Sci. Technol.* **16**, 716.

Volborth, A. (1969). "Elemental Analysis in Geochemistry." Elsevier, Amsterdam.
Wainerdi, R. E., and Uken, E. A. (1971). "Modern Methods of Geochemical Analysis." Plenum, New York.
White, M. L., Serpiello, J. W., Striny, K. M., and Rosenzweig, W. (1981). *Annu. Proc. Reliab. Phys. [Symp.]* **19,** 43–47.

CHAPTER 7

Processing of Gel Glasses

Jerzy Zarzycki

LABORATORY OF MATERIALS SCIENCE AND
CNRS GLASS LABORATORY
UNIVERSITY OF MONTPELLIER
MONTPELLIER, FRANCE

I. Introduction	209
II. Historical Outline	210
III. Gel Preparation	214
A. Gel Formation by Destabilization of Sols	214
B. Gel Formation From Organometallic Compounds	219
IV. Gel Drying	221
A. Phenomenological Approach	221
B. Structural Approach	223
C. Preparation of Monolithic Gels	228
V. Densification Process	231
A. Structure of Dried Gels (Xerogels and Aerogels)	231
B. Gel-to-Glass Transformation	234
C. Sintering by Viscous Flow	235
D. Devitrification Kinetics	240
VI. Special Features of Gel-Produced Glasses	244
VII. Conclusion	245
References	245

I. Introduction

Quenching a melt is not the only way to produce glasses. Besides this classic method other possibilities exist: condensation of a vapor on a cold substrate, disordering of a solid by irradiation, reaction in the liquid state followed by solvent elimination, etc. For a more detailed account, see, for example, the chapter on "Unusual Methods of Producing Glasses" (Schultz, 1983).

The gel route to glass formation is based on the possibility of forming the glass network by chemical polymerization of suitable compounds in the liquid state at low temperatures. In this way a "precursor" material (gel) is formed from which glass may be obtained by subsequent elimination of the unwanted residues (water, organic compounds, etc.) and collapse of the structure at temperatures much lower than those required for direct melt formation by fusion of the constituent oxides.

This is of interest in the case of many systems which are difficult or even impossible to prepare directly because of the excessive temperatures involved and/or high viscosity of the resulting melts which prevent satisfactory homogenization.

In particular, when oxides exhibit marked differences in volatility, chemical polymerization at low temperatures may be the only way to prevent heavy losses of some constituents.

By this method a high degree of homogeneity is directly obtained on a molecular scale and the absence of contamination during fusion makes it possible for high-purity glasses to be easily obtained.

From a theoretical standpoint, approaching the glassy state from lower temperatures instead of from higher ones may imply that some phase transitions (unmixing, crystallization) may be circumvented and new glasses impossible to prepare by quench may become accessible.

In contrast to "thermal" polymerization during melt formation, "chemical" polymerization at low temperatures offers more possibilities of building the glass network. This route to glass formation should give inorganic glasses some of the flexibility of the organic polymer synthesis which was lacking in the direct route.

Technically, the method is readily applicable to the formation of thin coatings and fibers; it requires special care, however, if *monolithic* pieces of glass are to be obtained. Apart from some applications confined to thin-layer technology, processing of glasses from gels is still in the laboratory stage.

II. Historical Outline

Even though the production of amorphous SiO_2 coatings by solution methods had already been developed in Germany during World War II (Geffcken and Berger, 1943), the general idea of using a low-temperature chemical mixing process to prepare mixtures of high-melting oxides homogeneous on a molecular scale is attributed to D. M. Roy (1952) and D. M. Roy *et al.* (1953). The traditional method of preparing mixtures for phase studies which consisted of melting oxides (usually) to a glass, crushing, remixing, and remelting was replaced by the preparation of coprecipitated gels which were then remelted to yield highly homogeneous glasses. In this way a wide variety of silicate and aluminosilicate systems was prepared.

R. Roy (1956) indicated various methods of preparing homogeneous glass batches: coprecipitation of salts, spray drying of solutions, and hydrolysis of alkoxide–salt solutions. R. Roy (1969) also reported that silica glass can be made at temperatures as low as 1200°C and easily at 1350–1400°C. The organic silica–nitrate method was found especially suitable for systems containing silica, alumina, titania, and zirconia. It was not restricted to glass-forming areas or to glass-forming oxides. It was also noticed that the formation of a stiff gel has the additional advantage of preventing any further gravity differentiation. Luth and Ingamells (1965) discuss gel preparation of starting materials for hydrothermal experimentation.

McCarthy et al. (1971) have demonstrated that it is possible to obtain glasses from gels by using hot-pressing techniques. Particulate silica gels were compacted at 650–1050°C and 270 atm to yield translucent products, but none, however, were of optical quality. They noticed that greater densification was achieved with gels (Ludox, Cab-O-Sil) than with premelted SiO_2 glass under the same conditions.

Dislich (1971a,b,c) indicated how multicomponent glasses can be prepared without going through the molten state. Borosilicate glass of the system SiO_2–B_2O_3–Al_2O_3–Na_2O–K_2O was prepared from gels obtained from metal alcoholates. Dried gel fragments were either remelted at 1600°C or hot pressed at 650–700°C under 100-ton pressure. In other experiments thin glass layers were obtained in the systems

SiO_2–Al_2O_3–MgO–P_2O_5–B_2O_3–CaO–BaO–As_2O_3,
SiO_2–PbO–Na_2O,
SiO_2–Na_2O–Al_2O_3.

The possibility of obtaining basic glasses for the glass–ceramic process was shown to be possible for the systems

SiO_2–Al_2O_3–P_2O_5–Li_2O–MgO–Na_2O–TiO_2–ZrO_2,
SiO_2–Al_2O_3–ZnO–Li_2O–TiO_2–ZrO_2–BaO–MgO–CaO–K_2O.

Fuji and Ishido (1965) have described how they obtained transparent SiO_2 bodies from gels derived from tetraethyl silicate sintered at normal pressure.

Konijnendijk and Groenendijk (1972) prepared homogeneous borosilicate glasses by wet-chemical techniques using a colloidal silica solution (Ludox) or by hydrolyzing tetraethyl orthosilicate.

Gels of the system SiO_2–B_2O_3–Al_2O_3–Na_2O–K_2O yielded, after drying, friable products which could be easily melted at 1200°C to a bubble-free glass. The authors noticed that the gain in the melting temperature is about 200°C. Macroscopic inhomogeneities were avoided and submicroscopic homogeneity improved (see also Konijnendijk et al., 1973).

Shoup (1976) described a method of producing multicomponent silica glasses directly from the gel. Here SiO_2–K_2O gels were leached to remove potassium, dried, and fired to give compact glasses, thanks to a suitable pore control obtained during gelling by the addition of a catalyst (formamide).

Kamiya *et al.* (1974) prepared SiO_2–TiO_2 and SiO_2–Al_2O_3 noncrystalline solids from metal alkoxide mixtures. Devitrification during heat treatment was studied. The specimens contained OH groups but no organic residues.

A systematic study has been undertaken in our laboratory since 1973 with the original aim of producing refractory glasses and glasses suitable for the glass–ceramic process but difficult to melt by usual methods. A comparative study of glasses obtained either from gels or from oxide mixtures was first reported for the SiO_2–La_2O_3, SiO_2–La_2O_3–Al_2O_3, and SiO_2–La_2O_3–ZrO_2 systems (Mukherjee *et al.*, 1976a). Improvement in the homogeneity of gel-produced glasses and a finer phase separation texture were accompanied by a marked increase in nucleation rates. The gels (or oxides) were remelted in a solar furnace facility. It was concluded that the structure of gel-produced glasses differed significantly from that of glasses melted from oxides. The influence of hydroxyl in the gel-produced glasses on the crystallization rate was investigated (Mukherjee *et al.*, 1976b; Mukherjee and Zarzycki, 1979).

This period marks a sudden increase in interest in the preparation of glasses by the gel route. Initial studies were based on *hot pressing* of gels. Investigations of Decottignies (1977) and Decottignies *et al.* (1977a,b, 1978) first on SiO_2 and SiO_2–La_2O_3 systems were followed by those on the systems SiO_2–B_2O_3, SiO_2–TiO_2, and SiO_2–P_2O_5 (Jabra, 1979; Jabra *et al.*, 1979, 1980), and demonstrated the ease with which excellent glasses could be obtained in this way. The properties of hot-pressed glasses were compared to those of glasses obtained by CVD (chemical vapor deposition) (Jabra *et al.*, 1980). Structural evidence was given of Si—O—Ti bond formation (Jabra *et al.*, 1981). A general survey of this work is given by Phalippou *et al.* (1981).

At the same time efforts continued in different laboratories to produce glasses from gels in the form of coatings, fibers, or hollow spheres. Attention was first focused on the production of refractory oxide fibers from gels: Wainer (1968) and Wintel (1968) patented the preparation of oxide fibers from aqueous colloidal solutions. The possibility of producing fibers of SiO_2, SiO_2–TiO_2, SiO_2–ZrO_2, and SiO_2–ZrO_2–Na_2O glasses was investigated by Kamiya *et al.* (1976, 1977, 1978, 1980; Kamiya and Sakka, 1977a). The hydrolysis conditions for producing bulk (monolithic) glasses were also examined by Kamiya and Sakka (1980). Review papers on these researches were given by Kamiya and Sakka (1977b) and Sakka and Kamiya (1980). The problem of "spinnability" is discussed in a recent paper by Sakka and Kamiya (1982). The production of *hollow glass spheres* for use as targets in inertial-confinement fusion was studied by Nogami et al. (1980, 1982) and Downs et al. (1981).

The conditions necessary for the production of *thin coatings* were intensively investigated. Schröder (1962, 1969) presented an interesting review of the properties and applications of oxide layers deposited on glass from organic solutions; Shimbo et al. (1975) reported on Al_2O_3–SiO_2 coatings suitable for optical waveguides.

The formation of porous transparent Al_2O_3 films was studied by Yoldas (1975a,b,c), resulting in the production of antireflective coatings for glass surfaces (Yoldas, 1980a).

Nogami and Moriya (1977a,b) studied the formation of noncrystalline films from solutions in the SiO_2–TiO_2, SiO_2–SnO_2, and SiO_2–Al_2O_3 systems, then the properties of SiO_2–ZrO_2 films, in particular their alkali resistance.

Independently Brinker and Mukherjee (1981a) studied the preparation of antireflective coatings for solar cells from gels, and Mukherjee and Lowdermilk (1982) the stability of gel-derived single-layer antireflection films for lasers (see also Mukherjee, 1982a,b, 1984).

For a general review of the possibilities and applications of the dip-coating technique see Dislich and Hussmann (1981a,b).

The general problem of the preparation of *bulk glasses* and *ceramics* from organometallic compounds was discussed by Yoldas (1977). For the obtention of nonparticulate glasses in the SiO_2–B_2O_3 and SiO_2–TiO_2 systems see Yoldas (1979, 1980b).

Yamane et al. (1978, 1979, 1982) have described the preparation of crack-free disks of dried SiO_2 gel using suitable hydrolysis conditions and the slow drying technique (see also Yamane and Okano, 1979).

After preliminary studies (Nogami and Moriya, 1979) the conditions for obtaining monolithic silica glass were formulated (Nogami and Moriya, 1980), the conversion to glass occurring between 700 and 1050°C.

Kamiya and Sakka (1980) obtained monolithic glasses in the TiO_2–SiO_2 system. The pieces of glass obtained were, however, of reduced dimensions (~1 cm^3).

The conditions for obtaining monolithic gels in the system SiO_2–B_2O_3–Al_2O_3–Na_2O–BaO were also investigated by Brinker and Mukherjee (1981b), and in the SiO_2–TiO_2 system by Gonzalez-Oliver et al. (1982). Klein and Garvey (1980, 1982, 1984) succeeded in preparing small crack-free disks of SiO_2 gels by the slow drying technique. The first monolithic pieces of gels of sizable dimensions (250 mm length) were obtained by using the method of hypercritical evacuation; see Prassas (1981), Zarzycki et al. (1982), Prassas et al. (1982b), and Zarzycki (1984).

Conditions of gel preparation in the SiO_2–CaO and SiO_2–CaO–Na_2O systems were investigated by Saito and Hayashi (1978) and Hayashi and Saito (1980). The SiO_2–SrO system was studied by Yamane and Kojima (1981), SiO_2–Fe_2O_3 by Gugielmi and Principi (1982).

After general investigations on SiO_2 and SiO_2–Na_2O gels, Puyané et al. (1980, 1982) described the production of glass rods for optical fibers from SiO_2–GeO_2, SiO_2–TiO_2, SiO_2–P_2O_5, and SiO_2–Y_2O_3 gels. A refined monitoring system was used to follow the contraction of the gel during firing to adjust the drying speed to avoid cracking. A detailed report on hydrolysis and gelling conditions leading to monolithic SiO_2 gels was given by Yu et al. (1982).

Review articles have been published by Mukherjee (1980), Klein (1981), and Partlow and Yoldas (1981). A recent review of the gel method for making glass was compiled by Sakka (1982). Mackenzie (1982, 1984) discussed the advantages of the gel method from an economical standpoint.

III. Gel Preparation

There are two ways of obtaining silica-based gels:

(1) Destabilization of silica *sol* (e.g., Ludox), pure or containing other metal ions added in the form of aqueous solutions of salts (method I).

(2) Hydrolysis and polycondensation of *organometallic compounds* dissolved in *alcohols* in the presence of a limited amount of water (method II).

Both methods lead to noncrystalline materials (precursors) containing substantial amounts of water and/or organic residues, which can be eliminated by suitable curing treatments. The dried and purified gels are essentially porous materials and a densification treatment is necessary to convert them into solid glasses devoid of residual porosity. The overall scheme of the process is presented in Fig. 1.

A. Gel Formation by Destabilization of Sols

1. Silica Sols

Silica sols can be prepared either by mechanical or electrical dispersion of the material or by chemical condensation methods, starting from solutions of Na-silicates, K-silicates, NH_4-silicates, or hydrolyzable products such as $SiCl_4$ or $Si(OR)_4$, where R is an alkyl group.

STARTING COMPOUNDS →(GEL FORMATION AND AGING)→ WET GEL →(DRYING)→ DRY GEL →(DENSIFICATION)→ GLASS

FIG. 1. Gel–glass process.

It has been shown (see, for example, Iler, 1979) that the formation of silicic acid in aqueous solutions is followed by polymerization of monomers Si(OH)$_4$ when its concentration exceeds 100 ppm (limiting solubility in water at 25°C). The polymerization reaction is based on the condensation of silanol groups:

$$\equiv\text{Si}-\text{OH} + \text{HO}-\text{Si}\equiv \rightarrow \equiv\text{Si}-\text{O}-\text{Si}\equiv + \text{H}_2\text{O}. \tag{1}$$

The following steps may be distinguished:

(1) formation of dimers and higher molecular species,
(2) condensation of these to form primary particles,
(3) growth of the particles,
(4) linking of the particles together in chains and then into three-dimensional networks.

Reaction (1) intervenes in the formation of primary particles, in their growth, and in their subsequent linking during gel formation. Above pH = 2 the rate of polymerization is proportional to the concentration of OH, below pH = 2 to that of the H$^+$ ions. The tendency is to produce a maximum of Si—O—Si bonds and a minimum of uncondensed Si—OH groups, resulting in the formation of ring structures, then linking these cyclic polymers to larger three-dimensional molecules. The condensation leads to a most compact state, the Si—OH groups being situated at the edge of the condensate.

These amorphous spheroidal groupings of about 1–2 nm are formed by a *nucleation* process similar to that which occurs in the formation of crystalline precipitates.

Because of size differences, *Ostwald ripening* then sets in. The smaller particles, which have a higher solubility, dissolve and the silica is redeposited on the larger ones, the total number of particles decreasing.

At low pH values, particle growth stops once a size of 2–4 nm is reached. Above pH = 7 particle growth continues at room temperature until particles about 5–10 nm in diameter are formed, then it slows down. At higher temperatures particle growth continues, especially for pH > 7.

For the pH range between 6 and 10.5 the silica particles are negatively charged and repel one another—growth continues without aggregation, resulting in the formation of stable *sols*. If, however, salts are present, aggregation and gelling occur.

At low pH the particles have little ionic charge; they can collide and form by aggregation continuous networks leading to *gels* (Fig. 2). This process may involve primary particles of different sizes according to the pH level and the presence or absence of salts.

Condensation can be controlled and even stopped when the particles reach the required size. The addition of stabilizing ions prevents their further condensation (step 4).

Commercial silica hydrosols (e.g., Ludox) are stable sols with 20–50 wt. % SiO_2. They are made up of dense silica particles with an average diameter of between 7 and 21 nm. The pH is between 9 and 11.

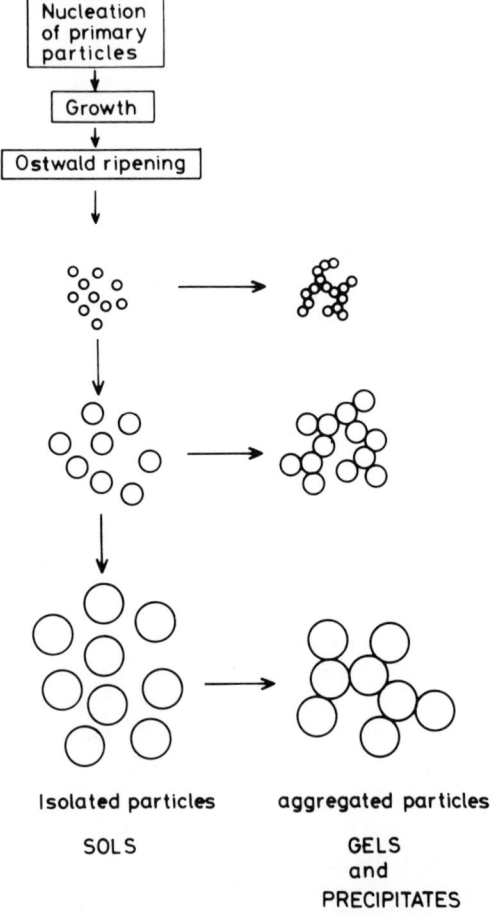

FIG. 2. Polymerization steps leading to the formation of sols and gels. [From Zarzycki *et al.* (1982).]

2. Destabilization

To obtain a *gel* from a stable sol this latter must be destabilized either by temperature increase or by the addition of an electrolyte.

Increase of temperature reduces the amount of intermicellar liquid by evaporation and increases thermal agitation, which induces collisions between particles and their linking by condensation of surface hydroxyls.

By electrolyte addition the pH of the sol may be modified in order to reduce the electric repulsion between the particles, (depending on the ζ potential). This is accomplished by adding an acid to diminish pH to 5–6 and to induce gel formation by *aggregation*. This conversion of sol into gel is progressive, the growing aggregates (microgel) invading progressively the whole volume of the sol. The local concentration of silica and water remains, however, the same. A gel has the same density and the same refractive index as the sol. When about half of the silica has entered the gel phase a rapid increase in viscosity is noted.

The mechanism of interparticle bonding leading to microgels and gels involves the attachment of two neighboring silica particles via the formation of Si—O—Si bonds [reaction (1)].

Neutral Si—OH and ionized Si— groups on the surface condense to form Si—O—Si linkages by the same mechanism as in primary-particle formation. However, the presence of soluble silica or monomer near the points of contact may contribute to the cementing together of the particles.

3. Aging

A further step is the *strengthening* of the network of particles by a mechanism involving the partial coalescence of the particles. The negative radius of curvature at the *neck* joining the two particles implies that the local solubility is less there than near the surface of the particle. Transport and deposition of silica occur there preferentially, leading to a thickening of the neck (Fig. 3).

This occurs in aging treatments in which chains of particles may be

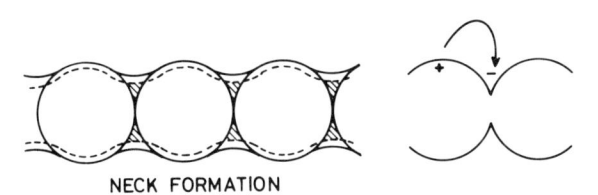

NECK FORMATION

FIG. 3. Strengthening of particulate chains by the deposition of silica at the necks. [From Zarzycki *et al.* (1982).]

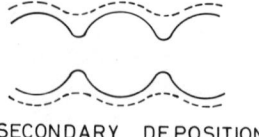

SECONDARY DEPOSITION

FIG. 4. Strengthening of chains during aging. [From Zarzycki *et al.* (1982).]

converted into more or less "fibrillar" structures by this rearrangement mechanism. Particle size in sols may be increased by adding "active" silica in the form of particles less than 2 nm or even smaller polymer species; they redissolve in the presence of larger particles and redeposit on them. This "nourishment" of particles is at the base of the so-called "buildup" process (Fig. 4).

The sol–gel transition should be distinguished from a *precipitation* (or flocculation) mechanism in which separate aggregates are formed in contrast to the continuous three-dimensional particle networks (Fig. 5).

Colloidal particles will form gels only if there are no active forces which would promote coagulation into aggregates with a higher silica concentration than the original sol. Metal cations, especially the polyvalent ones, may lead to precipitation rather than gelling.

SOL

GEL

PRECIPITATE

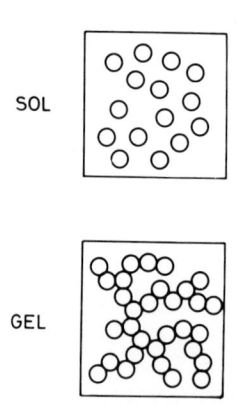

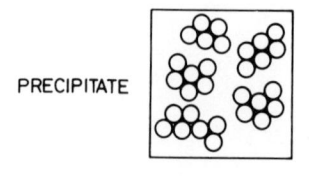

FIG. 5. Difference between gelation and precipitation. [From Zarzycki *et al.* (1982).]

4. Multicomponent Gels

At least one gellifying constituent (generally silica sol) is required. Other constituents may be added in the form of soluble salts (nitrates, sulfates, etc.) or organometallic compounds.

By adjusting the temperature, concentration, and especially the pH of the resulting sol, a homogeneous solution is obtained which may then be gelled in a controlled way in order to avoid precipitation.

Rabinovich et al. (1982) presented an alternative way of preparing glasses from colloidal sols. Their method consists in using a "fumed" silica (Cab-O-Sil) which is mixed with water in a high-shear blender to form a uniform sol (slip) which can be shaped by "gel molding." Gelled castings are dried and, to avoid cracking, a (unspecified) treatment is used. Sintering of these bodies leads either to nontransparent materials or to translucent glass. To obtain optically transparent glasses the addition of 1–4 wt. % of B_2O_3 was found to be beneficial.

B. Gel Formation From Organometallic Compounds

Metal alcoholates, also called metal alkoxides, $M(OR)_n$, where M is a metal and R an alkyl group, react with water and undergo hydrolysis and polycondensation reaction, leading to the formation of metal oxide. The overall reaction scheme consists, at least formally, of two steps:

$$M(OR)_n + nH_2O \rightarrow M(OH)_n + nR(OH), \quad (2)$$
$$pM(OH)_n \rightarrow pMO_{n/2} + \tfrac{1}{2}pnH_2O. \quad (3)$$

The resulting oxide is produced in the form of extremely small particles (~2 nm) which may form a gel.

In reality the situation is more complex; reactions (2) and (3) proceed simultaneously and are generally incomplete. Hydrolysis may be achieved by using a quantity of water less than the stoichiometric one and a number of radicals R remain unreacted. The polycondensation is arrested and the final product rather corresponds to the formula

$$(MO)_x(OH)_y(OR)_z.$$

When several different compounds, e.g., $M(OR)_n$, $M'(OR)_{n'}$, are reacted, a complexation step may precede reactions (2) and (3). This first step has received much less attention so far than the subsequent one (Dislich and Hinz, 1982).

In this way complex networks involving one or several different cations M, M', e.g.,

$$-M-O-M'-O-M-,$$

may be produced. The use of alkoxides of Si, B, Ti, Zr, etc., leads to the formation of complex gels formed of small particles which prefigure the network of corresponding oxide glasses.

Table I gives the list of organometallic compounds most frequently used in the synthesis of glasses by this method. The reagents are dissolved in alcohol, generally methyl or ethyl alcohol, and the water necessary for the hydrolysis is either taken from the atmosphere or added to the starting solution to accelerate the process. Some of the cations may also be introduced in the form of alcoholic or aqueous solutions of salts (nitrates, acetates, etc.). A carefully controlled amount of acid acting as a catalyst (e.g., HCl, glacial acetic acid) is added. The gelling time depends on the pH, temperature, amounts of H_2O and catalyst added, etc.

Little is known about the detailed reactions which take place during gel formation, especially in the case of multicomponent systems.

Hydrolysis of pure silicon tetraethoxide was studied by Aelion et al. (1950) and Schmidt and Kaiser (1981).

There are only a few studies of systems with several constituents. Carturan et al. (1978) investigated the system SiO_2–Al_2O_3–Na_2O and Schmidt et al. (1982) the system SiO_2–B_2O_3–Na_2O using IR spectroscopy. Yamane (1980), Brinker et al. (1982), and Brinker and Scherer (1984) used small-angle x-ray scattering (SAXS) to follow the gelling process.

TABLE I

ALKOXIDES USED IN GEL SYNTHESIS

Cation	$M(OR)_n$	
Si	$Si(OCH_3)_4$	Tetramethylorthosilicate / Tetramethoxysilane (TMS)
	$Si(OC_2H_5)_4$	Tetraethylorthosilicate / Tetraethoxysilane (TES)
Al	$Al(O\text{-}iso\text{-}C_3H_7)_3$	Aluminum isopropoxide
	$Al(O\text{-}sec\text{-}C_4H_9)_3$	Aluminum secondary butoxide
Ti	$Ti(O\text{-}C_2H_5)_4$	Titanium ethoxide
	$Ti(O\text{-}iso\text{-}C_3H_7)_4$	Titanium tetraisopropoxide
	$Ti(O\text{-}C_4H_9)_4$	Titanium tetrabutoxide
	$Ti(O\text{-}C_5H_7)_4$	Titanium tetramyloxide
B	$B(OCH_3)_3$	Trimethylborate
Ge	$Ge(O\text{-}C_2H_5)_4$	Germanium ethoxide
Zr	$Zr(O\text{-}iso\text{-}C_3H_7)_4$	Zirconium isopropoxide
	$Zr(O\text{-}C_4H_9)_4$	Zirconium tetratertiary butoxide
Y	$Y(O\text{-}C_2H_5)_3$	Yttrium ethoxide
Ca	$Ca(O\text{-}C_2H_5)_2$	Calcium ethoxide

IV. Gel Drying

A. Phenomenological Approach

The freshly prepared gel is formed of a network of particles holding an interstitial liquid—the solvent trapped during the gelling step, water in the case of *hydrogels* prepared by method I, mixtures of alcohols and water for the alcogels prepared by method II.

Elimination of these liquid phases leads to dry gels, namely, *xerogels*.

When a "wet" gel is dried the following sequence of events is generally observed on a macroscopic scale:

(1) progressive shrinkage and hardening,
(2) stress development,
(3) fragmentation.

The drying of thin gel films on solid substrates usually presents no particular problems owing to the large surface of evaporation compared to the thickness of the layer. Excessive shrinkage may, however, produce cracking. The same is true for drying of fibers.

The chief difficulty is encountered when massive pieces of gel are required. This is the central problem with *monolithic gels,* and one to which much research has recently been devoted. Monolithic pieces are required if the subsequent consolidation into glass is to be accomplished without the use of hot pressing, by means of which a particulate material may still be compressed into a solid body.

Cracking during the drying stage is the result of nonuniform shrinkage of the drying body. This is a well-known problem in ceramic technology for which, however, quantitative treatments are still scarce.

Cooper (1978) has proposed a phenomenological theory based on an analogy of stress generation due to drying and that due to nonuniform temperature distribution.

This correspondence is based on the similarity observed between the expansion curves of the ceramic systems and those of thermal expansion during the glass transition. The theory of thermomechanical stresses can then be directly applied to the case of nonuniform water distribution in ceramic bodies.

For a prismatic slab of thickness $2w$ the maximum surface stress σ is found to be of the form

$$\frac{\sigma}{A} = \frac{1}{9} j \frac{w}{D}$$

with $A = E/(1 - \nu)$, where E is Young's modulus, ν Poisson's ratio, j the water flux, and D the diffusion constant. Similar results were obtained for other geometries. From a knowledge of the rupture stress σ_r, the constants E and ν, as well as the diffusion constant D, it is then in principle possible to calculate the admissible flux j to avoid the fracture of a body of a given geometry and size. In any case the optimum rate of drying is seen to be inversely proportional to the linear dimension of the object.

The flow of water through systems of this kind has been extensively studied in soil mechanics, where refined mathematical treatments have been developed to account for permeation in porous systems (Parlange, 1973; Philip, 1974).

In practice the necessary times for drying gels in order to preserve monolithicity are considerable. Figure 6 shows, for example, the drying curves of a SiO_2 gel produced by method II from 60% tetramethoxysilane solution in methyl alcohol cast in circular dishes (Prassas, 1981; Prassas et al., 1982b). Curve A corresponds to evaporation of a specimen covered by a plastic sheath. Before gelation the specimen loses about 8–10% at a constant rate. During gelation (15–18 h) the rate diminishes and the average rate (obtained for 10 specimens) is about 0.15 ± 0.03 g/h for an evaporation surface of 8 cm^2.

After gelation the weight loss slows down and is practically constant and equal to 0.037 g/h for 200 h. When the loss is about 60% of the initial gel weight the rate still decreases until no further loss is detected after 400 h. The gel remains monolithic.

Curve B represents conditions of accelerated drying, which permit monolithic gels to be obtained in about 200 h. The rate also drops when the weight loss attains 60% of the initial weight of the gel.

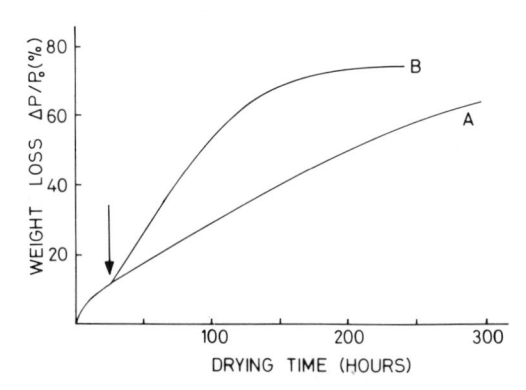

FIG. 6. Drying curves of a SiO_2 gel. A, evaporation of a specimen covered with a plastic sheath; B, accelerated drying. The arrow marks the gelling point. [After Prassas (1981).]

For evacuation speeds not exceeding 0.08 g/h monolithic specimens are obtained with 100% certainty. Their final density is 1.74–1.76 g/cm^3 and they have undergone a linear shrinkage of 45–47% independent of the speed of solvent evacuation.

B. STRUCTURAL APPROACH

The preceding phenomenological approach supposes that the ceramic is an isotropic continuum; no attempt is made to correlate the stresses with the texture of the structure of the material.

In fact the stresses arise not only from the differences in expansion coefficient due to variable water content, but, in the first case, from the *action of capillary forces* which become operative when the pores start to empty and a liquid–air interface is present in the form of menisci distributed in the pores of the drying gel.

Two steps may be distinguished: at the beginning the volume decrease of the material is equal to the volume of evaporated liquid. There is enough liquid to fill the pores and no liquid–air interface is present—no capillary forces are operating. During the second stage the volume is reduced by an amount smaller than the volume of water lost, numerous menisci are formed in the pores, and capillary attraction presses the particles together (Fig. 7). For particles in contact the volume reduction may result first in an elastic deformation of the system and then when the system becomes rigid an irreversible collapse may occur during drying. The resulting stresses generally produce fragmentation of the gel (cracking) unless special precautions are taken in order to preserve monolithicity.

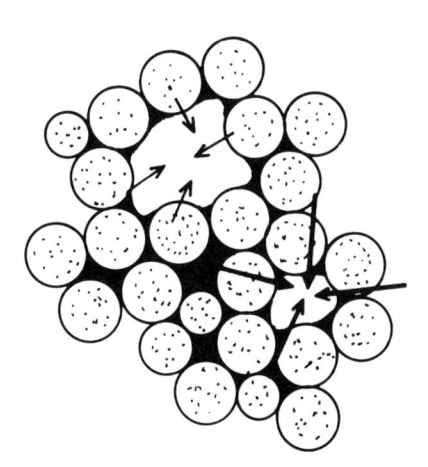

FIG. 7. Capillary forces during drying of a wet particulate material. [From Zarzycki *et al.* (1982).]

1. Capillary Forces and Capillary Potential

When a liquid evaporates from a porous material the solid phase is subjected to forces due to capillary phenomena at the liquid–gas–solid interfaces. The capillary pressure Δp developed across a curved interface with principal radii R_1 and R_2 is given by Laplace's formula

$$\Delta p = \gamma(1/R_1 + 1/R_2), \tag{4}$$

where γ is the surface tension of the liquid.

For a cylindrical capillary of radius r and a liquid having a wetting angle θ this pressure is

$$\Delta p = (2\gamma \cos \theta)/r. \tag{5}$$

The behavior of a liquid in different capillaries can be characterized by the *capillary potential* defined as the potential energy of the field of capillary forces per unit mass of liquid.

By analogy with the potential of the field of gravity, the capillary potential ψ_c for a cylindrical capillary is

$$\psi_c = (2\gamma \cos \theta)/r\rho_L, \tag{6}$$

where ρ_L is the density of the liquid.

A wetting liquid will tend to occupy the position with the highest capillary potential and a nonwetting liquid the position with the lowest potential. A wetting liquid moves spontaneously from a wide into a narrow capillary.

In a drying gel there are capillary pores and gaps of different sizes and shapes; the value of the capillary potential differs at different points of the system. The liquid will tend to occupy positions ensuring the minimum energy of the system as a whole.

A wetting liquid preferentially forms cups (menisci). Capillary forces act on the curved liquid–gas surfaces and on the three-phase liquid–solid–gas contact lines pulling or pushing apart neighboring particles.

The magnitude of capillary forces depends on the size of the capillaries in the system; they can generate considerable stress. Stresses developed by the capillary forces depend, in the first place, on the capillary pressure Δp. Figure 8 shows variations of Δp as a function of the radius r for (a) water ($\gamma = 0.073$ N/m) and (b) methyl alcohol ($\gamma = 0.022$ N/m). The wetting angle taken was $\theta = 0°$.

The local variation of Δp across the structure and stress concentration effects due to porosity make a detailed calculation impractical, but an

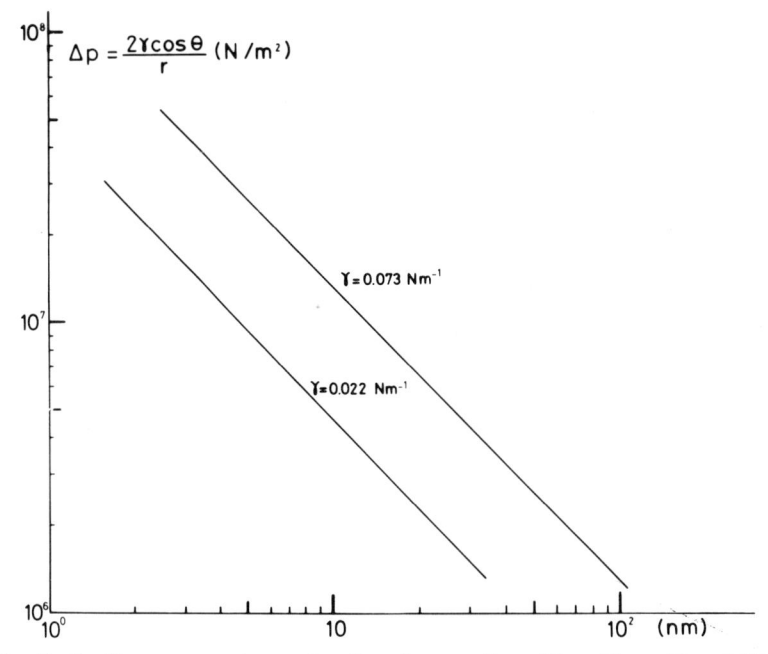

FIG. 8. Capillary pressure Δp as a function of pore radius r. [From Zarzycki et al. (1982).]

estimate of the admissible pore size for a given rupture stress can be obtained from the curves of Fig. 8.

2. Moisture Stress

In reality, during the drying of gels many mechanisms influence the process: not only capillarity but also adsorption, osmotic pressure, changes in the structure of pore water, electric double layer, etc. It is not easy to separate these individual contributions. However, the problem can be treated globally, by introducing the concept of "moisture stress," which is a measure of the energy content of the water–gel system (Schofield, 1935).

Moisture stress can be defined as the work done per unit mass of water by the water when a small quantity of it is transported from the water–gel system to a free water surface at the same temperature and height as the gel. Thermodynamically it is the partial specific Gibbs free energy (or partial specific thermodynamic potential) of the water in the gel. (It is always a negative quantity, but in comparisons the negative sign is usually ignored, so a larger value means that which is numerically greater.)

If p is the pressure of water in the gel and ρ the density of water, the moisture stress ψ is given by

$$\psi = \int_{p_0}^{p} \frac{dp}{\rho}, \tag{7}$$

p_0 being, for example, the atmospheric pressure. For $\rho = 1$ g/cm^3, ψ is numerically equal to the *pressure deficiency* of the water in the gel. For this reason ψ is often expressed in pressure units, although it is an energy quantity. Expressing pressure suction head h in centimeters of water, we find that its common logarithm defines the pH scale (Schofield, 1935) in common use in soil mechanics:

$$pF = \log_{10} h \quad \text{(cm)}. \tag{8}$$

At high moisture stress the energy of attraction of water to gel lowers the water vapor pressure and the moisture stress is given by

$$\psi = RT \ln(e/e_0) \tag{9}$$

where R is the gas constant per gram and e/e_0 the relative humidity of the gel water.

The importance of the moisture stress lies in the fact that it is ψ and not the water content c which is the determining parameter in problems involving moisture changes. This has been recognized in soil engineering but not sufficiently so in ceramic technology (Packard, 1967).

Water will move only in response to a gradient of ψ, not of c. Because of inhomogeneities, gradients of moisture stress will occur where no gradients of moisture content exist, and this will induce permeation and possible fracture.

The general form of the curves relating ψ to c is shown in Fig. 9, which has three essential parts: an approximately linear portion relating ψ to c for lower values of ψ, a more or less horizontal elbow, and a steep portion for higher ψ. The beginning of the elbow marks the point of air entry into the pores (shrinkage limit in the case of ceramic bodies). Before this point is reached, for lower ψ, the suction in the pore water is equivalent to the pressure on the solid and the shear strength of the material is proportional to the moisture stress acting on it. This portion of the curve could be used to assess and compare the mechanical strength of *moist*, deformable gels.

The more or less flat plateau corresponds to the emptying of the pores of the *rigid* gel. As pF increases, the increasing suction causes the pores to be emptied, the larger ones first, then progressively those of decreasing size.

The steep portion for high pF corresponds to adsorption phenomena and coincides with the usual adsorption isotherm of the gel.

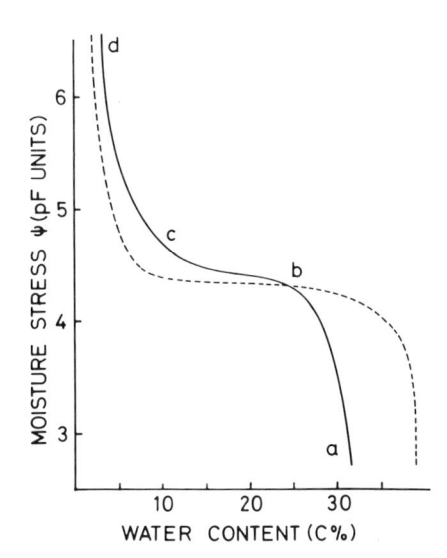

FIG. 9. Moisture stress as a function of water concentration for two different gels (schematic).

The first two portions of this curve could be used to assess the capabilities of a given gel to withstand the drying procedure.

The steeper the first portion, the faster the increase of mechanical resistance for a given loss of water content; the beginning of the plateau marks the beginning of possible fracture. A flatter plateau would correspond to a greater homogeneity in pore distribution and an early tendency of the elbow to larger pores.

The $\psi(c)$ curve thus permits quantitative characterization of the given material; its application to gel–glass technology should be investigated (Zarzycki, 1984).

3. Rheological Behavior of the Gel

During the gelling and subsequent drying processes the mechanical characteristics of the material undergo very significant changes. The initial solution is a Newtonian liquid which changes during gelling to a Bingham material with a steady increase of viscosity. The resulting gel is first a viscoelastic liquid and then a viscoelastic solid which is progressively transformed into a purely elastic solid during the drying process.

The stiffening process is irreversible in SiO_2-containing gels, the dehydration establishing new Si—O—Si bonds between particles and thus cementing them in new positions, resulting in a collapse and overall shrinkage of the skeleton of the gel.

The porosity of the system diminishes, but at the same time the pore size distribution is changed, the dry gel being thus a modified version of the initial hydrogel. For pores of different sizes lying near one another differential stresses are generated from capillary forces, etc., and maximum differentials arise when a small pore lies in the vicinity of a large (empty) pore. Fractures are initiated if these stress differences are greater than the tensile resistance of the material. As the point of air entry is passed on, larger pores start emptying, and if the sample is placed in an atmosphere the relative humidity e/e_0 of which is progressively decreased, smaller and smaller pores will empty according to Thomson's relation:

$$\ln(e/e_0) = -2\gamma V/RTr \tag{10}$$

where V is the molecular volume of water, T the absolute temperature, R the gas constant, and r the radius of the pore.

The capillary stress Δp will thus affect smaller and smaller pores and, if the surrounding material is deformed viscoelastically, the radius of a pore will simultaneously decrease, delaying its emptying until the partial water pressure e has lessened. At the same time, however, as the water content is decreased, the system hardens, the viscoelastic behavior being replaced by a purely elastic deformation. In this competition between closure of pores by capillary forces and increased rigidity, the smaller pores may undergo complete collapse provided the rigidity has not reached an excessive value. This depends on the nature of the gel.

C. Preparation of Monolithic Gels

If massive pieces of glass are to be obtained without the use of hot-pressing techniques, which have the disadvantage of limiting specimen size, these pieces cannot be in particulate form. The obtention of *monolithic* dry gels is thus an essential prerequisite. Very few indications exist in the literature as to how this can be done. Essentially empirical, the majority concern pure silica.

By very slow drying (about two weeks) of the gels prepared by hydrolysis and polycondensation of tetramethoxysilane (TMS), Yamane *et al.* (1978, 1979) obtained disks free from fissures and of small dimension (∼1 cm). The dry gels had an apparent density of 1–1.5 g/cm^3; the average pore diameter was 2–5 nm.

Similarly, Nogami and Moryia (1979, 1980) obtained monoliths of small dimensions. Their densities were 1.8 and 0.8 g/cm^3 and their pore radii between 2 and 7 nm.

Shoup (1976) prepared, by the sol–gel process, silica monoliths of appreciable dimensions. The process uses sols containing Na$^+$ and K$^+$ ions. Gelling is obtained at pH > 10 upon addition of polar solvents such as formamide or ethyl acetate. To eliminate traces of alkali ions the humid gel is washed with acid solution. The dry gels have a uniform open porosity with a pore diameter of ~200 nm. Monolithicity was attained if pore diameter exceeded 60 nm.

Yoldas (1975c) obtained monolithic transparent aluminas in three stages:

(1) hydrolysis of aluminum alkoxide, $Al(OR)_3$;
(2) peptization of the hydroxide and formation of a transparent sol;
(3) gelation of the sol.

Not enough details are given about the drying stage. The author has determined a critical acid concentration in stage 2 necessary to retain monolithicity of 0.03–0.1 moles per mole alkoxide. There is no discussion of the causes of fracture.

The preceding analysis of the phenomena accompanying drying has shown the importance of capillary forces and of differential stresses which operate during shrinking.

All actions which tend to minimize these stresses and increase the mechanical resistance of the network should enhance the probability of monolithic gel formation. The following are possible:

(1) strengthening the gel by reinforcement,
(2) enlarging the pores,
(3) reducing the surface tension of the liquid,
(4) making the surface hydrophobic,
(5) operating under hypercritical conditions where the liquid–vapor interface vanishes,
(6) evacuating the solvent by freeze drying.

The use of solvents with small surface tension is only efficient when the gel presents large pores; it is complicated by the necessary compatibility of the liquids in the initial mixture to ensure homogeneous polymerization. The role of such solvents in the porosity of silica gels has been investigated by Neimark and Scheinfain (1953).

Tensioactive substances which lower the surface tension of the liquid may be used as well as substances which diminish the wetting of the solid phase. Ammonia and its organic derivatives have been proposed to decrease the wetting of silica particles (Iler, 1979).

An increase in the radius of the pores may be obtained by varying the conditions of hydrolysis or by the addition of foreign substances.

Enhancement of the mechanical resistance of the gel may be obtained before drying by an aging process or by the addition of "active silica" during the gelling process. Numerous examples exist of the favorable results thus obtained (Okkerse, 1960; Yates, 1971).

The increase in the mechanical resistance of the gel is continuous during drying. For a silica gel with 60% TMS the solid phase represents 10 vol. % of the initial solution. In the first instants of evaporation the solid phase is in the form of an open network and presents very slight mechanical resistance. The capillary forces then draw the particles together and new Si—O—Si bonds strengthen the solid network. For this humid gel the stresses originating in micropores can bring about fragmentation during the first drying stages; it is necessary to evacuate the solvent trapped in the larger pores without provoking evaporation in the smaller ones. This can be done only if a pressure of the solvent is maintained above that of the gel, a pressure slightly lower than the saturation pressure given by Thomson's relation [see Eq. (10)]. The most efficient way to eliminate the destructive action of the surface tension of the liquid is to suppress the liquid–vapor interface by operating under *hypercritical conditions*.

This method consists in treating the gel in an autoclave under hypercritical conditions for the solvent. Kistler (1932) had already suspected that the liquid phase in a gel was independent of the skeleton and was able to prepare uncollapsed gels the pores of which were filled with air, or *aerogels*. The substitution of water by alcohol and subsequent evacuation of alcohol under hypercritical conditions had to be done in order to avoid the peptizing action of water at high temperatures.

Nicolaon and Teichner (1968) applied the method of hypercritical evacuation directly to alcogels obtained from organometallic compounds. These studies led to the production of a very light material with high porosity suitable for applications in catalysis. No attention was, however, directed toward *monolithicity* and the obtention of materials suitable for densification into glass.

Studies performed in our laboratory (Prassas, 1981; Zarzycki *et al.*, 1982) have led to the conclusion that monolithicity can be attained in this way but that many practical variables must be optimized in order to obtain monoliths with 100% certainty.

The process is schematized in Fig. 10, which shows the equilibrium curve between the liquid and gas phase of the solvent. In order to ensure the continuity of the liquid → gas transition, the path of the thermal treatment must not cross the equilibrium curve AC. To circumvent the critical point C a path such as abde may theoretically be used. The liquid is first compressed beyond the critical pressure, then its temperature is increased at constant pressure. Isothermal expansion would then bring the liquid within the gas domain without loss of continuity.

In practice this path is modified in the following way: The open container containing the gel is placed inside an autoclave and, to obtain hypercritical conditions, a given quantity of solvent (methanol) is added

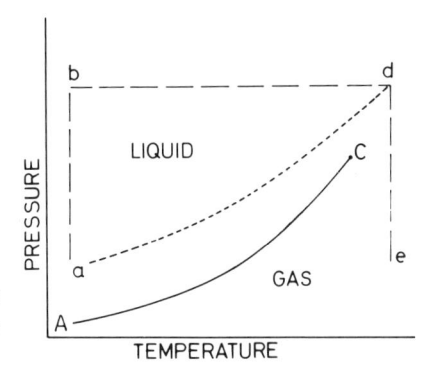

FIG. 10. Paths for hypercritical solvent evacuation (schematic). [After Prassas (1981).]

to the autoclave. This is closed and electrically heated. When the critical temperature of methanol is exceeded (path ad), successive flushings with dry argon eliminate the last trace of alcohol. The autoclave is then cooled down and the gel removed at ambient temperature.

A number of runs have shown that monolithicity depends on many variables, namely,

the speed of heating,
the proportion of additional solvent,
the concentration of TMS and water of hydrolysis,
the geometry and size of the sample,
previous aging of the gel.

Optimizing these variables, we obtained monolithic samples with 100% certainty. Figure 11 shows as an example cylindrical samples of monolithic silica aerogels up to 40 mm in diameter and 250 mm in length. The porosity of the samples can be controlled by varying the TMS concentration.

The gels are hydrophobic and contain an appreciable percentage of organic adsorbed radicals. Their mechanical resistance is, however, sufficient for these to be eliminated by subsequent thermal treatment without loss of monolithicity and finally converted into a clear glass of excellent optical quality.

V. Densification Process

A. STRUCTURE OF DRIED GELS (XEROGELS AND AEROGELS)

The final structure of the dry gel will depend on the structure of the wet gel originally formed in solution; it will be a contracted or a distorted version of the latter. The models proposed comprise (Fig. 12) (a) aggregates of particles of approximately the same size, essentially massive in

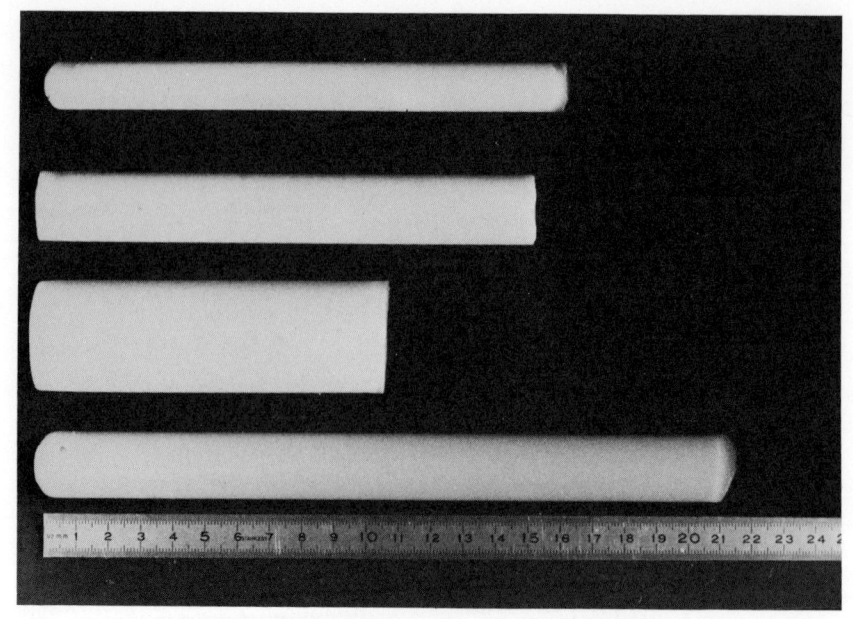

FIG. 11. Examples of monolithic aerogels obtained by hypercritical solvent evacuation.

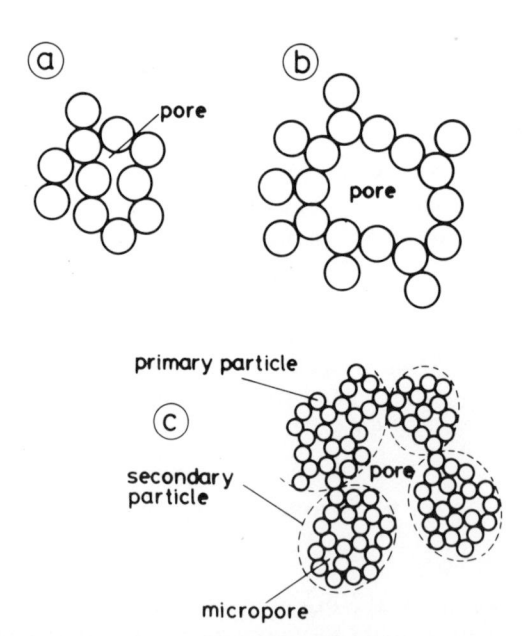

FIG. 12. Different structural models of dried gels: (a) massive aggregates, (b) aggregates with macropores, and (c) secondary aggregates of primary particles. [From Zarzycki et al. (1982).]

FIG. 13. Texture of gels (schematic): (a) dense agglomerate of particles with closed pores, and (b) lattice of particles with open pores. [From Zarzycki (1982). North-Holland Publishing Company, Amsterdam, 1982.]

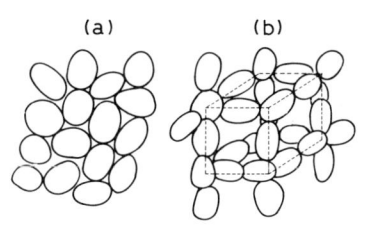

nature; (b) aggregates of particles formed by primary particles (with ultrapores); (c) aggregates of a more complex nature in which three levels of particles can be distinguished as well as micropores and macropores.

The dried amorphous gel differs from a glass by its *texture*. The gel is essentially an *agglomerate* of elementary particles, the size of which may be of the order of 10 nm, arranged more or less compactly. The porosity may vary considerably according to the method of preparation.

The residual space represents the *pores,* which may be closed for dense arrangement of particles (Fig. 13a) or *open* when the texture consists of more or less regular "lattices" of particles leaving large interstices (Fig. 13b).

Small-angle scattering of x rays (SAXS) and of neutrons (SANS) constitute convenient techniques for distinguishing a gel from a homogeneous glass; in the case of a glass (devoid of phase separation) only a very weak scattering due to frozen-in thermal fluctuations may be detected; a strong small-angle intensity characterizes a gel.

The constituent particles are coated with residual OH and OR groups which are partly eliminated during the transition from a particulate texture to a continuous solid; these groupings may be detected and analyzed by conventional infrared spectroscopic techniques.

According to the packing geometry, the systems present different porosity and specific surface; the final characteristics of a dried gel are determined by the physicochemical conditions at every step of the preparation:

(1) the size of primary particles at the moment of aggregation;
(2) the concentration of particles in solution;
(3) the pH, salt concentration, temperature, and time of aging or other treatment in the wet state;
(4) mechanical forces present during drying;
(5) the temperature, pH, pressure, salt concentration, and surface tension of the liquid medium during drying;
(6) the temperature, time, and atmosphere during drying.

B. Gel-to-Glass Transformation

To transform the particulate structure of a dried gel into continuous glass the elementary particles must weld together, which results in progressive pore elimination. This is achieved by heating the gel in order to promote diffusion phenomena and viscous flow. During this heat treatment the residual OH and OR groups will tend to be eliminated in the form of H_2O and ROH, a process which is accompanied by an additional polymerization of the system:

$$\equiv\!Si\!-\!OR + HO\!-\!Si\!\equiv \longrightarrow \equiv\!Si\!-\!O\!-\!Si\!\equiv + ROH. \tag{11}$$

The escape of residual products from *closed* pores can be a problem: organic residues may carbonize at sufficient temperature and thus bring about discoloration of the gel and leave carbonaceous particles in the glass. Trapped products may also produce at a later stage the loss of monolithicity of the gel. It is therefore important to induce the escape of residues before closure of the pores; oxidation treatments are often necessary to eliminate certain organic groups.

In almost all studies the progressive elimination of residues is followed by DTA and TGA analyses. Losses of H_2O and organic volatiles during heat treatment and the influence of oxygen-containing atmospheres have been extensively investigated in order to determine the optimal heat treatment schedule which leads to a dense and clear glass. Escaping products may also be monitored by gas chromatography.

Oxidation treatments were found necessary in some cases to eliminate carbon precipitates.

Infrared transmission spectroscopy, particularly in the 3400–3600 and 1300–400 cm^{-1} regions, are commonly used to identify the remaining groups and follow the gel into glass conversion. Hench *et al.* (1982) and Prassas and Hench (1984) used IR reflection spectroscopy and Bertoluzzo *et al.* (1982) Raman spectroscopy.

In general during thermal treatment the following temperature zones are found (Fig. 14):

50–265°C: departure of OH_2 and of ROH;
265–300°C: oxidation of —OR groups;
300–1000°C: departure of —OH.

It is important to define in each particular case the heating schedule in order to eliminate unwanted residues without impairing monolithicity before viscous flow phenomena start.

Occluded —OH and H_2O may provoke bloating upon heating of the gel. This will often be the case with gels dried slowly at low tempera-

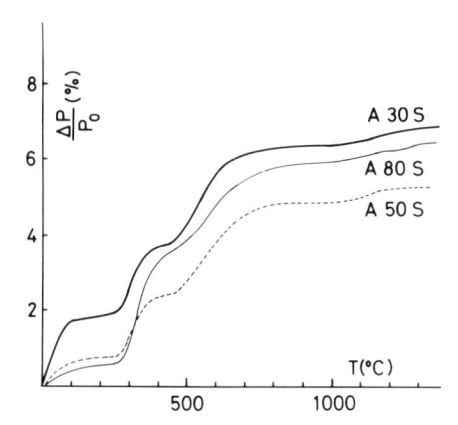

FIG. 14. Weight losses of silica aerogels heated at 10°C/min. [From Prassas (1981).]

tures—even if they have remained monolithic so far. Remaining —OH groups may be eliminated by chlorination treatments (Hair and Hertl, 1973; Phalippou *et al.*, 1984) if very low OH levels are required in the final glass.

On the other hand, occluded water may be used in foaming processes, e.g., in blowing gel particles into microballoons. The further addition of agents releasing CO_2 such as urea was used by Nogami *et al.* (1982) to increase the amount of the gas phase used for expansion.

Leaving these particularities aside, we may describe densification as essentially a *sintering* process by which the pores of a dry gel are eliminated and the material converted into clear, massive glass. After the elimination of residues the driving force in this process is essentially supplied by the *surface energy* of the porous gel. This tends to decrease the interface, thus eliminating the pores, their collapse being governed in the case of glasses by *Newtonian viscous flow*. Extra pressure, as in hot-pressing techniques, may be applied to speed up the process.

C. Sintering by Viscous Flow

The densification of gels into glasses was discussed by Zarzycki (1982). To study this transformation a *model* must be adopted to represent the texture of the porous solid. Two models have been used in the case of gels:

(1) The model of *closed pores* proposed by Mackenzie and Shuttleworth (1949) in their theory of sintering.

(2) The model of *open pores* devised by Scherer (1977a) to study less dense, "latticelike" textures encountered in the sintering of "soots" in optical-fibers technology.

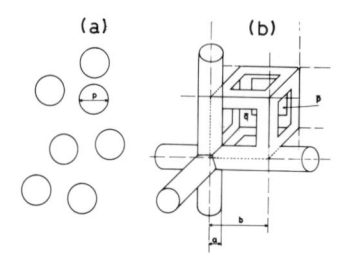

FIG. 15. Models of gel texture (schematic): (a) closed pores, and (b) open pores. [From Zarzycki (1982). North-Holland Publishing Company, Amsterdam, 1982.]

These two models, which idealize the situations depicted in Figs. 13a and 13b, are represented in Figs. 15a and 15b, respectively.

1. Closed-Pore Model

Mackenzie and Shuttleworth assume that the pores are identical spheres of initial radius r_i; their number n per unit volume of *solid* phase is supposed to remain constant during densification.

If the reactive density $D = \rho/\rho_s$ is defined as the ratio of the apparent density of the porous solid (gel) to the density of the solid phase (glass), the relation between n and r_i is

$$n \tfrac{4}{3}\pi r_i^3 = (1 - D)/D \tag{12}$$

or

$$r_i = (3/4\pi)^{1/3}[(1 - D)/D]^{1/3} n^{-1/3}. \tag{13}$$

To evaluate n it is most convenient to consider the total surface area $S = 4\pi r_i^2 n$ of the pores per *unit volume* of *solid* phase:

$$S = 4\pi (3/4\pi)^{2/3} n^{1/3} [(1 - D)/D]^{2/3}. \tag{14}$$

Experimentally, by using various techniques (BET, SAXS, etc.), a specific surface $S_{sp} = S/\rho_s$ is measured (expressed generally in m²/g).

In the case of sintering of gels the solid phase is a glass which has a Newtonian viscosity η independent of the rate of strain and a surface energy γ_g. The kinetic equation of sintering is then

$$\frac{dD}{dt} = \frac{3}{2}\left(\frac{4\pi}{3}\right)^{1/3} \frac{\gamma_g n^{1/3}}{\eta} (1 - D)^{2/3} D^{1/3}, \tag{15}$$

which may be integrated to

$$\frac{\gamma_g n^{1/3} \Delta t}{\eta} = \frac{2}{3}\left(\frac{3}{4\pi}\right)^{1/3} \int_0^D \frac{dD}{(1 - D)^{2/3} D^{1/3}} \tag{16}$$

where Δt is the time necessary to reach the reduced density D.

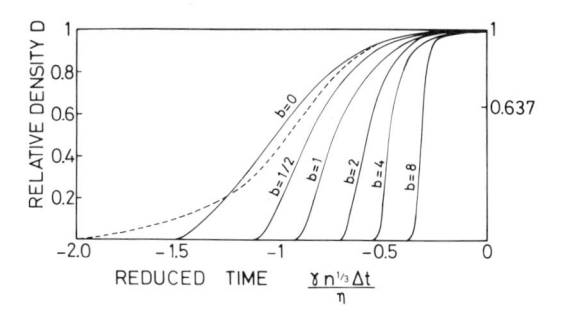

FIG. 16. Kinetics of sintering for different values of parameter b (see text). The full curves represent results for the Mackenzie–Shuttleworth closed-pore model; the broken curve represents the Scherer's open-pore model for $b = 0$. [From Zarzycki (1982). North-Holland Publishing Company, Amsterdam, 1982.]

A *reduced time* t_r is defined by the relation

$$t_r = \Delta t \gamma_g n^{1/3} / \eta. \tag{17}$$

If an external pressure P is applied during sintering, γ_g must be replaced by the expression

$$\gamma_g(1 + Pr_i/2\gamma_g) \tag{18}$$

or

$$\gamma_g\{1 + b[(1 - D)/D]^{1/3}\}, \tag{19}$$

where

$$b = (3/4\pi)^{1/3} P/2\gamma_g n^{1/3}. \tag{20}$$

the parameter b is proportional to the pressure difference between the pores and the outside of the compact.

This slightly complicates the evaluation of the integral, but in both cases it can be shown that sintering to $D = 1$ occurs in a *finite* time. Figure 16 shows the results of these calculations. (Sintering without applied pressure corresponds to $b = 0$.) It can be seen in particular that the *reduced* sintering time from $D = 0.5$ to $D = 1$ is nearly equal to unity.

2. Open-Pore Model

For gels with an open "latticelike" texture which implies an open porosity, the Mackenzie–Shuttleworth (MS) model is no longer applicable. Scherer (1977a) proposed a model which consists of a regular cubic lattice of intersecting cylinders of radius a, the edge of the lattice being l.

This model idealizes the situation of Fig. 13b, the two parameters a and l being related respectively to the radius of elementary particles and to a "spacing" of the pseudolattice. It is valid for the values of the ratio $x = a/l < 0.5$. For $x = 0.5$ the pores are closed and the MS model is again applicable.

It has been shown that the densification kinetics derived from Scherer's model follow rather closely the results of the MS model for values of $D > 0.3$ (Fig. 16). The model presents an advantage for small values of D; the (arbitrary) choice of a cubic lattice by Scherer has little effect on the final results. In this two-parameter model the relative density D is

$$D = 3\pi x^2 - 8\sqrt{2}\, x^3, \tag{21}$$

and the specific surface

$$S_{sp} = (1/\rho_s)(6\pi - 24\sqrt{2}\, x)/lx(3\pi - 8\sqrt{2}\, x) \tag{22}$$

In principle knowledge of D and S_{sp} is sufficient to determine the parameters $x = a/l$ and l. In practice, additional determinations of the equivalent diameter of the interconnecting pores (situated on the sides of the cell) are made by means of Hg penetration porosimetry as well as electron microscopical observations to determine the parameter a. The model may be developed further by assuming pore size distributions (Scherer, 1977b). The connection between the Scherer model and the Mackenzie–Shuttleworth model shows that

$$n^{1/3} \simeq (1/l_0)D^{-1/3},$$

where l_0 is the initial length of the cell edge.

This makes it possible to use the same reduced time t_r as for the MS model. Scherer's model thus constitutes a convenient extension toward the low-D region, where the significance of the MS model becomes doubtful. It has been successfully used by Scherer and Bachman (1977) for the evaluation of densification of SiO_2 "soots" and gels.

Wenzel (1982) discussed the validity of pore volume data obtained by different techniques.

3. Hot-Pressing: The Murray–Rodgers–Williams (MRS) Approximation

Murray et al. (1954) have simplified the MS model in the case in which $P \gg 2\gamma_g/r_i$, i.e., for $b \gg 1$, the equation becoming

$$dD/dt = (3P/4\eta)(1 - D), \tag{23}$$

which may be integrated into

$$\ln(1 - D) = -(3P/4\eta)t + \ln(1 - D_i). \tag{24}$$

The consequence, however, is that the time t required for complete densification is *no longer finite;* it is necessary to specify the final density D_f.

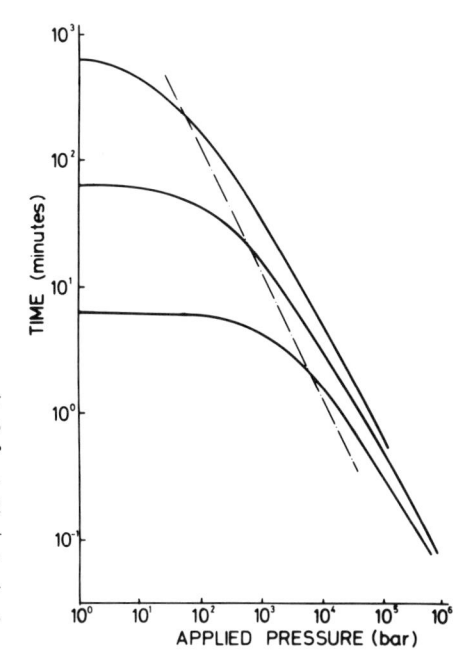

FIG. 17. Sintering time t necessary for compaction of a silica gel from $D = 0.5$ to $D = 1$ for different applied pressures P with $\eta = 10^{11}$ P and $\gamma_g = 280$ erg/cm^2 calculated from the model of Mackenzie and Shuttleworth. Broken line: Murray–Rodgers–Williams approximation for $D_f = 0.999$. [From Zarzycki (1982). North-Holland Publishing Company, Amsterdam, 1982.]

For example, for $D_f = 0.999$ and $D_i = 0.5$, the time t_p it takes to densify a gel with $\eta = 10^{11}$ P is

$$t_p = [1.37 \times 10^4/P \text{ (bar)}] \quad \text{min.} \tag{25}$$

The plot in Fig. 17 shows the approximation introduced. The slope is correct for higher P but the position of the line depends on the D_f adopted. For higher P a better fit is obtained by taking, for example, $D_f = 0.99999$.

The discrepancy between the simplified theory and the complete MS treatment may explain the initially faster compaction rates observed in some experiments by Decottignies *et al.* (1978). Evidently the MRW approximation does not take into account the variations of porosity ($n^{1/3}$ factor).

This treatment is valuable for hot-pressing techniques. It enables a quick determination of the $\eta(T)$ relationship in the pressing temperature interval by a method of linear temperature incrementation (Decottignies *et al.*, 1978).

In the case of "flash" pressing, if it is assumed that the temperature varies linearly with time in the temperature interval over which densification occurs, the basic MRW equation (23) may be written

$$\frac{d \log(1 - D)}{dT} = \frac{3P}{4\eta \times 2.3 v_0}, \tag{26}$$

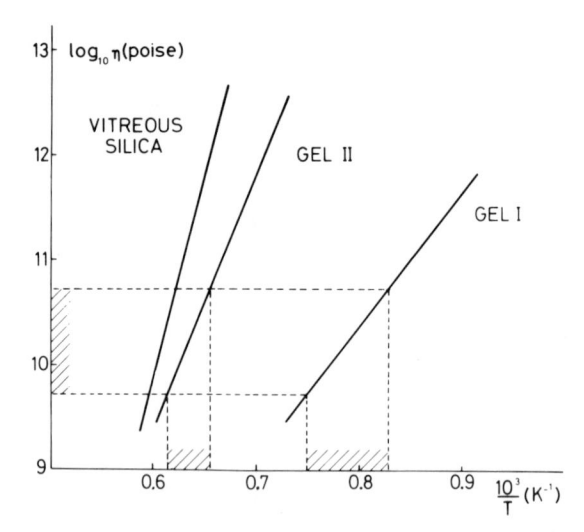

FIG. 18. Viscosity–temperature relationships for vitreous SiO$_2$ and silica gels prepared by methods I and II. Hot-pressing intervals are indicated by cross hatching. [From Zarzycki (1981).]

where v_0 is the rate of temperature increase. From the slope of the $\log(1 - D) = f(T)$ curves it is then possible to calculate the viscosity η of the gel within the $10^{9.5}$–$10^{11.5}$ P viscosity interval. The results of such a determination for SiO$_2$ and silica gels are given in Fig. 18. They show the marked difference between gels prepared by methods I and II and vitreous silica.

D. Devitrification Kinetics

1. Use of TTT Diagrams

The gel being densified will at the same time tend to devitrify. The successful conversion of gel into glass therefore depends on a competition between phenomena which lead to densification and those which promote crystallization. It should be noted that in some cases devitrification of the gel results from atmospheric attack during the drying stage, in particular for alkali-rich compositions (Prassas et al., 1982a). Devitrification of the gel will be most likely, however, at high temperatures during the last densifying stage. It is therefore necessary to take into account the crystallization kinetics of the gel.

These are best presented by means of TTT (time–temperature–transformation) diagrams, which are a convenient way of studying the devitrification versus compaction problem (Zarzycki, 1981). The TTT diagrams show the time t_y to reach a determined crystallized fraction y as a function

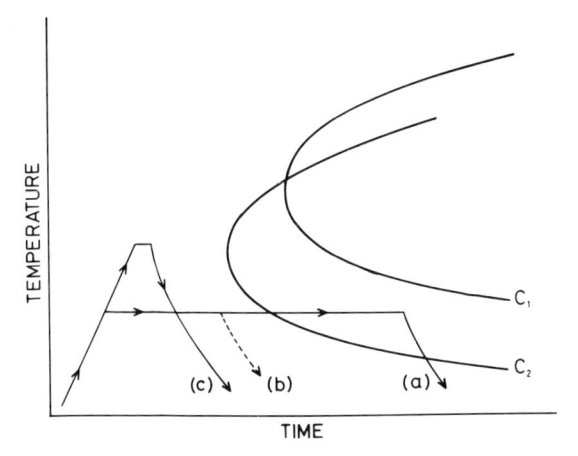

FIG. 19. Time-temperature-transformation (TTT) diagrams and thermal paths for compaction. [From Zarzycki (1982). North-Holland Publishing Company, Amsterdam, 1982.]

of the temperature T. Treating y as a parameter, one obtains a set of C_y curves which represent the kinetic behavior of the system. In particular, if y_0 corresponds to the smallest crystallized fraction detectable by analytical techniques, the curve C_{y_0} represents a frontier not to be crossed during a thermal treatment schedule if crystallization is to be avoided. (The value generally adopted for y_0 is 10^{-6}, but other criteria are possible.) Using this method, one can evaluate the sintering time of a given gel when the characteristics of the material and the thermal treatment program are specified.

The relative positions of this thermal treatment path during densification and of the C_{y_0} curve of the gel determine the possibility of obtaining either glassy or crystallized material at the end of the compaction program (Fig. 19). If, for example, for a given gel corresponding to C_1 there is no danger of devitrification using path (a), this is no longer true for a gel corresponding to the curve C_2 and the same path would lead to crystallized material. The solution would then be either to shorten the time of sintering, e.g., by applying a suitable pressure P [path (b)] or to increase the temperature for a short time by using the technique of "flash pressing" [path (c)].

In the case of gels the position of the curves C strongly depends on the impurities of the material, principally on water content, which influences the viscosity η as well as the surface tension γ_g of the material.

It is thus essential to recognize the factors that determine the position of the C_{y_0} curves in the TTT diagrams; they depend on the nucleation–growth characteristics of a given system.

2. Nucleation–Growth Conditions

From a knowledge of the nucleation rate I (number of nuclei formed per unit volume and unit time) and the crystal growth rate u it is possible to evaluate the time t_y it will take to obtain a fraction y of crystalline phase in order to construct the TTT diagram (C_y curve). The crystallized fraction y is related to I, u, and the time t_y by one of the expressions

$$y = \tfrac{1}{3}\pi I u^3 t_y^4, \tag{27}$$

for interface-controlled growth, or

$$y = B(\tilde{D})^{2/3} I (t_y)^{5/2}, \tag{28}$$

for diffusion-controlled growth, where B is a constant depending on supersaturation and $\tilde{D}$ the diffusion constant. It can be shown (Zarzycki, 1981) that the crystallized fraction is

$$y \propto (t_y/\eta)^4 T^3 \exp(-W^*/kT) \tag{29}$$

in the first case, and

$$y \propto (t_y/\eta)^{5/2} B \tag{30}$$

in the second case, where W^* is the thermodynamic barrier for nucleation.

These expressions show clearly that, whatever the mechanism of crystallization, at a given temperature T the identical fraction y is obtained provided the ratio t_y/η remains constant. The position of the C_y curve depends on this ratio. On the other hand, as shown previously, both the reduced time required for sintering and the hot-pressing time depend on the same factor. The time/viscosity ratio thus appears as a determining factor in *both* the densification and devitrification processes occurring during gel → glass conversion, which makes the use of TTT diagrams particularly suitable.

3. Viscosity–Time Equivalence

It is possible to take into account the influence of the viscosity in the TTT diagram. At a given temperature, if the viscosity η of the material has been modified, e.g., by impurities, the same fraction y will be obtained in a time t'_y such that

$$t'_y/\eta' = t_y/\eta, \tag{31}$$

η' being the new viscosity, provided that this does not change significantly the proportionality constant of Eq. (29) or (30). This is a plausible hypothesis since all transport phenomena are included in the η factor.

It is most convenient to plot both the TTT diagram and the $\eta = f(T)$ curves by using plots of $\log t_y$ versus $1/T$ and $\log \eta$ versus $1/T$. A change in viscosity characteristics then corresponds at each temperature to a transformation of C_y into C_y' according to the relationship

$$\Delta \log t_y = \Delta \log \eta. \tag{32}$$

This is interesting since it enables predictions to be made for a given system which has undergone small compositional changes influencing η. In practice, it is much easier to obtain rapidly the viscosity–temperature characteristics than a C_y curve of the TTT diagram.

A further advantage of plotting the TTT diagrams in this system of coordinates is that for lower temperatures the C_y curves tend to become linear with a slope corresponding to that of the $\log \eta$ versus $1/T$ curves, which is useful for extrapolation.

If C_y is a known limit not to be transgressed for the original material, C_y' will constitute the new limit for the modified material and the pressing schedule may be fixed accordingly.

4. Influence of Heterogeneous Nucleation

Figure 20 shows as an example the TTT diagrams of vitreous SiO_2 and silica gels obtained respectively by methods I and II. The application of

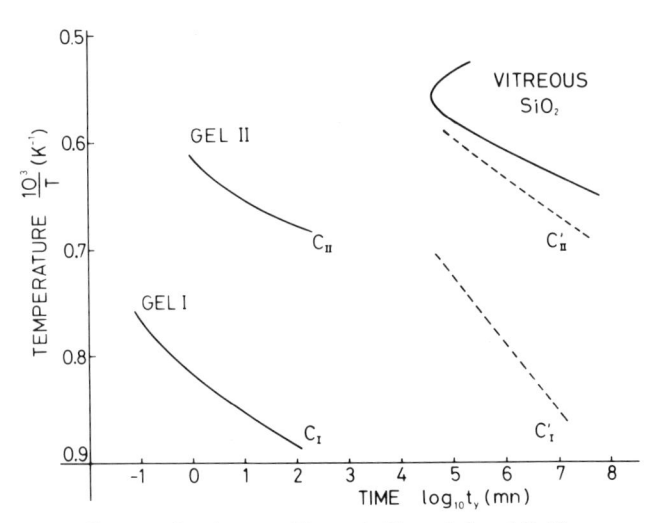

FIG. 20. TTT diagrams for vitreous silica and silica gels I and II. The curves correspond to the crystallized fraction $y = 10^{-6}$. The broken curves represent the theoretical curves for gels I and II deduced from the curve of vitreous silica to be expected if the viscosity variations alone were responsible for the differences observed in devitrification behavior. [From Zarzycki (1981).]

the viscosity-time equivalence leads to curves C'_I and C'_{II}, which should be correct if devitrification behavior is influenced by variations of viscosity only. It can be seen that these curves are on the same temperature level as the curves C_I and C_{II} derived from experimental results, but the latter are shifted considerably to the left. This indicates that other factors must also be taken into account. It has been suggested (Zarzycki, 1981) that this effect is due to heterogeneous nucleation, possibly resulting from the large interface of the gel and from surface impurities, in this case Na^+ ions.

For a gel of a given composition it thus appears that the possibility of converting it into glass depends on its mode of preparation and the final impurity level.

5. Devitrification Behavior

Comparative studies of glasses obtained either by fusion of gels or directly, by melting the mixtures of oxides carried out by Mukherjee et al. (1976a) for the SiO_2-La_2O_3 and SiO_2-La_2O_3-ZrO_2 systems indicate that gel-produced glasses devitrify almost ten times more rapidly. This was ascribed to the OH content of these glasses.

The crystalline phases were investigated in detail by Mukherjee and Zarzycki (1979).

The devitrification behavior of SiO_2-La_2O_3 glasses obtained by hot pressing was investigated by Decottignies (1977) and Decottignies et al. (1977b, 1978). The influence of thermal treatment and of pressure on the precipitation of quartz and cristobalite was determined.

The compositional limits of obtaining noncrystallized glasses were obtained for SiO_2-Al_2O_3 and SiO_2-TiO_2 systems by Kamiya et al. (1974) and for the SiO_2-ZrO_2 system by Kamiya et al. (1980).

Phalippou et al. (1978) compared the kinetics of transformation of silica gels obtained either from sols or from organometallic compounds.

Phalippou et al. (1982) determined the influence of various additions (Na_2O, B_2O_3) to SiO_2 gels on their crystallization behavior.

Höland et al. (1982) studied the system SiO_2-Al_2O_3-MgO, Pancrazi et al. (1982) the system SiO_2-Al_2O_3-CaO, and Woignier et al. (1982) the systems SiO_2-Al_2O_3-X_2O with X = Li, Na, K.

VI. Special Features of Gel-Produced Glasses

The differences in devitrification behavior have been noted above.

A systematic comparison of the physical constants of glasses obtained either by hot pressing of gels or by direct fusion or CVD methods has been made by Jabra (1979) and Jabra et al. (1980) for SiO_2-B_2O_3, SiO_2-TiO_2, and SiO_2–P_2O_5 glasses. Only minor differences were detected; for example, the glass transition temperature is slightly lower for gel-produced glasses. This can be explained by the rather high temperature required for the hot-pressing process, which may obliterate structural differences.

Weinberg and Neilson (1978) compared the phase separation behavior of SiO_2–Na_2O glasses obtained either from gels or by melting of oxides. They found a miscibility temperature about 100°C higher for the gel glass and a finer texture. The same authors recently reported (1983) an elevation of the liquidus in a gel-derived SiO_2–Na_2O glass at about 40°C. Both effects were interpreted as being due to higher OH content.

Yoldas (1982, 1984) claims, on the contrary, that a structural effect may be introduced by varying the polymerization conditions of the original network from organometallic compounds. This would permit the introduction into the synthesis of inorganic glasses some of the flexibility hitherto restricted to organic polymers.

VII. Conclusion

From the preceding discussion of the gel → glass process it can be seen that the different steps are not equally well understood. In particular, the detailed mechanism of gel formation in multicomponent systems has only begun to be investigated. The drying conditions have recently received considerable attention, particularly with regard to ensuring monolithicity; decisive progress has been made with the hypercritical solvent evacuation technique.

The mechanism of rheological changes of the gels and their fracture behavior remain, however, practically unknown.

The sintering conditions are better understood but the final obtention of glasses is conditioned by heterogeneous nucleation which complicates theoretical estimates.

Even at this stage, however, the gel process has already shown great potential. It is hoped that before too long the gel route will impart to inorganic glasses some of the flexibility already available in the synthesis of organic polymers.

References

Aelion, R., Loebel, A., and Eirich, F. (1950). *J. Am. Chem. Soc.* **72,** 5705–5712.
Banks, W. H., and Barkas, W. W. (1946). *Nature (London)* **158,** 341–342.
Bertoluzzo, A., Fagnano, C., Morelli, M. A., Gottardi, V., and Guglielmi, M. (1982). *J. Non-Cryst. Solids* **48,** 117–128.
Brinker, C. J., and Scherer, G. W. (1984). *In* "Ultrastructure Processing of Ceramics, Glasses, and Composites" (L. Hench and D. R. Ulrich, eds.), pp. 43–59. Wiley, New York.

Brinker, C. J., and Mukherjee, S. P. (1981a). *Thin Solid Films* **77**, 141–148.
Brinker, C. J., and Mukherjee, S. P. (1981b). *J. Mater. Sci.* **16**, 1980–1988.
Brinker, C. J., Keefer, K. D., Schaeffer, D. W., and Ashley, C. S. (1982). *J. Non-Cryst. Solids* **48**, 47–64.
Carturan, G., Gottardi, V., and Graziani, J. S. (1978). *J. Non-Cryst. Solids* **29**, 41–48.
Cooper, A. R. (1978). *In* "Ceramic Processing before Firing" (G. Y. Onoda and L. L. Hench, eds.), pp. 261–276. Wiley, New York.
Decottignies, M. (1977). Synthèse de verres par pressage à chaud des gels; application à l'élaboration de matériaux vitrocéramiques. Ph.D. thesis, Univ. of Montpellier, Montpellier, France.
Decottignies, M., Phalippou, J., and Zarzycki, J. (1977a). *C.R. Hebd. Seances Acad. Sci.*, **285**, 265–268.
Decottignies, M., Mukherjee, S. P., Phalippou, J., and Zarzycki, J. (1977b). *C.R. Hebd. Seances Acad. Sci., Ser. C* **285**, 289–292.
Decottignies, M., Phalippou, J., and Zarzycki, J. (1978). *J. Mater. Sci.* **13**, 2605–2618.
Dislich, H. (1971a). *Angew. Chem.* **83**, 428.
Dislich, H. (1971b). *Angew. Chem., Int. Ed. Engl.* **10**, 363–370.
Dislich, H. (1971c). *Glastech. Ber.* **44**, 1–8.
Dislich, H., and Hinz, P. (1982). *J. Non-Cryst. Solids* **48**, 11–16.
Dislich, H., and Hussmann, E. (1981a). *Thin Solid Films* **77**, 129.
Dislich, H., and Hussmann, E. (1981b). *Sprechsaal* **114**, 361–366.
Downs, R. L., Ebner, M. A., Homyk, B. D., and Nolen, R. L. (1981). *J. Vac. Sci. Technol.* **18**, 1272–1275.
Fuji, K., and Ishido, Y. (1965). *Yogyo Kyokaishi* **73**, 113–116.
Geffcken, W., and Berger, E. (1943). D.R.P. 736,411.
Gonzalez-Oliver, C. J. R., James, P. F., and Rawson, H. (1982). *J. Non-Cryst. Solids* **48**, 129–152.
Gugielmi, M., and Pincipi, G. (1982). *J. Non-Cryst. Solids* **48**, 161–175.
Hair, M. L., and Hertl, W. (1973). *J. Phys. Chem.* **77**, 2070–2075.
Hayashi, T., and Saito, H. (1980). *J. Mater. Sci.* **15**, 1971–1977.
Hench, L. L., Prassas, M., and Phalippou, J. (1982). *J. Non-Cryst. Solids* **53**, 183–193.
Höland, W., Plumat, E. R., and Duvigneaud, P. H. (1982). *J. Non-Cryst. Solids* **48**, 205–217.
Iler, R. K. (1979). "The Chemistry of Silica." Wiley (Interscience), New York.
Jabra, R. (1979). Synthèse par voie des gels des verres des systèmes SiO_2–TiO_2, SiO_2–P_2O_5, SiO_2–B_2O_3 et leur caractérisation. Ph.D. thesis, Univ. of Montpellier, Montpellier, France.
Jabra, R., Phalippou, J., and Zarzycki, J. (1979). *Rev. Chim. Miner.* **16**, 245–266.
Jabra, R., Phalippou, J., and Zarzycki, J. (1980). *J. Non-Cryst. Solids* **42**, 489–498.
Jabra, R., Phalippou, J., Prassas, M., and Zarzycki, J. (1981). *J. Chim. Phys.* **78**, 777–780.
Kamiya, K., and Sakka, S. (1977a). *Yogyo Kyokaishi* **85**, 308–309.
Kamiya, K., and Sakka, S. (1977b). *Res. Rep. Fac. Eng., Mie Univ.* **2**, 87–104.
Kamiya, K., and Sakka, S. (1980). *J. Mater. Sci. Lett.* **15**, 2937.
Kamiya, K., Sakka, S. and Yamanaka, I. (1974). *Int. Congr. Glass, 10th, 1974,* Part II, Vol. 13, pp. 44–48.
Kamiya, K., Sakka, S., and Tashiro, N. (1976). *Yogyo Kyokaishi* **84**, 614–618.
Kamiya, K., Sakka, S., and Ito, S. (1977). *Yogyo Kyokaishi* **85**, 599–605.
Kamiya, K., Sakka, S., and Mizutani, M. (1978). *Yogyo Kyokaishi* **86**, 552–559.
Kamiya, K., Sakka, S., and Tatemishi, Y. (1980). *J. Mater. Sci.* **15**, 1765–1771.
Kistler, S. S. (1932). *J. Phys. Chem.* **36**, 52–64.

Klein, L. C. (1981). *Glass Ind.* **62,** 14–16.
Klein, L. C., and Garvey, G. J. (1980). *J. Non-Cryst. Solids* **38-39,** 45–50.
Klein, L. C., and Garvey, G. J. (1982). *J. Non-Cryst. Solids* **48,** 97–104.
Klein, L. C., and Garvey, G. J. (1984). *In* "Ultrastructure Processing of Ceramics, Glasses, and Composites" (L. Hench and D. R. Ulrich, eds.), pp. 88–99. Wiley, New York.
Konijnendijk, W. L., and Groenendijk, H. (1972). *Klei Keram.* **22,** 7–13.
Konijnendijk, W. L., Van Duuren, M., and Groenendijk, H. (1973). *Verres Refract.* **27,** 11–13.
Luth, W. C., and Ingamells, C. O. (1965). *Am. Mineral.* **50,** 255–258.
McCarthy, G. J., Roy, R., and McKay, J. M. (1971). *J. Am. Ceram. Soc.* **54,** 637–638.
Mackenzie, J. D. (1982). *J. Non-Cryst. Solids* **48,** 1–10.
Mackenzie, J. D. (1984). *In* "Ultrastructure Processing of Ceramics, Glasses, and Composites" (L. Hench and D. R. Ulrich, eds.), pp. 15–26. Wiley, New York.
Mackenzie, J. K., and Shuttleworth, R. (1949). *Proc. Phys. Soc., London* **62,** 833–852.
Mukherjee, S. P. (1980). *J. Non-Cryst. Solids* **42,** 477–488.
Mukherjee, S. P. (1982a). *In* "Materials Processing in the Reduced Gravity Environment of Space" (G. Rindone, ed.), pp. 321–331. North-Holland Publ., Amsterdam.
Mukherjee, S. P. (1982b). *J. Phys. C-9* **43,** 265–270.
Mukherjee, S. P. (1984). *In* "Ultrastructure Processing of Ceramics, Glasses, and Composites" (L. Hench and D. R. Ulrich, eds.), pp. 178–188.
Mukherjee, S. P., and Lowdermilk, W. H. (1982). *J. Non-Cryst. Solids* **48,** 177–184.
Mukherjee, S. P., and Zarzycki, J. (1979). *J. Am. Ceram. Soc.* **62,** 1–4.
Mukherjee, S. P., Zarzycki, J., and Traverse, J. P. (1976a). *J. Mater. Sci.* **11,** 341–355.
Mukherjee, S. P., Zarzycki, J., Badie, J. M., and Traverse, J. P. (1976b). *J. Non-Cryst. Solids* **20,** 455–458.
Murray, P., Rodgers, E. P., and Willliams, A. E. (1954). *Trans. Br. Ceram. Soc.* **53,** 474–510.
Neimark, I. E., and Sheinfain, R. Y. (1953). *Kolloidn. Zh.* **15,** 145.
Nicolaon, G. A., and Teichner, S. J. (1968). *Bull. Soc. Chim. Fr.* pp. 1900–1906.
Nogami, M., and Moriya, Y. (1977a). *Yogyo Kyokaishi* **85,** 59–65.
Nogami, M., and Moriya, Y. (1977b). *Yogyo Kyokaishi* **85,** 448–454.
Nogami, M., and Moriya, Y. (1979). *Yogyo Kyokaishi* **87,** 43.
Nogami, M., and Moriya, Y. (1980). *J. Non-Cryst. Solids* **37,** 191–201.
Nogami, M., Moriya, Y., Hayakawa, J., and Komiyama, T. (1980). *Yogyo Kyokaishi* **88,** 712.
Nogami, M., Hayakawa, J., and Moriya, Y. (1982). *J. Mater. Sci.* **17,** 2845–2849.
Okkerse, C. (1960). Submicroporous and macroporous silica. Ph.D. Thesis, Univ. of Delft, Delft, Netherlands.
Packard, R. O. (1967). *J. Am. Ceram. Soc.* **50,** 223–229.
Pancrazi, F., Phalippou, J., Sorrentino, F., and Zarzycki, J. (1982). *J. Phys. C-9* **43,** 279–284.
Parlange, J. Y. (1973). *Soil Sci.* **116,** 1–7.
Partlow, D. P., and Yoldas, B. E. (1981). *J. Non-Cryst. Solids* **46,** 153–161.
Phalippou, J., Zarzycki, J., and Lalanne, J. F. (1978). *Ann. Chim. Sci. Mater.* **3,** 99–105.
Phalippou, J., Prassas, M., and Zarzycki, J. (1981). *Verres Refract.* **35,** 975–982.
Phalippou, J., Prassas, M., and Zarzycki, J. (1982). *J. Non-Cryst. Solids* **48,** 17–30.
Phalippou, J., Woignier, T., and Zarzycki, J. (1984). *In* "Ultrastructure Processing of Ceramics, Glasses, and Composites" (L. Hench and D. R. Ulrich, eds.), pp. 70–87.
Philip, J. R. (1974). *Soil Sci.* **117,** 257–262.

Prassas, M. (1981). Synthèse des gels du système SiO_2–Na_2O et des gels monolithiques de silice. Etude de leur conversion en verre. Ph.D. thesis, Univ. of Montpellier, Montpellier, France.
Prassas, M., and Hench, L. L. (1984). *In* "Ultrastructure Processing of Ceramics, Glasses, and Composites" (L. Hench and D. R. Ulrich, eds.), pp. 100–125. Wiley, New York.
Prassas, M., Phalippou, J., Hench, L. L., and Zarzycki, J. (1982a). *J. Non-Cryst. Solids* **48**, 79–95.
Prassas, M., Phalippou, J., and Zarzycki, J. (1982b). *J. Phys.* C-9 **43**, 257–260.
Puyané, R., James, P. F., and Rawson, H. (1980). *J. Non-Cryst. Solids* **41**, 105–115.
Puyané, R., Harmes, A. L., and Gonzalez-Oliver, C. J. R. (1982). *Proc. Eur. Conf. Opt. Commun., 8th, Cannes 1982*, 623.
Rabinovich, E. M., Johnson, D. W., McChesney, J. B., and Vogel, E. M. (1982). *J. Non-Cryst. Solids* **47**, 435–439.
Roy, D. M. (1952). Phase equilibria in the system MgO–Al_2O_3–SiO_2–H_2O. Ph.D. thesis, Pennsylvania State University, University Park, Pennsylvania.
Roy, D. M., Roy, R., and Osborn, E. F. (1953). *Am. J. Sci.* **251**, 337–361.
Roy, R. (1956). *J. Am. Ceram. Soc.* **39**, 145–146.
Roy, R. (1969). *J. Am. Ceram. Soc.* **52**, 344.
Saito, H., and Hayashi, T. (1978). *Prepr. Annu. Meet. Jpn. Ceram. Soc., Okayama*, p. 90.
Sakka, S. (1982). *In* "Treatise on Materials Science and Technology" (M. Tomozawa and R. H. Doremus, eds.), Vol. 22, pp. 129–167. Academic Press, New York.
Sakka, S., and Kamiya, K. (1980). *J. Non-Cryst. Solids* **42**, 403–421.
Sakka, S., and Kamiya, K. (1982). *J. Non-Cryst. Solids* **48**, 31–46.
Scherer, G. (1977a). *J. Am. Ceram. Soc.* **60**, 236–239.
Scherer, G. (1977b). *J. Am. Ceram. Soc.* **60**, 243–246.
Scherer, G., and Bachman, D. L. (1977). *J. Am. Ceram. Soc.* **60**, 239–243.
Schmidt, H., and Kaiser, A. (1981). *Glastech. Ber.* **54**, 338–342.
Schmidt, H., Scholze, H., and Kaiser, A. (1982). *J. Non-Cryst. Solids* **48**, 65–77.
Schofield, R. K. (1935). *Trans. Int. Congr. Soil Sci., 3rd, 1935*, Vol 2, pp. 37–48.
Schröder, H. (1962). *Opt. Acta* **9**, 249–254.
Schröder, H. (1969). *Phys. Thin Films* **5**, 87–141.
Schultz, P. (1983). *In* "Glass Science and Technology" (D. R. Uhlmann and N. J. Kreidl, eds.), Vol. 1, pp. 49–103. Academic Press, New York.
Shimbo, S., Tanzawa, K., and Yokota, R. (1975). *Abstr. Annu. Meet. Glass Div. Jpn., Ceram. Soc. Tokyo, 15th*, p. 18.
Shoup, R. D. (1976). *In* "Colloid and Interface Science" (M. Kerker, ed.), Vol. 3, pp. 63–69. Academic Press, New York.
Wainer, E. (1968). Patentschrift, 1,249,832.
Weinberg, M. C., and Neilson, G. F. (1978). *J. Mater. Sci.* **13**, 1206–1216.
Wenzel, J. (1982). *J. Mater. Sci.* **17**, 3380–3382.
Wintel, G. (1968). Japanese patent 48-17615.
Woignier, T., Phalippou, J., and Zarzycki, J. (1982). *J. Phys.* C-9 **43**, 261–264.
Yamane, M. (1980). *Yogyo Kyokaishi* **88**, 589.
Yamane, M., and Kojima, T. (1981). *J. Non-Cryst. Solids* **44**, 181.
Yamane, M., and Okano, S. (1979). *Yogyo Kyokaishi* **87**, 434–438.
Yamane, M., Aso, S., and Sakaino, T. (1978). *J. Mater. Sci.* **13**, 865–870.
Yamane, M., Aso, S., Okano, S., and Sakaino, T. (1979). *J. Mater. Sci.* **14**, 607–611.
Yamane, M., Inoue, S., and Nakazawa, K. (1982). *J. Non-Cryst. Solids* **48**, 153–159.
Yates, P. C. (1971). Canadian patent 8,784,444.
Yoldas, B. E. (1975a). *Am. Ceram. Soc. Bull.* **54**, 286–288.

Yoldas, B. E. (1975b). *Am. Ceram. Soc. Bull.* **54,** 289–290.
Yoldas, B. E. (1975c). *J. Mater. Sci.* **10,** 1856–1860.
Yoldas, B. E. (1977). *J. Mater. Sci.* **12,** 1203–1208.
Yoldas, B. E. (1979). *J. Mater. Sci.* **14,** 1843–1849.
Yoldas, B. E. (1980a). *Appl. Opt.* **19,** 1425–1429.
Yoldas, B. E. (1980b). *J. Non-Cryst. Solids* **38-39,** 81–86.
Yoldas, B. E. (1982). *J. Non-Cryst. Solids* **51,** 105–121.
Yoldas, B. E. (1984). *In* "Ultrastructure Processing of Ceramics, Glasses, and Composites" (L. Hench and D. R. Ulrich, eds.), pp. 60–69. Wiley, New York.
Yu, P., Liu, H., and Wang, Y. (1982). *J. Non-Cryst. Solids* **52,** 511–520.
Zarzycki, J. (1981). *Adv. Ceram.* **4,** 204–217.
Zarzycki, J. (1982). *J. Non-Cryst. Solids* **48,** 105–116.
Zarzycki, J. (1984). *In* "Ultrastructure Processing of Ceramics, Glasses, and Composites" (L. Hench and D. R. Ulrich, eds.), pp. 27–42.
Zarzycki, J., Prassas, M., and Phalippou, J. (1982). *J. Mater. Sci.* **17,** 3371–3379.

CHAPTER 8

Coatings on Glass*

H. Dislich

SCHOTT GLASWERKE
MAINZ, FEDERAL REPUBLIC OF GERMANY

I. Introduction and Overview	252
A. Why Coatings on Glass	253
B. How to Coat Glass	254
C. Driving Forces for Coating Development	255
II. Vacuum Processes and Products	256
A. Thermal Evaporation	256
B. Sputtering	257
C. Plasma Polymerization	259
III. Spray Processes and Products	259
A. Deposition on Flat Glass	260
B. Coatings on Containers	261
C. Coatings of Lamps	261
D. Possible Extensions of the Process	261
IV. CVD Processes and Products	262
A. Flat Glass Coatings	263
B. Interior Coatings of Tubing	264
C. Possible Extensions of the Process	265
V. Wet Reduction Processes and Products	265
A. Products	266
VI. Dip Coating Processes and Products	267
A. Process	267
B. History	268
C. Products	269
D. Process Chemistry	272
E. Extension of Process Chemistry to Multicomponent Oxide Systems	274
F. A Possible Future Product: The Heat Mirror	276
VII. Leaching Processes and Products	277
A. Present Trends	278
VIII. Specialties	279
A. Optical Coatings	279
B. Anticorrosion Coatings	280

* The trade names Duran, Irox, Thermex, and Zerodur are registered to Schott Glaswerke.

    C. Chemical Variation of the Glass Surface   280
    D. Coloration and Decoration   281
 IX. Concluding Remarks   281
    References   282

I. Introduction and Overview

Considering the many beneficial effects of coatings on glass, the inclusion of a chapter in the present treatise hardly needs justification. The chapter will concentrate on essentials in the widely scattered pertinent literature. The need for such concentration was clearly felt at the Geneva Forum "Coatings on Glass 1980," which presented an excellent overview of the present state of the art (Geneva Forum, 1981).

The principal focus of the present chapter on highly transparent oxide coatings from gels reflects the facts that such coatings are closest in kind to glasses, frequently are glasses, and have their genesis in the glassy state. Network formation and modification are typical features, although they result not from a melt process but from a sol–gel process at a much lower temperature. This process is today the subject of many investigations and is discussed by Zarzycki in Chapter 7. This and other extensions of glass-forming processes have led to the abandonment of the ASTM definition of glass as "an inorganic product of fusion which has cooled to the solid state without crystallizing" in favor of Mackenzie's more general one: "Glass is a non-crystalline solid." In the area of coating technology, it has become apparent that glass is much more versatile, even as to its preparation, than had been commonly assumed.

An evenhanded treatment of all coating processes—which differ enormously from one another—would require an entire book. For most processes and products of concern in the present work, much more industrial experience exists than can be gathered from publications and patents. Hence overgeneralizations by the author are inevitable and should be accepted within the context of his presentation.

In this connection, it should be stressed that vacuum deposition processes for window panes and optical substrates today command the largest market. Deposition by wet reduction is applied to window panes in the United States. Spray and CVD processes have considerable potential for use in direct deposition during the float process, and the old leaching process (although not a coating process, closely related to it) appears to be at the beginning of a renaissance. Diffusional processes (staining, chemical strengthening, embedding of colloidal metal particles in an electrical field) will not be treated in this chapter.

Although market impact cannot be treated in this chapter, trends in industrial development cannot be disregarded, since they often determine emphasis on or neglect of entire areas. For this reason, contemporary trends and thrusts will be treated, but with no claim for completeness.

A. Why Coatings on Glass

The multimillennial history of glass—even today only rarely replaced by more recent materials—demonstrates that it possesses a remarkable combination of properties. A desire prevails, however, to vary and improve its properties—often by the use of coatings.

The extensive literature on thin films (less than a few micrometers in thickness) prior to 1969 has been reviewed by Mayer (1972). The extension of the use of thin films since then may be summarized as follows.

1. Changes in Chemical Quality

Glasses can be made more resistant by anticorrosion coatings (extreme optical glasses, communication fibers). The hydrophilicity of glass surfaces can be increased (antifogging ski goggles). Hydrophilicity can be attained by siliconizing (quantitative emptying of ampules containing medical solutions requiring exact doses). Even biocompatibility can be obtained through coating.

2. Changes in Mechanical Quality

Improvement in scratch resistance by addition of antifriction layers (e.g., SnO_2 or TiO_2 coatings) with additional organic coatings on containers (decreasing the incidence of fracture by contact in fast bottling lines).

3. Changes in Electrical Quality

Increase in the small room-temperature surface conductivity by several orders of magnitude through oxidic semiconductor coatings (removal of static charge, antifogging).

4. Changes in Optical Quality

Transmission and reflection of glass in various wavelength regions can be altered dramatically by coatings. Antireflection layers increase transmission. Layers of low refractive index on fiber cores of higher index constitute fiber light guides.

Window glass is presently changing its old function of admitting light and excluding rain and wind, and is becoming a functional element in the energy balance of buildings. Coatings can do the job, and must do it by simple means for economical reasons.

Coated float glass has a bright future in building and transportation (Hinder, 1981). The increase in cost as well as the scarcity of energy calls for more heat conversion (heat mirrors) and solar control. The standard two-pane window without coating will be replaced. Opportunities also exist in the areas of thermal and photovoltaic solar energy. As to transportation, deicing and demisting appear to be front runners.

B. How to Coat Glass

Almost all layers in present use are transparent, thus maintaining one of the essential properties of glass. Some coatings consist of only few molecular layers (siliconizing, antifriction coatings). Others are prepared with a thickness of $\frac{1}{4}\lambda$ (TiO_2-interference sun-protection layers; antireflection layers). Optical claddings of light guides are usually several micrometers thick, hydrophilic layers often 10 μm.

The deposition process often determines optical quality. For very thin coatings it usually makes little difference to the optical quality which of the processes mentioned above has been used. For fibers, geometry and the drawing process usually permit only the dipping process.

For the geometry of ophthalmic lenses, the vacuum process is most rational. If the substrate has many corners and edges, dipping is not applicable; here vacuum, chemical, and liquid spray deposition are more suitable.

In the case of large glass panes, especially of window glass, all processes are competitive. Chemical variability; optical, chemical, and mechanical quality of the coatings; and rate of deposition and cost of investment may all be decisive, as may location, history, scientific and technological background, and risk acceptable to the entrepreneur.

In what follows, the typical performance characteristics of each process are cited, with detailed discussion reserved for selected cases. The principal methods of producing coatings are physical vapor deposition (PVD), liquid spray deposition (LSD), chemical vapor deposition (CVD), wet reduction deposition (WRD), dip coating deposition (DCD), and leaching process (LP). The last three abbreviations are not in general use and are introduced here for the purpose of uniformity. With the exception of PVD, all the processes involve a significant chemical component. Physical vapor deposition WRD, and LP can be applied at low temperatures, down to ambient temperature, although sometimes higher temperatures are chosen. Liquid spray deposition, CVD, and DCD require temperatures of a few hundred degrees Celsius. If oxides are formed, 500°C is typical.

PVD yields metals, oxides, halides, nitrides, and other substances (although with varying degrees of difficulty); it thus is the most versatile and has been developed the most fully. It is only in this process that it is usual to measure thickness during the coating process.

LSD is conventionally used for forming oxides, as is CVD—although the latter may also yield metals, halides, nitrides, etc., in its more costly variations. Metals are usually produced by WRD, and oxides by DCD. The dipping process (DCD) is, however, also suited for sulfides, nitrides, doping with halides, and embedding of noble metals in an oxide matrix. The leaching process (LP) differs from all other processes by its subtractive character. It "leaves behind" in the surface of the substrate porous oxides which differ from the bulk composition.

All processes require the glass surface being coated to be extremely clean and well conditioned. Cleaning processes and the nature of the glass surface are treated in Holland's text (Holland, 1964) and many later publications (e.g., Guenther, 1981).

C. Driving Forces for Coating Development

The development of glass coating on a broad basis started essentially after World War II, driven by a variety of possibilities and technological demands. For instance, vacuum coatings were prominently developed at Balzers, the OCLI, and in the American optical industry; dip coatings at Schott in Germany; antifriction coatings in the container industry; wet reduction coatings at PPG; etc.

As the technology was increasingly mastered, the coating of large areas of flat glass was achieved. Windows are transparent for solar and heat radiation. In summer, impinging solar energy causes a considerable warming up of the interior of a building, which must be counteracted by costly air conditioning. In winter and at night, heat escapes through the window. Both problems can be controlled by glass coatings.

Early solar control coatings were developed which either reflected part of the impinging solar energy (300–2500 nm) (the better solution) or absorbed it (which is not so good, because the absorbed heat is emitted by secondary radiation). Examples of the former include noble-metal or TiO_2-interference dipped coatings, and of the latter, vacuum metal coatings of Co, Fe, Ni, Cr, and alloys.

With the dramatic increase in the cost of energy in the 1970s, interest in heat mirror coatings increased dramatically. These coatings had to permit solar energy to enter the building to a large extent, and to prevent by reflection the longer-wavelength heat radiation (3–50 μm, maximum at 9.5 μm) from escaping the building. This need provided the main driving force for recent developments, directed to both composition and process improvement.

In principle, there are two solutions to the problem:

(1) oxidic semiconductor layers, as of Sn and In,
(2) very thin metal layers (Au, Ag, Cu) with additional oxide layers providing adhesion, protection, and antireflection.

Metal layers representing compromise solutions have been on the market in the form of vacuum coatings. Thicker Au, Ag, or Cu layers (~20 nm) provide good sun protection with additional heat control. Thin gold layers (~10 nm) have a good heat-control effect but do not have optimal transmissivity for solar energy. Hence the present trend is toward silver and semiconductor layers, using various processes (see below).

Since it is not possible here to treat the entire window problem, reference is made to the excellent publication by Glaeser (1980), which also is credited for parts of the description of the vacuum processes in this chapter.

II. Vacuum Processes and Products

In vacuum coating, materials are transformed to the vapor phase in a vacuum and condensed onto the substrate. Individual processes differ—as will be shown—by the manner of transformation to the vapor phase. Since the advantages and disadvantages of each process are strongly determined by technological detail, reference is again made to Glaeser (1980). Installations in operation today have capacities between 50,000 and 1,000,000 m^2/year. Vacuum-deposited coatings for sunglasses are described by Groth (1977).

A. Thermal Evaporation

Thermal evaporation is achieved either by treating filled evaporation boats or by direct electron bombardment (which has the advantage of avoiding reactions with the wall material). The evaporation rate is high for electron-beam evaporation, an advantage for rather thick layers; but there is a large loss of deposition material and it is quite difficult to obtain uniform layers. Both variants permit the preparation of multilayer systems which, in the case of electrical heating, exhibit rather uniform layer thicknesses under proper control.

In this manner, Ni, Cr, and Ti, sun-protection layers and Au, Ag, and Cu sun-protection layers with heat control have been produced on flat glass. By means of electron-beam evaporation, chromium and chromium oxide layers have been produced which are mechanically and chemically stabilized for external use by SiO_2 protective coatings. Without such protective layers, the coatings can only be used in the interior space of hermetically sealed insulating glazing. Protective layers >10 μm thick consisting of a low-alkali borosilicate glass are deposited by electron-beam evaporation (Dutz *et al.*, 1968).

B. Sputtering

In cathodic sputtering the material to be deposited is bombarded in an electrical dc field with argon ions from a gas discharge. Atoms or molecules are released from the target and condense on the substrate. The uniformity of the coatings is good and material loss is small.

The rather old process of cathodic sputtering has proven satisfactory for coatings of Au, Ag, and Cu in connection with, e.g., bismuth oxide. Bismuth is sputtered as a metal in an oxygen-containing atmosphere and condensed reactively as the oxide. Sulfides, nitrides, and carbides may be obtained in an analogous manner. Sun- and heat-protection coatings on flat glass prepared in this manner can only be used in the protected interior of insulating glazing. The disadvantage of the relatively low deposition rate is, in part, compensated for by a capacity for simultaneous deposition on several glass panes. The technically and economically attractive goal of producing a pure heat mirror as a triple-layer system (oxide–Ag–oxide) requires three deposition chambers in series because of the different sputtering gases—a costly process. The sputtering rates for the metals most important for window coatings (such as Au, Ag, Cu) have tolerable sputtering rates, but those for the oxides (or halides, etc.) are quite low.

In the magnetron sputtering process, the concentration of gas ions can be increased to the point that the sputtering rate rises by a factor of 10 or more. This is a decisive step for the economy of the process. Indium–tin oxide (ITO) and cadmium stannate coatings have been developed using reactive planar magnetron sputtering, starting from alloys which deposit relatively rapidly in an Ar–O_2 atmosphere at room temperature (Howson and Ridge, 1981). The technical implementation of these and similar developments is of considerable interest.

At present, pure sun-protection coatings which are stable in the atmosphere are produced on flat glass from high-grade steel and titanium with titanium oxide layers. Pure heat mirror coatings (e.g., ITO) with good electrical conductivity, high IR reflection, and high transmission of solar energy have not yet been produced on large areas, perhaps because they cost too much (Glaeser, 1980).

For this reason, ITO coatings do not yet play a role in the window market. They have, however, conquered the LCD display market, replacing SnO_2, which is harder to etch and more absorptive. Donnelly, the OCLI, and Balzers are the leading producers. Nichols (1982) gives a detailed description of the ITO sputtered film. The sputtering process is preferred because the defined control of the In:Sn ratio (90:10) is difficult in the thermal process. The data on uniformity given by Nichols (1982) for a 6 × 6-in. ITO film with oxygen deficit (100–80 Ω/□ and 72–64% transmission) indicate, however, that there remain opportunities for improvement—and possibly the introduction of other processes. Heating in air for 2 min to 550°C increases the transmission to 95%.

1. The Heat Mirror Problem in Vacuum Coatings

In looking at the requirements of a heat mirror—now applicable both to windows and solar collectors, and presented above as a driving force for development—the following profile results: (1) maximum transmission for solar radiation, (2) maximum reflection for heat radiation, (3) excellent adhesion to glass, (4) good long-term stability.

Cu and Au—suitable in principle—suffer from absorption of solar radiation. Indium–tin oxide and SnO_2 have been discussed. Cadmium stannate, an *n*-type semiconductor with oxygen defects, does have good IR reflection and low visible light absorption, but it does not seem easy to shift its absorption edge toward the UV. Silver seems to be the present favorite, as probably first shown in the combination TiO_2–Ag–TiO_2 (Bachner *et al.*, 1974).

The first product appeared on the market in 1982 (Firmenprospekt Interpane, 1982) and more products seem sure to follow. Adhesion to glass might be improved by means of oxide coatings. Problems with the long-term stability of Ag have been known for a long time. These include solarization stability. A competitive product might be ITO, which will be discussed further in Section VI.

2. Optical Coatings

In the area of high-performance eyeglasses and photo-optic devices, antireflection films have played an important role. The materials used include oxides, fluorides (chiefly MgF_2 with its low index of refraction), and cermets (mixture of insulators such as MgF_2 or SiO_2 with metals).

These coatings on glass lenses have a high value. For eyeglasses, the situation is quite different in different countries. Few eyeglasses with antireflection or colored coatings are sold in the United States, while in Japan the percentage is 70%, and in Europe 30%.

For measuring technology and analytical processes, interference filters produced by the vacuum process are widely used. These consist of many thin coatings (on glass) whose optical thickness is mostly $\frac{1}{4}\lambda$ or a multiple thereof, where λ is the wavelength. A manifold of bandwidths with high transmission or reflection can be obtained (Firmenprospekt Schott Glaswerke, 1981a).

C. Plasma Polymerization

In this process—also taking place in a vacuum—vapors of organic (or organometallic molecules) are split in a high-frequency glow discharge or under electron bombardment, often with the formation of free radicals, and a high-polymer coating, usually in a highly cross-linked form, is produced on the substrate. Not only conventional monomers, but also highly saturated molecules may be used as starting compounds.

A characteristic of this process is the extensive freedom from defects and pores of the polymer films, which predestines them for the passivation of surfaces. Fluoropolymers produced in this manner are recommended as anticorrosion coatings for sensitive optical materials (not just glasses). Extremely low-loss organic polymer coatings for communication fibers have been described, as have antireflection and multilayer interference coatings. For details, see Schroeder (1975) and Pulker (1981).

III. Spray Processes and Products

Compared to the costly and highly sophisticated vacuum apparatus, the equipment used in spray processes is quite robust. The coating material is sprayed from a solution by means of moving nozzles in the form of a fog of fine droplets and deposits on a heated glass substrate. In most cases, the coatings are oxide coatings originating from halides or acetylacetonates by hydrolytic or pyrolytic processes (Fig. 1).

Preferably organic solvents or alcohol are used as the carrier. The temperatures of the substrates are mostly between 400 and 700°C. The rate of film formation is high. Uniformity in thickness is limited, as can be seen in the thickness range of interference. For this reason, the process is used in areas where thickness is not critical or where color variations are permissible. The oldest use (40 years ago) is the deicing of airplane windows by transparent heating coatings. Even today, these coatings are characterized by irregularities, which do not disturb viewing in transmission but cause multicolored reflections.

$$SnCl_4 + 2H_2O \rightarrow SnO_2 + 4HCl\uparrow$$

Fig. 1. Oxides from halogenides. $\quad 2InCl_3 + 3H_2O \rightarrow In_2O_3 + 6HCl\uparrow$

A. Deposition on Flat Glass

In the case of windows, the process is used to increase reflection or absorption. A schematic representation of a spray apparatus is given in Fig. 2. Properly cut borosilicate glass panes hanging vertically are heated to >500°C and sprayed with an alcoholic $SnCl_4$ solution containing F^- ions. A transparent F-doped SnO_2 layer with 70–80% IR reflection and 70% transmission in the visible is obtained. The resulting product, e.g., Thermax 32 (Firmenprospekt Schott Glaswerke, 1982), is used as a household stove viewing window. It reflects the heat of the stove back into the stove, keeping the external temperature of the pane low. The coatings are robust against the environment.

Many attempts have been made to utilize the heat of a freshly formed glass ribbon for, e.g., the conversion of $SnCl_4$ to SnO_2. The glass surface is virginal and of a high temperature uniformity. Glaverbel thus coats Libbey Owens Ford–type ribbon with SnO_2 to obtain the product Glaverplus Comport (Firmenprospekt Glaverbel). Subsequently the glass may be strengthened thermally without destroying the coating.

The most attractive goal, however, is coating the fast-moving float ribbon with its excellent surface quality. Mixed oxide layers, 70 nm thick, of Co, Cr, Fe, Ni, Ti, and V, have been deposited as sun-protection coatings. These are predominantly absorbing coatings without heat control. Acetylacetonates are the starting materials. The coatings are stable enough to be used externally.

What has not yet been achieved at this writing is the large-scale deposition of a 300-nm-thick SnO_2 coating on the float ribbon. This would represent a heat mirror of considerable importance in glazing buildings, since it would have, with about 80% reflection, sufficient transmission. There have been a flood of patents and a few publications, but nothing yet on the market in this area. It seems that to obtain the homogeneity of coating necessary for a window, the CVD process is superior (see Section IV).

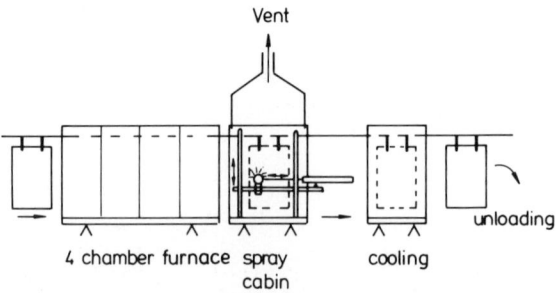

Fig. 2. Schematic drawing of a spray deposition unit.

B. Coatings on Containers

In the glass container industry, the spray process has been used for a long time: very thin (50–100 molecules) layers of TiO_2 and SnO_2 are deposited at the hot end of the annealing conveyor. Depending on the type of equipment, this kind of deposition might also be classified as CVD.

On the cold end, the glass is sprayed with organic media, e.g., polyoxyethylene stearate. This combination of (antifriction) coatings helps to prevent the initiation of microfissures (Griffith flaws) during transportation when the glass surfaces rub each other. The resulting reduction in breakage in rapidly running bottling lines is an essential economic advantage. By this increase in strength, lightweight bottles could be introduced (weight reduction ~30%).

The starting materials are mostly $SnCl_4$ and titanium tetraisopropylate. Several papers in the Geneva Forum (1981) give the latest state of this art as well as easily understandable illustrations of the equipment. Important topics today are the compliance of cold end coatings with food and drug legislation, and the resistance of the coating combination against washing.

C. Coatings of Lamps

Tin dioxide spray coatings of sodium vapor lamps reflect heat radiation back into the discharge tube, while sodium light is transmitted. Still better is the effect for ITO spray coatings (Frank *et al.*, 1981; Fillard and Manifacier, 1981; Groth and Van Boort, 1968).

D. Possible Extensions of the Process

The chemistry described so far—starting from halides and acetylacetonates, terminating in oxides—is very simple. The spray process might, however, make use of a more sophisticated chemistry, leading to very special glass coatings (Garcia and Tomar, 1981). In addition to simple and semiconducting oxides, it should be possible to produce sulfides and selenides. Sulfur and selenium can be introduced via N,N-demethyl-thio (or seleno-, respectively) urea. Examples include not only binary compounds with Cd, Zn, Cu, In, Pb, Ga, Sn, S, Se, or Ti (such as CdS) but also films of $Zn_xCd_{1-x}S$, CdS_xSe_{1-x}, $CuInSe_2$, $AgIn_2S_8$, Ce_2SnO_4, $CdSnO_3$, and many others. The aim is various solar-cell applications, including antireflection films and gas sensors. This author cannot assess the potential problem of maintaining stoichiometry; in the dipping process, to be discussed later, this has been achieved in the case of oxides.

In solar collectors, the covering pane may be coated with heat mirror layers on the basis of the oxides of Sn and In. However, they have met strong competition from selective coatings on the absorber sheets.

IV. CVD Processes and Products

The boundary between spray and CVD processes is not well defined, as seen in the case of the pyrosol process (Viguié, 1974) to be described later. In contrast to the other processes treated here, the CVD process has so far not found its chief applications on glass substrates. However, great expectations prevail for float glass coatings.

The deposition material is usually transformed to the gas or vapor state at elevated temperatures and is generally transported to the reaction chamber by a carrier gas. In this chamber, the glass to be coated is kept at the higher temperature at which deposition onto its surface occurs. The remaining reaction products are withdrawn continuously from the reaction chamber in the form of gases. The principles of operation are demonstrated in Figs. 3 and 4.

From the abundance of possible inorganic coatings, carbides, nitrides, oxides, borides, and silicides may be mentioned. Some essential reaction mechanisms are shown in Fig. 5 (according to Erben *et al.*, 1981). For further details, see Geneva Forum (1981), Schroeder (1975), Erben *et al.* (1981), The National Technical Information Service (1975), and particularly Blocher (1981) and Kalbskopf (1981).

The CVD process can be influenced by high-frequency fields, high-energy radiation, and glow discharge. Coatings otherwise requiring unduly high temperatures may be obtained at lower temperature. The number of substances that can be deposited is very large, and codeposition is

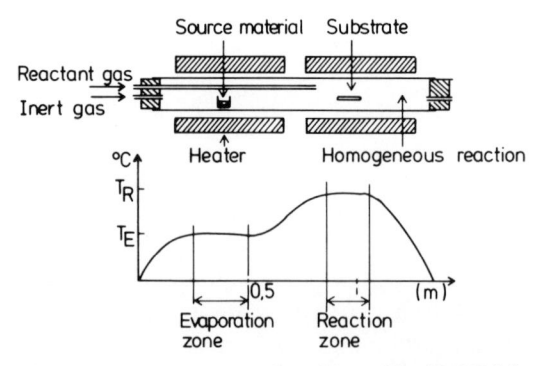

FIG. 3. Basic CVD operation. [From Viguié (1974).]

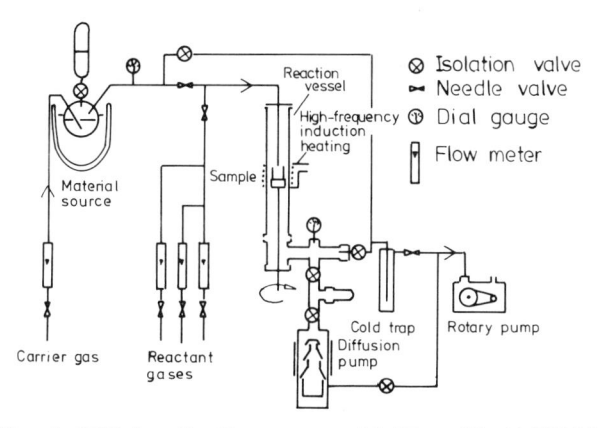

FIG. 4. CVD from liquid source material. [From Viguié (1974).]

possible. Often the density of the film corresponds to theoretical density, and adherence to the substrate is mostly good. The thickness of the coating can be controlled well and may reach 200 μm at deposition rates of 0.05–25 μm/min. According to Blocher (1981), the following coatings have been deposited on glass (in part, just experimentally): SnO_2, In_2O_3, TiO_2, Al_2O_3, SiO_2, Fe_2O_3, BeO, Ni, Cr, V, Mo, W, C, Si, Ge, B, Al, Be, Ce, Os, Re.

A. Flat Glass Coatings

The process is used industrially in connection with the float glass process, utilizing the heat of the freshly formed glass ribbon. With the equipment illustrated in Fig. 6, 30-nm-thick Si layers are prepared from SiH_4 in the reducing part of the float process to form sun-protection coatings without heat blocking effects (Glaeser, 1980; Pilkington Brothers Ltd., 1975). They are stable under atmospheric influences.

$$Ni(CO)_4 \longrightarrow Ni + 4CO \uparrow \quad \text{Pyrolysis}$$

$$ZrCl_4 + 2H_2O \longrightarrow ZrO_2 + 4HCl \uparrow \quad \text{Hydrolysis}$$

$$WF_6 + 3H_2 \longrightarrow W + 6HF \uparrow \quad \text{Reduction}$$

$$TiCl_4 + CH_4 \longrightarrow TiC + 4HCl \uparrow \quad \text{Reduction + Reaction}$$

FIG. 5. CVD reaction mechanisms.

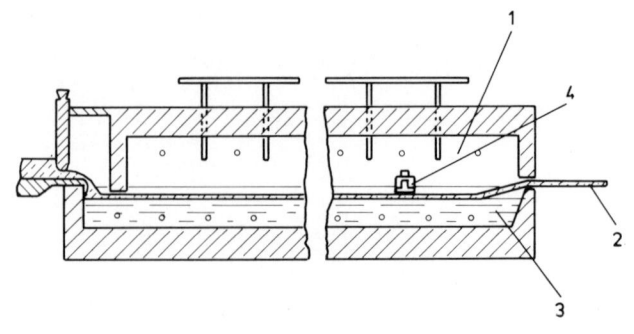

FIG. 6. Schematic drawing of a CVD unit for coating float glass on-line. 1, Bath tank with molten metal and floating glass; 2, ribbon of glass; 3, molten metal bath; 4, gas distributor. [Courtesy of Pilkington Brothers P.L.C. (1974); Bernard James Kirkbride, Joseph Earl Lewis, Robert Andrew Downey, and Charles Victor Thomasson; and the Comptroller of Her Majesty's Stationery Office.]

A transfer of this technology to heat mirror coatings for the ~300-nm doped SnO_2 coatings must have been tried by nearly every float glass producer, considering the large number of patents in the area (e.g., Bauberger and Kalbskopf, 1980; Ternew and Van Cauter, 1979). In spite of this considerable economic interest, however, no product has reached the market. This might be due to the difficulty of coupling two quite complex processes, the float ribbon and the CVD process. More specifically, it may be due to the fact that the float ribbon speed does not correlate with the CVD rate, as shown by Kalbskopf (1981). This difficulty can only be met by several deposition stations, separated by gas curtains which would have to be controlled most carefully (Kalbskopf, 1981). There is also, however, a chemical problem. If one works, in the case of SnO_2, with the relatively inexpensive $SnCl_4$, a reaction with the alkali of the glass takes place at the necessary temperature of $>500°C$, which results in the formation of alkali halides, causing turbidity in the coating. This might have to be prevented by a blocking coating to be applied previously, or by the choice of much more expensive or otherwise less suitable halogen-free Sn compounds. Whether and how this problem will be solved will have a considerable affect on the prospects for other coating processes and products in the heat mirror domain.

Silicon dioxide and SnO_2 (and other) coatings of much smaller dimensions are deposited on glass and used in LCD displays using the CVD equipment of Watkins-Johnson (Firmenprospekt Watkins-Johnson Co., 1978).

B. Interior Coatings of Tubing

The techniques used to deposit glassy layers on the inside of SiO_2 and doped-SiO_2 tubing in the manufacture of optical waveguides are described by Scherer and Schultz in Volume 1 (Chapter 2) and by Klein in Chapter 9 of the present volume of this treatise (see also Mauver, 1972; MacChesney, 1977).

For the interior coating of long tubing of small diameter (but not for optical waveguide preforms), the pyrosol process (Blandenet *et al.*, 1981) is generally preferred. This process can be classified as intermediate between spray and CVD processes. The starting material need not have a high vapor pressure since it is fogged from the solution by ultrasound and then transported by a carrier gas to the substrate, which is held at a higher temperature, where it is converted to a coating. Compared to compressed-air atomizing, smaller droplets and a narrower droplet size distribution are obtained. Both features depend on several parameters and can be controlled to obtain improved homogeneity. Here, too, acetylacetonates are preferred, but cyclopentadienyles (ferrocene), tin alkyls, alcoholates and halides are also used. To date, oxides, sulfides, selenides, tellurides, and noble metals (Pd, Pt, Ru) have been described. In the center of interest are SnO_2 and ITO. Here again one sees the goal of coating the float ribbon, but this goal lies far away. The difficulties mentioned in connection with other processes are also applicable here, to a greater or lesser extent.

C. Possible Extensions of the Process

Fluoroacetonates have higher vapor pressures than the conventional simple acetylacetonates, but are much more expensive. They do allow low-temperature CVD processes (Bloos and Lukas, 1973–1974). Aluminoalkyls, from which Al coatings are obtained, have still higher vapor pressures. The availability of organometallic products with high vapor pressure is growing steadily. Some of these compounds are being produced in large quantities, and thus less expensively. This field seems not yet to have been exploited, but an example has been provided by the multicomponent oxide glass coatings prepared by Hitachi on glass and Si substrates using $ZnEt_2$, BEt_3, and $Si(OEt)_4$ (Kosugi *et al.*, 1976a,b).

V. Wet Reduction Processes and Products

Probably the oldest kind of glass coating is the production of optically dense metal (e.g., silver) mirrors. A metal salt solution is first sprayed on the surface, followed by a reducing solution which precipitates the metal at ambient or moderately elevated temperature. This reduction starts spontaneously at various points unrecognizable in the optically dense coatings of the final product.

To obtain transparent metal coatings for windows, much more insight into this spontaneous reaction is needed, since variations in thickness would become visible in the required very thin coatings. A controlled influence on the course of this reduction process will be possible only on the basis of exact knowledge of the complicated chemical reactions.

One example of many related developments will now be given to show how the present state of knowledge was developed. Apparently starting from the old pyrolytic coating of drinking glasses with Ag soaps, Donley (1966) was able to overcome their frequent insolubility by reacting them to form an Ag–amino compound. Around 1960, a series of more than 20 patents appeared in rapid sequence, mostly by Miller and Franz, one of which may be cited as an example (Franz, 1971). These disclose the formation first of Cu, Ag, and Au films, later also Ni, Co, and Fe films. Essential features were the sensitization of the glass surface with Sn salts and $PdCl_2$, their contact with Ag salt solutions, the use of chelate formers, the variation of reducing agents (e.g., $NaBH_4$), and the control of the reduction rate to obtain a uniform establishment of the desired transmission. For this purpose, solutions with useful pot life were developed, e.g., a hydrous alkaline Ag solution containing amine and stabilized by a persulfate compound. Along with sequential spraying, simultaneous spraying of Ag and reducing solutions was also tried. Equipment for controllable spray coatings was likewise developed. Next, oxide coatings such as Ni + CuO (Cu subsequently oxidized by dichromate) and the combination Ni + Ag were included; Ni is precipitated first, then treated with an Ag solution containing a complexing agent for Ni, leading to the replacement of part of the Ni in the film by Ag (Greenberg, 1975).

This abundant chemistry of complexing, here only outlined, was necessary to attain the present results in the wet reduction process. There is an analogy to the dipping process in this mobilization of chemistry. From a technological viewpoint, it should be noted that the glass panes move on a horizontal roller conveyor through serially arranged spray zones whose number is determined by that of the coatings to be applied. Finally the panes are washed and dried.

A. Products

Gold, Cu, Ag, and Ni coatings are produced. They are relatively thick and have corresponding low transmission (<30%) in the visible. It is interesting that this low transmission (for European demands) is now in the process of being abandoned also in the United States (Gillery, 1982).

The metal coatings (e.g., Au, Ag, Cu) are laminated between coatings of the oxides of In, Sn, Bi, Zn, Ti. With thin metal coatings prepared by classical wet reduction processes, the achievement of transmissions >50% will not be simple because of the above-mentioned problems of homogeneity. The Au, Ag, and Cu coatings produced by the wet reduction process provide sun protection and heat blocking properties, while the Ni coatings provide only sun protection. They are stable enough for use in an insulating glass system, but not for exterior use.

VI. Dip Coating Processes and Products

The dipping process supplies by its very nature products coated on both sides. This is often advantageous in reinforcing the effect of the coating or in diminishing the number of coatings required. It is a process typically used for oxide coating of large areas. Its chemistry—i.e., the mode of layer formation—corresponds to that of the sol–gel process, now so much discussed because it yields not only coatings (its first products) but also bulk glass (Dislich *et al.*, 1969a,b). See Chapter 7 by Zarzycki in this volume. An extensive survey of the state of the art in 1969 was published by Schroeder (1962, 1969).

A. Process

Large (to 3 × 4-m) window panes are cleaned carefully, generally by using several baths in series. This is essential to assure uniform wetting by the solution to be applied. Since glass surfaces may be in various conditions as received, the cleaning operation may be considered a type of standardization of the glass surface. After cleaning, the glass panes are dipped into a solution containing hydrolyzable metal compounds, as shown in Fig. 7. Next the panes are withdrawn very uniformly into an atmosphere of fixed H_2O content. It is here that the chemistry of the coating process begins with condensation including (this is essential) gel formation. This will be discussed in more detail below. The panes are then subjected to a heating schedule up to 400–500°C, during which time the reactions continue until a transparent metal oxide coating is formed.

The thickness d of the coating is primarily determined by the concentration and viscosity of the solution as well as by the velocity v of withdrawal ($d \sim v^{2/3}$).

The film, which is fluid at first, runs down along the glass pane, remains fixed in part, and solidifies through the above-mentioned reactions. The disturbed edge sections, which are typically 5–10 mm wide, are simply cut off. These coating procedures may be repeated for the formation of multiple layers. The process lends itself well to automation, as exemplified by IROX sun-protection glass of Schott Glass Works. In principle, for small substrates or special cases, spinning or solution layering could be used.

Typical advantages of the process and resulting products are (1) two-

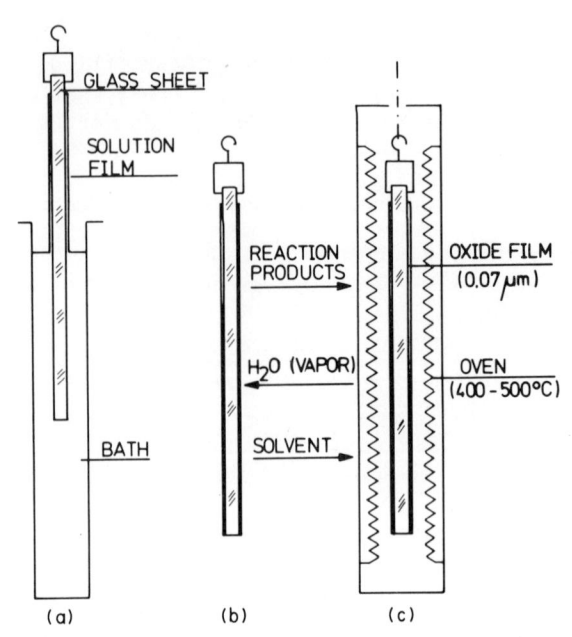

FIG. 7. Production in dipping process: (a) drawing, (b) chemical reaction, and (c) baking.

sided coating in one operation, (2) good homogeneity, (3) good stability against the environment (exterior use possible), (4) much chemical variability in the oxides.

These advantages depend on knowing and controlling a series of interdependent parameters, since, unlike the vacuum processes, the layers are not controlled during their formation. This may explain why this process is as yet used commercially only at Schott Glaswerke. For large-scale production, it is restricted to oxides. Metal layers cannot be produced in this manner, but metals can be embedded colloidally in an oxide matrix.

In addition to flat glass, tubing may be coated advantageously. Very small substrates or those with many corners and edges are not suited for the dipping process.

B. History

The history of this process is briefly presented in Fig. 8, in order to dispel the impression that dipping coatings originated only in the 1970s with the wide discussion of the sol–gel process. Evidently it was rather the multicomponent oxide coatings that initiated the recent sol–gel discussion.

1939:	W. Geffcken and E. Berger Dtsch. Reichspatent 736 411 Jenaer Glaswerk Schott & Gen., Jena Single oxide coatings by sol-gel process
1952– 1979:	H. Schroeder Thin-film physics with dip coatings
1959:	Car rear-view mirrors TiO_2-SiO_2-TiO_2 Production at Schott
1964:	Antireflection coatings TiO_2/SiO_2-TiO_2-SiO_2 Production at Schott
1969:	IROX^R, Sunshielding windows TiO_2 Production at Schott, (>1 Mio m^2)
1969:	H. Dislich, P. Hinz, R. Kaufmann F.R.G. - Patent 19 41 191 Jenaer Glaswerk Schott & Gen., Mainz Multicomponent oxide coatings (Glass, glass ceramics, crystalline substances)

FIG. 8. History of dip coating process.

C. Products

Dip coatings are predominantly used as optical interference coatings and only few have a protective function (Dislich and Hussmann, 1981). The refractive indices attainable are between $n_D = 1.455$ (SiO_2) and 2.25 (TiO_2) [see Table I (after Schroeder 1969)]. By mixing the starting solution, any desired refractive index within the range mentioned can be obtained (Fig. 9), an additional advantage of the dipping process. Coatings can also be produced which have pronounced absorptions in the visible [see Table II (after Schroeder 1969), which gives optical constants n, k at ~550 nm].

The resulting optical possibilities are discussed by Schroeder (1969). Firmenprospekt Schott Glaswerke (1981b) list a variety of products such as antireflection coatings for picture and instrument glazing, heat mirror films for tungsten halogen lamps, cold light mirrors, selectively reflecting mirrors, achromatic light dividers, radiation damping filters, color effect filters, and color conversion filters. In all cases the deposition method is the same as that described above, but the panes are not so large and in most cases TiO_2 and SiO_2 are coated onto each other. In what follows, two large-volume products are treated as examples.

TABLE I

Characteristics of Metal Oxide Coatings which are Nonabsorbing in the Visible Region[a]

Oxide	n	Upper absorption limit (nm)	Structure
Al_2O_3	1.62	≈250	Amorphous, crystalline
CeO_2	2.11	400	Crystalline
HfO_2	2.04 ($\lambda = 400$)	≈200	Crystalline
In_2O_3	1.95	420	Crystalline
La_2O_3	1.78	220	
PbO		≈380	Amorphous
SiO_2	1.455	≈205	Amorphous
SnO_2		350	Crystalline
Ta_2O_5	2.1	310	
ThO_2	1.93	≈220	Crystalline
TiO_2	2.3	380	Crystalline
Y_2O_3	1.82	≤300	
ZrO_2	1.98		Crystalline

[a] From Dislich and Hussmann (1981).

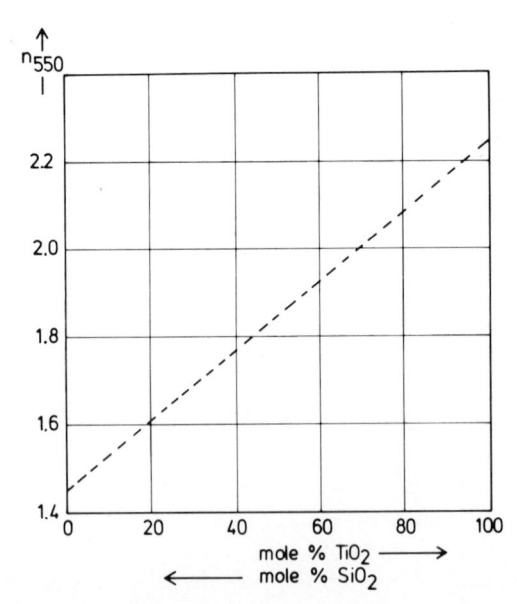

FIG. 9. Refractive index of SiO_2–TiO_2 films as a function of molar proportion of TiO_2. [From Dislich and Hussmann (1981).]

TABLE II

CHARACTERISTICS OF METAL OXIDE COATINGS WITH PRONOUNCED
ABSORPTIONS IN THE VISIBLE REGION[a]

Oxide	n	k	Color (transmitted light)
Cobalt oxide	≈2.0	≈0.16	Brown
Chromium oxide			Yellow-orange
Copper oxide			Brown
Ferric oxide	2.38	0.14	Yellow-red
Nickel oxide			Gray
Rhodium oxide		≈0.2	Gray-brown
Ruthenium oxide			Gray
Uranium oxide	1.95	0.015	Yellow
Vanadium oxide	≈2.0	0.01	Greenish yellow

[a] From Dislich and Hussmann (1981).

1. Sun Shielding Glass (IROX)

The sun shielding glass (IROX) has TiO_2 coatings on either side, in which Pd has been embedded colloidally to obtain the desired absorption. Figure 10 shows the characteristics of IROX AO, and Fig. 11 a building glazed with this product. One of the chief advantages of IROX is color neutrality in transmission and reflection.

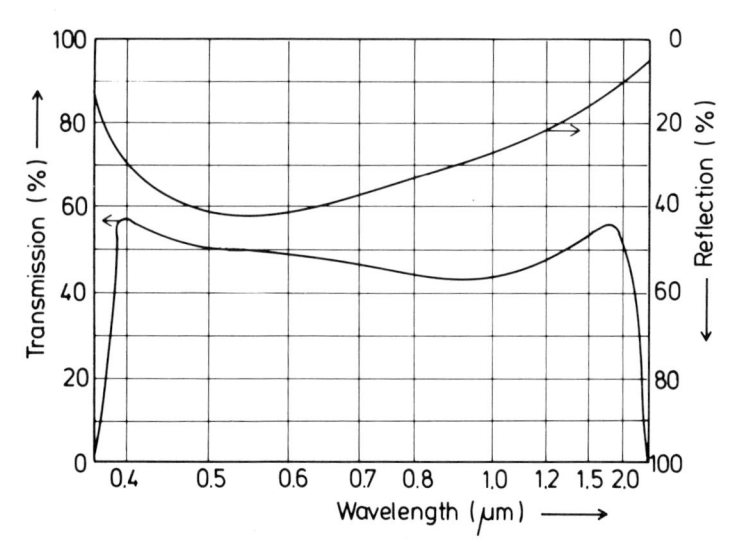

FIG. 10. Reflection, transmission, and absorption characteristics of the sunshielding glass IROX AO. [From Dislich and Hussmann (1981).]

FIG. 11. A high-rise climatized building glazed with IROX. [From Dislich and Hussmann (1981).]

Because of their high scratch resistance and environmental stability, single panes coated in this manner may be used as sun-protection windows in warmer climates. This is impossible with many vacuum metal coatings. On the other hand, the coating has no increased IR reflection, and so in cooler climates two- or three-pane insulating glass is used, filled if necessary with a poorly conducting gas. Such Ar/SF_6-filled three-pane windows have a k value of 1.8 W/m² K.

2. Interference Rear-View Mirrors

Another large European sol–gel dip coating product is the much-used car rear-view mirror, consisting of a coating combination TiO_2–SiO_2–TiO_2 (manufactured by Deutsche Spezialglas A.G.). Its essential properties are better contrast and reduced glare; its reflection curve is shown in Fig. 12.

D. PROCESS CHEMISTRY

The starting materials are always hydrolyzable metal compounds, often alkoxides, whose hydrolysis products at elevated temperature are

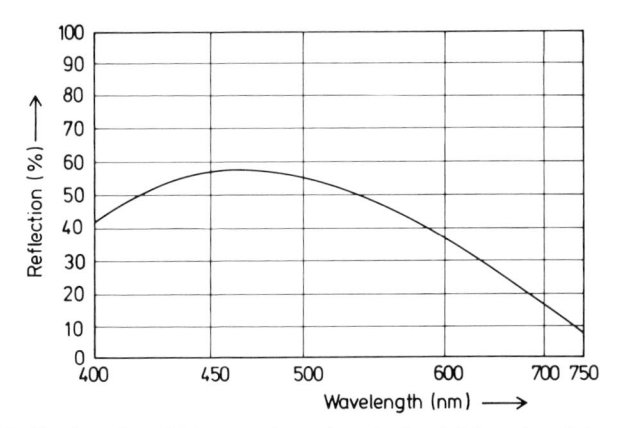

FIG. 12. Reflection of a vehicle rear-view mirror in the visible region. [From Dislich and Hussmann (1981).]

converted into transparent metal oxides by polycondensation. A vitreous SiO_2 layer is produced according to Fig. 13.

During this process, a series of controllable steps are passed (Dislich and Hussmann, 1981) which will not be discussed more closely here. A chemical anchoring of the coating on the glass is achieved because the Si—OH groups in the glass surface coreact as shown in Fig. 14.

In this way, the oxide structure of the substrate glass effectively continues into the coating. Oxides of Al, In, Si, Ti, Zr, Sn, Pb, Ta, Cr, Fe, Ni, Co, and some rare-earth elements are produced in this manner. The coatings may, according to their chemistry and thermal treatment, be vitreous or crystalline.

1. Phosphate Coatings

According to Rothon (1981), coatings of phosphates, particularly those of aluminum, may be produced in a similar way. They form thick layers at lower temperature than simple oxides. These coatings are said to have considerable stability against F and fluorides. At this writing no significant industrial uses are known.

$$Si(OCH_3)_4 + 4H_2O \longrightarrow Si(OH)_4 + 4CH_3OH\uparrow$$

$$Si(OH)_4 \longrightarrow SiO_2 + 2H_2O\uparrow$$

FIG. 13. Synthesis of SiO_2 from $Si(OCH_3)_4$ (schematic). [From Dislich and Hinz (1982). North-Holland Publishing Company, Amsterdam, 1982.]

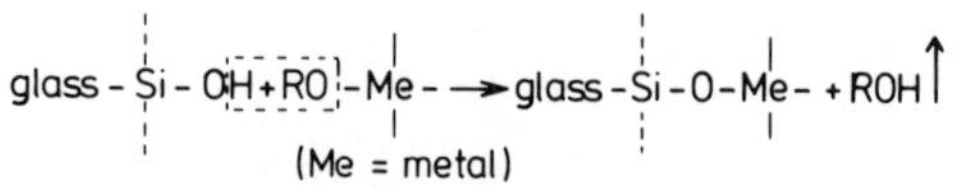

$$\text{glass} - \underset{|}{\overset{|}{\text{Si}}} - \overline{\text{OH} + \text{RO}} - \underset{|}{\overset{|}{\text{Me}}} - \longrightarrow \text{glass} - \underset{|}{\overset{|}{\text{Si}}} - \text{O} - \underset{|}{\overset{|}{\text{Me}}} - + \text{ROH} \uparrow$$
(Me = metal)

FIG. 14. Adhesion of a metal oxide film on glass. [From Dislich and Hinz (1982). North-Holland Publishing Company, Amsterdam, 1982.]

2. Sulfide Coatings

One of the few publications (Acharya *et al.*, 1982) describes the preparation of photosensitive Sb_2S_3 films from a methanolic $SbCl_3$–thiourea solution. It is essential that the films be obtained free of oxygen.

E. Extension of Process Chemistry to Multicomponent Oxide Systems

If one wants to produce not just single or mixed oxide coatings but well-defined multicomponent oxide coatings, e.g., a definite borosilicate glass, one may make use (in addition to what has been described) of two principles (Dislich, 1971). These are (1) most metal alkoxides react with each other, forming heteropolar or homopolar complexes or associates; and (2) all metal alkoxides, also complexes, are more or less hydrolyzable and the hydrolysis products polycondense to form oxidic end products.

1. Preparation of Multicomponent Oxide Coatings

The first synthesis was that of borosilicate layers (Dislich, 1971) (see Fig. 15). A silicate–phosphate glass is used as an anticorrosion coating on hydrolytically sensitive special optical glasses for a laser-protection filter by Deutsche Speziaglas A.G. (see Fig. 16).

$$m\,Si(OR)_4 + n\,B(OH)_3 + o\,Al(OR)_3 + p\,NaOR + q\,KOR$$

$$\xrightarrow{\text{Complexation}} (Si_m B_n Al_o Na_p K_q)(OR)_{4m+3o+p+q}(OH)_{3n}$$

$$\xrightarrow{\text{Hydrolysis}} (Si_m B_n Al_o Na_p K_q)(OH)_{4m+3o+p+q+3n}$$

$$\xrightarrow[-H_2O]{\text{Heat}} Si_m B_n Al_o Na_p K_q O_{(4m+3o+p+q+3n)/2}$$
Condensation

FIG. 15. Synthesis of a borosilicate glass layer. [From Dislich and Hussmann (1981).]

$$m\ Si(OMe)_4 + n\ Al(O\text{-sec. Bu})_3 + oP_2O_5 + pMg(OMe)_2 \xrightarrow[\text{temp.}]{H_2O}$$

$$Si_m\ Al_n\ P_o\ Mg_p\ O_{\frac{4m+3n+5o+2p}{2}}$$

FIG. 16. Synthesis of silicate–phosphate glass film. [From Dislich and Hinz (1982). North-Holland Publishing Company, Amsterdam, 1982.]

2. Preparation of Glass Ceramic Coatings

In an analogous manner, Schott's glass–ceramic Zerodur may be obtained as a coating (see Fig. 17).

An eight-component system, this is an example of the efficiency of the synthesis according to Dislich *et al.* (1969a,b; Dislich, 1971). It forms at 620°C, first as a glass and subsequently crystallizes partially at 680–830°C. The crystals are high-quartz solid solution and ZrO_2 containing nucleating phases. For details on glass ceramics see Chapter 7, by Beall and Duke, in Volume 1 of this treatise.

3. Preparation of Crystalline Coatings

Magnesium aluminum spinel can be produced analogously as a coating (Fig. 18). It is important that the bonds existing in spinel (—O—Al—O—Mg—O) be preformed in the solution. This permits much lower synthesis temperatures (Dislich, 1971).

$$m\ Si(OMe)_4 + n\ Al(Osec.\ Bu)_3 + oP_2O_5 + p\ LiOEt + q\ Mg(OMe)_2 + r\ NaOMe + s\ Ti(OBu)_4 + t\ Zr(OPr)_4$$

$\xrightarrow{\text{complexation}}$

$$[(Si_m\ Al_n\ P_o\ Li_p\ Mg_q\ Na_r\ Ti_s\ Zr_t)(OR)_{4m+3n+p+2q+r+4s+4t}(OH)_{30}]$$

$\xrightarrow{\text{hydrolysis}}$

$$[(Si_m\ Al_n\ P_o\ Li_p\ Mg_q\ Na_r\ Ti_s\ Zr_t)(OH)_{4m+3n+p+2q+r+4s+4t+30}]$$

$\xrightarrow{\text{condensation}}$

$$[Si_m\ Al_n\ P_o\ Li_p\ Mg_q\ Na_r\ Ti_s\ Zr_t\ O_{\frac{4m+3n+30+p+2q+r+4s+4t}{2}}]$$

FIG. 17. Synthesis of a glass–ceramic. [From Dislich and Hinz (1982). North-Holland Publishing Company, Amsterdam, 1982.]

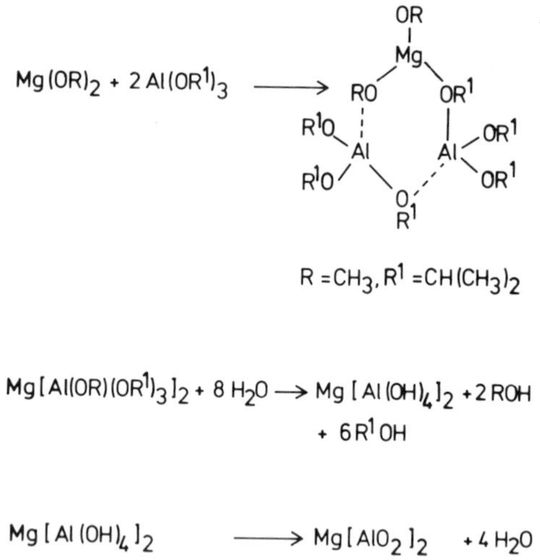

$$Mg(OR)_2 + 2\,Al(OR^1)_3 \longrightarrow$$

[structure diagram of Mg[Al(OR)(OR¹)₃]₂ with bridging OR groups]

$$R = CH_3,\ R^1 = CH(CH_3)_2$$

$$Mg[Al(OR)(OR^1)_3]_2 + 8\,H_2O \longrightarrow Mg[Al(OH)_4]_2 + 2\,ROH + 6\,R^1OH$$

$$Mg[Al(OH)_4]_2 \longrightarrow Mg[AlO_2]_2 + 4\,H_2O$$

FIG. 18. Spinel synthesis. [From Dislich and Hussmann (1981).]

These syntheses were the basis for a general extension in the direction of stoichiometrically defined multicomponent systems which can be tailored.

F. A Possible Future Product: The Heat Mirror

The number of possibilities for defined multicomponent oxide coatings is so large that a restriction to what has technological–economical meaning is indicated. For this reason, many possibilities will be omitted (e.g., nitrides). An extensive presentation of the state of the art for inorganic network polymers which can be produced in this way (not only as coatings) is given in Dislich (1983).

The problem of greatest technological–economical interest appears to be whether the heat mirror can be produced by the dipping process, a process typically suitable for large areas.

After finding that the results described in Subsection VI.E may be transferred to the area of oxidic semiconductor coatings (Fig. 19), ITO appeared to be the technically most advantageous system (Dislich *et al.*, 1977; Dislich and Hinz, 1982), because the following properties are possi-

$$2\,Cd(OAc)_2 + Sn(n\text{-}OBu)_4 \xrightarrow[\text{temp.}]{H_2O}$$

$$Cd_2SnO_4 + 4\,AcOH + 4n\text{-}BuOH$$

FIG. 19. Synthesis of cadmium stannate. [From Dislich and Hinz (1982). North-Holland Publishing Company, Amsterdam, 1982.]

Film thickness	80 - 100 nm
Carrier density	$5 - 6 \times 10^{20}$ cm^{-3}
Mobility	60 - 70 cm^2 V^{-1} sec^{-1}
Conductivity	5000 - 6000 Ω cm
Sheet resistance	~25 $\Omega/\square$
Light transmittance	> 90 %
IR reflection	~80 %

Deposited in

— an industrial climate	- 2 years without damage
— a sunny nonindustrial climate	- 2 years ″ ″
— a 10% NaCl solution 50 °C	- 1 years ″ ″
— boiling water	-100 hours ″ ″

FIG. 20. Characteristics of ITO coatings.

Light transmittance	83 %
Total energy transmittance	74 %
U value*	0.28
k value	1.6

*Standard ASHRAE winter test conditions: 0°F (-18°C) outdoor temperature, 70°F (21°C) indoor temperature, 15 mph (6.7 m/sec) wind velocity, no sun. For all glass: 1/2" (12.7 mm) air gap, 1/4" (6 mm) glass. English U values represent Btu/h ft^2 °F. Metric k values represent W/(M^{2}K).

FIG. 21. Characteristic properties of ITO coatings.

bly attainable: (1) high transmission for solar radiation and in the visible range, (2) high IR reflection, (3) high scratch resistance and stability against the environment.

Laboratory experiments confirm the suitability of the sol–gel–dipping process for the preparation of defined ITO layers (Arfsten et al., 1984). The characteristics of ITO coatings are shown in Figs. 20 and 21.

If such a window could be realized technically, it would represent the best solution for the window as a passive solar collector, because of its excellent total energy transmission and high light transmission paired with heat blocking, not far from the best achieved to date.

VII. Leaching Processes and Products

The leaching process is subtractive, in contrast to all the other (additive) processes. This means that something is taken out of the glass surface and a "coating" with utilizable properties is left behind. Utilizable means that the surface has a lower density, and therefore a lower refractive index, which has an antireflection effect.

The first observation of such phenomena can be traced back to Lord Rayleigh and to Fraunhofer. The first investigations leading to technological applications are connected with the names of Taylor, Kolimorgen, Amy, Jones, Homer, and Schroeder (see Schroeder, 1974).

Schroeder (1975) briefly describes these processes as follows.

> From highly refractive optical glasses, acidic baths dissolve Pb^{2+}, Ba^{2+}, etc. (as well as the alkali), replacing them by hydrogen or H_3O^+ ions respectively. The resulting pore sizes are between 1 and 5 nm.
>
> Alkali-containing silicate glasses are leached selectively by nearly neutral electrolyte solutions (in the presence of inhibitors, mostly Al-compounds). Pores are >10 nm in size.
>
> Conventional flat glasses are selectively etched by vapors of SiF_4 or solutions of H_2SiF_6 with excess SiO_2, and heat treated subsequently. Pore sizes are noticeably larger than in the aforementioned cases [Libbey-Owens-Ford, 1943; Radio Corporation of America, 1946].
>
> Sodium borosilicate glasses are phase-separated by heat-treating; the precipitated alkali-borate-rich phase is leached by acids. A porous SiO_2-rich skeleton remains.

For a deeper understanding, see Schroeder (1974). Until recently, the use of this process was restricted to the leaching of large-dimension radiation-shielding glasses and of parts of complex shape less suitable for additive processes.

A. Present Trends

When it became known (Minot, 1976) that very durable surface layers can be obtained by leaching, coatings were produced on alkali borosilicate glass (Corning code 7740), following heat treatment and phase separation by subsequent treatment with ammonium bifluoride solution and distilled water (Elmer and Martin, 1979). The result was a porous SiO_2-rich structure with a refractive index gradient which noticeably decreased reflection losses, not only in the visible but also in the near-IR range. According to Elmer and Power (1980), these coatings are quite weather resistant. A disadvantage of these coatings is the appearance of transmission losses at short wavelengths caused by scattering due to phase separation.

At present there is a noticeable interest in antireflection coatings produced by leaching for glasses of high optical quality which are not phase separated. Such a process exists (Schroeder, 1952), and is based on leaching with neutral salt solutions, particularly Na_2HAsO_4. Recently this process has been further developed for borosilicate glasses (Cook *et al.*, 1981). The surface reflection of BK7 and similar borosilicate glasses is reduced from 4 to 0.1% per side, and maximum transmission can be obtained for any desired wavelength. To this end Al^{3+}-containing Na_2HAsO_4 solutions in distilled water were used at 87°C. Large BK7 lenses used for laser-induced nuclear fusion at Lawrence Livermore National Laboratory were so treated (Cook *et al.*, 1982). It is important that these lenses are about twice as resistant to laser damage at 12 J/cm² (1-ns, 1.06-μm pulse) as those treated in a vacuum process. By a type of preexposure below the critical threshold—whose effect is not yet understood—the resistance to laser damage is doubled once more (Lowdermilk *et al.*, 1981).

A combination of the dipping sol–gel process and leaching for the same purpose has been described (Lowdermilk and Mukherjee, 1982). First a coating of SiO_2–Na_2O–B_2O_3 is deposited; subsequently it is phase separated and leached. The resulting coatings also have a good resistance against laser damage.

VIII. Specialties

In this section even less completeness can be attempted. Rather, an attempt is made to indicate the manifold of possibilities and the occasional simplicity of the solution of a special problem by coating.

A. Optical Coatings

From the chemical structures of SiO_2 glass and of monomethylpolysiloxane (Fig. 22), the latter may be regarded as a monomethylated SiO_2. Its light transmission, including the UV, resembles that of SiO_2 glass, but it has, owing to the influence of the methyl groups, a noticeably lower refractive index. These two properties are utilized to produce optical claddings on SiO_2 glass rods and fibers. These are prepared in a dipping process by hydrolysis and polycondensation of CH_3–$Si(OR)_3$ (Dislich and Jacobsen, 1973) according to the principles described in Section VI.

The smallest refractive index of any technically utilized solid is Teflon FEP, with $n_D = 1.34$. Its structure is shown in Fig. 23. Silicon dioxide glass fibers can be coated with Teflon, also by a dipping process. Using this product, the first UV fiber optics were achieved by Dislich and Jacobsen (1966, 1973).

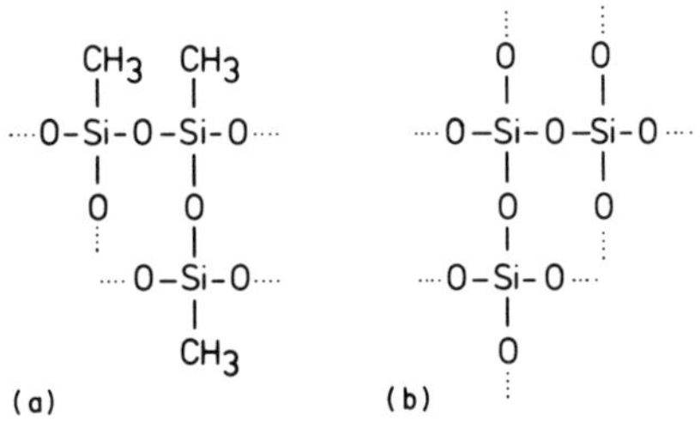

FIG. 22. Structures of (a) monomethylpolysiloxane and (b) silica glass (schematic). [From Dislich (1979).]

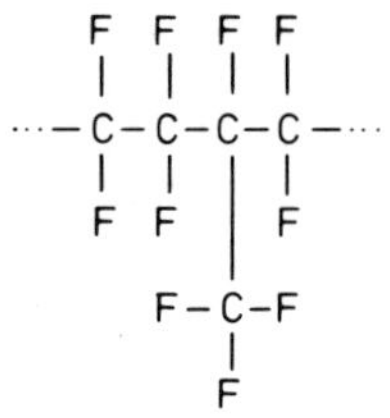

$$
\begin{array}{c}
\text{F} \quad \text{F} \quad \text{F} \quad \text{F} \\
| \quad\; | \quad\; | \quad\; | \\
\cdots -\text{C}-\text{C}-\text{C}-\text{C}-\cdots \\
| \quad\; | \quad\; | \quad\; | \\
\text{F} \quad \text{F} \quad\; | \quad \text{F} \\
\text{F}-\text{C}-\text{F} \\
| \\
\text{F}
\end{array}
$$

FIG. 23. Structure of Teflon FEP. [From Dislich (1979).]

B. Anticorrosion Coatings

This same transparent Teflon FEP is also extremely inert chemically against media corroding glass. It thus may serve to protect glass against corrosion, e.g., protecting sight glasses for pressure boilers against alkali attack (Dislich and Hinz, 1973).

The virgin surfaces of freshly drawn fibers (which must remain stable for decades) are at least partially preserved by plastic coatings—often by UV polymerization.

Glass–ceramic stove tops can be made stable against the (rare) attack of thermally decomposing caramelizing sugar by a special heteropolysiloxane coating (Steinbach et al., 1979).

C. Chemical Variation of the Glass Surface

The glass surface is characterized by, among other factors, its Si—OH groups. If one allows these to react at elevated temperatures with, e.g., $(CH_3)_2SiCl_2$ or silicone oils, the surface becomes siliconized, and thus hydrophobic. Glass ampules or flasks may thus be emptied quantitatively of hydrous medicine solutions, e.g., penicillin–Na.

If fluoroalkyl groups in large numbers are fixed on the glass surface, it becomes not only hydrophobic, but also—like Teflon—oleophobic. Small glass spheres thus treated for reflecting traffic signals are able to float on organic solvents (lacquer solutions) in spite of their high specific gravity, and so can be inserted in a dense monolayer sphere packing.

The hydrophilicity of a glass can be augmented to an antifogging function by coating with water-swelling plastics, e.g., on the basis of hydroxyethyl methacrylate.

The inner surface of porous glass can be controlled as to polarity by coating with organic residues. In the development of an artificial kidney from porous hollow fibers, separating effects are controlled, and the glass surface is also made biocompatible: see the survey by von Baeyer et al. (1980).

Adhesion between glass and plastics is improved by coating with adhesion-enhancing groups (such as $\equiv$Si—CH$_2$=CH$_2$). This is illustrated by combination glass–plastic filters (Dislich and Wiesner, 1970), or plastic coatings on bottles whose contents will not run off after fracture and which can be reused after sterilization.

D. Coloration and Decoration

This area can only be touched on briefly, although it is quite important. It extends from the classical gilding and silvering (also other metallizing) of drinking vessels to enameling. Metal soaps, pigments, organic compounds, and the wide range of enamels can be used, in a wide range of processes. Decisive for the selection of a process is frequently resistance temperature and environment. See Chapter 4 by Eppler in Volume 1 of this treatise.

IX. Concluding Remarks*

The fascinating feature of "coatings on glass" is their manifold chemical and operational possibilities corresponding to a multitude of demands. There is usually—as has been shown—a solution to each problem.

Difficult and requiring considerable effort is the path toward economical production. Different approaches are tried in different places, and only few prove themselves over the long run. Large products, such as windows, deserve particular attention.

It is important—and perhaps this chapter will serve this purpose—to continue to compare the strengths and weaknesses of the various processes.

Even today, consideration of the glass surface has remained a bit superficial. Many coatings will stand or fall with the quality of the glass surface to which they are applied. There is much experience in individual companies, but less exact knowledge, for the glass surface is "alive" and subject to continuous change (Guenther, 1981).

In the near future, the technological–economical realization of energy-controlling window coatings is bound to be crucial for the area of coatings on glass because of the large market they command; this will contribute to the advance of other areas.

Not included in this treatment because they are still in their infancy are the areas of controlled electrochromic or photochromic coatings, mixed organic–inorganic coatings, and catalytically active coatings. These should give the entire area new impulses for the more remote future.

---

* This manuscript was completed in December, 1982.

References

Acharya, H. N., Nayak, B. B., Chandhuri, T. K., and Mitra, G. B. (1982). *Thin Solid Films* **92**, 309–314.
Arfsten, N. J., Kaufmann, R., and Dislich, H. (1984). In "Ultrastructure Processing of Ceramics, Glasses and Composites" (L. L. Hench and D. R. Ulrich, eds.), pp. 189–196. Wiley, New York.
Bachner, F. J., Fan, J. C. C., Foley, G. H., and Zarracky, P. M. (1974). *Appl. Phys. Lett.* **25**, 693.
Baumberger, O., and Kalbskopf, R. (1980). DOS 30 05 797, Societa Italiana Vetro SIV S. p. A., Sal Salvo, Chieti, Italy.
Berger, E. (1934). *Glastech. Ber.* **12**, 189.
Berger, E. (1936a). *Glastech. Ber.* **14**, 351.
Berger, E. (1936b). *J. Soc. Glass Technol.* **20**, 257.
Blandenet, G., Court, M., and Lagarde, Y. (1981). *Thin Solid Films* **77**, 81–90.
Blocher, J. M., Jr. (1981). *Thin Solid Films* **77**, 51–53.
Bloos, K. H., and Lukas, H. (1973–1974). *Electrodeposition Surf. Treat.* **2**, 47–64.
Cook, L. M., Mader, K. H., and Schnabel, R. (1981). U.S. patent application 309, 149.
Cook, L. M., Lowdermilk, W. H., Milam, D., and Swain, J. E. (1982). *Appl. Opt.* **21**, 1482–1485.
Dislich, H. (1971). *Angew. Chem., Int. Ed. Engl.* **10**, 363–370.
Dislich, H. (1979). *Angew. Chem. Int. Ed. Engl.* **18**, 49–59.
Dislich, H. (1983). *J. Non-Cryst. Solids* **57**, 371–388.
Dislich, H., and Hinz, P. (1973). G.B. Patent 1,400,664, Jenaer Glaswerk Schott & Gen., Mainz, Federal Republic of Germany.
Dislich, H., and Hinz, P. (1982). *J. Non-Cryst. Solids* **48**, 11–16.
Dislich, H., and Hussmann, E. (1981). *Thin Solid Films* **77**, 129–139.
Dislich, H., and Jacobsen, A. (1966). U.S. patent 3,623,903, Jenaer Glaswerk Schott & Gen., Mainz, Federal Republic of Germany.
Dislich, H., and Jacobsen, A. (1973). *Angew. Chem., Int. Ed. Engl.* **12**, 439.
Dislich, H., and Wiesner, K. H. (1970). *Optik* **30**(4), 1.
Dislich, H., Hinz, P., and Kaufmann, R. (1969a). U.S. patent 3,759,683.
Dislich, H., Hinz, P., and Kaufmann, R. (1969b). U.S. patent 3,847,583, Jenaer Glaswerk Schott & Gen., Mainz, Federal Republic of Germany.
Dislich, H., Hinz, P., and Wolf, G. (1977). U.S. patent 4,229,491, Jenaer Glaswerk Schott & Gen., Mainz, Federal Republic of Germany.
Donley, H. E. (1966). U.S. patent 3,528,845, PPG Industries, Inc.
Dutz, H., Mulfinger, O., and Krolla, G. (1968). DBP 16 96 110, Jenaer Glaswerk Schott & Gen., Mainz, Federal Republic of Germany.
Elmer, T. H., and Martin, F. W. (1979). *Am. Ceram. Soc. Bull.* **58**, 1092–1097.
Elmer, T. H., and Power, J. M. (1980). *Am. Ceram. Soc. Bull.* **59**, 1124–1126.
Erben, E., Muehlratzer, A., and Zeilinger, H. (1981). *Metall (Berlin)* **35**, 1253–1257.
Fillard, J. P., and Manifacier, J. C. (1981). *Thin Solid Films* **77**, 67–80.
Firmenprospekt Glaverbel "Glaverplus Comfort, 881AL-AL5-0,5." Glaverbel, Belgium.
Firmenprospekt Interpane (1982). "Licht & Wärme, Tiplus® Neutral." Firmenprospekt Interpane, Lauenförde, Federal Republic of Germany.
Firmenprospekt Schott Glaswerke (1981a). For Precision Measurements Interference Filters and Special Filters by SCHOTT." Firmenprospekt Schott Glaswerke, Mainz, Federal Republic of Germany.
Firmenprospekt Schott Glaswerke (1981b). "Produktinformation Gläser mit Oberflächenschichten." Firmenprospekt Schott Glaswerke, Mainz, Federal Republic of Germany.

Firmenprospekt Schott Glaswerke (1982). "Thermax® 32 'A Hot Problem.'" Firmenprospekt Schott Glaswerke, Mainz, Federal Republic of Germany.
Firmenprospekt Watkins-Johnson Co. (1978). "Conveyorized Atmospheric Pressure CVD." Firmenprospekt Watkins-Johnson Co., Scotts Valley, California.
Frank, G., Kauser, E., and Koestlin, H. (1981). *Thin Solid Films* **77**, 101–107.
Franz, H. (1971). U.S. patent 3,723,138, PPG Industries, Inc.
Garcia, F. J., and Tomar, M. S. (1981). *Prog. Cryst. Growth Charact.* **4**, 221–248.
Geneva Forum (1981). *Thin Solid Films* **77**(1–3).
Gillery, F. H. (1982). *J. Non-Cryst. Solids* **47**, 21–26.
Glaeser, H. J. (1980). *Glastech. Ber.* **53**, 245–258.
Greenberg, C. B. (1975). U.S. patent 3,978,271, PPG Industries, Inc.
Groth, R. (1977). *Glastech. Ber.* **50**, 239–247.
Groth, R., and Van Boort, H. J. J. (1968). *Philips Tech. Rundsch.* **29**, 47–48.
Guenther, K. H. (1981). *Thin Solid Films* **77**, 239–251.
Hinder, A. B. (1981). *Thin Solid Films* **77**, 272.
Holland, L. (1964). "The Properties of Glass Surfaces." Chapman & Hall, London.
Howson, R. P., and Ridge, M. J. (1981). *Thin Solid Films* **77**, 119–125.
Kalbskopf, R. (1981). *Thin Solid Films* **77**, 65–66.
Kosugi, T., Wakni, Y., and Yamazaki, T. (1976a). Japan-Kokai 76 83,615, Hitachi Ltd., Tokyo.
Kosugi, T., Wakni, Y., and Yamazaki, T. (1976b). Japan-Kokai 77 90,513, Hitachi Ltd., Tokyo.
Libbey-Owens-Ford (1943). U.S. patent 2,490,263.
Lowdermilk, W. H., and Mukherjee, S. P. (1982). *Appl. Opt.* **21**, 293–296.
Lowdermilk, W. H., Swain, J. E., and Milam, D. (1981). Internal information.
MacChesney, J. B. (1977). U.S. patent 4,217,027, Bell Telephone Laboratory.
Maurer, R. D. (1972). U.S. patent 3,737,293, Corning Glass Works.
Mayer, H. (1972). "Physik dünner Schichten," 2 vols. Wiss. Verlagsges., Stuttgart.
Minot, M. J. (1976). *J. Opt. Soc. Am.* **66**, 515–519.
National Technical Information Service (1975). *Proc. Conf. Chem. Vap. Deposition, Int. Cmf., 5th, 1975,* NIT-Search, PS-75/74.
Nichols, D. R. (1982). *Photonics Spectra* **57**, 57–60.
Pilkington Brothers P.L.C. (1974). GB-Patent 1,507,465. Pilkington Brothers P.L.C., St. Helens, England.
Pulker, H. K. (1981). *Thin Solid Films* **77**, 203–212.
Radio Corporation of America (1946). U.S. patent 2,490,662.
Rothon, R. N. (1981). *Thin Solid Films* **77**, 149–153.
Schroeder, H. (1952). German Patent 821,826, Jenaer Glaswerk Schott & Gen., Mainz, Federal Republic of Germany.
Schroeder, H. (1962). *Opt. Acta* **9**, 249–254.
Schroeder, H. (1969). *Phys. Thin Films* **5**, 87–141.
Schroeder, H. (1974). *Int. Congr. Glass, 10th, 1974,* No. 8, pp. 118–130.
Schroeder, H. (1975). *Ullmanns Encykl. Tech. Chem.* **10**, 260–262.
Steinbach, H., Schnurrbusch, K., and Rieder, M. (1979). DOS 29 52 756, Bayer AG, Leverkusen, Federal Republic of Germany.
Terneu, R., and Van Cauter, A. (1979). DOS 29 41 843, BFG Glassgroup, Paris.
Viguié, J. C. (1974). *In* "Science and Technology of Surface Coatings" (B. N. Chapman and J. C. Anderson, eds.), pp. 361–368. Academic Press, New York.
von Baeyer, H., Schnabel, R., Vaulont, W., and Koczmarczyk, G. (1980). *Trans.—Am. Soc. Artif. Intern. Organs* **26**, 309–313.

CHAPTER 9

Optical Fiber Waveguides

Richard M. Klein

OPTICAL FIBER AND COMPONENTS DEPARTMENT
GTE LABORATORIES, INC.
WALTHAM, MASSACHUSETTS

I. Introduction	285
II. Fiber Characteristics	286
A. Optical Characteristics	286
B. Mechanical Characteristics	296
III. Fiber Processing	298
A. Vapor-Phase Techniques	299
B. Non-Vapor-Phase Techniques	323
IV. Summary	334
References	335

I. Introduction

Optical fibers are strands of glass which can be used to transmit information in the form of optical rather than electrical signals. Their rapidly expanding use in communication systems is based on the ability of optical fibers to transmit vast quantities of data over exceedingly long distances.

As might be deduced from this unusual application for glass, the processing of optical fiber waveguides is very different from that for conventional glasses. This is a consequence of the stringent demands placed on these components in service. For example, optical fibers may have to transmit optical signals for distances up to 100 km at data rates which can approach 1 GHz. In addition, each fiber must be able to be mated to similar fibers by means of either permanent splices or rematable connectors without a significant loss in signal intensity. Finally, optical fibers must be able to withstand the intermittant mechanical stresses of cabling and deployment, as well as the long-term stresses expected in service, and survive for periods of 30 years or more; optical properties have little relevance if the fiber breaks.

The object of this chapter is to describe the methods by which optical fiber waveguides can be fabricated. However, before this is done, the optical and mechanical characteristics of fibers will be reviewed, with an emphasis on showing how the performance criteria are related to fabrication requirements. With this background, various methods of processing fiber waveguides will be described, concentrating on the vapor-phase techniques that have dominated the commercial optical fiber market.

II. Fiber Characteristics

Since the means by which optical fiber waveguides are processed are so intimately bound to their performance specifications, we will first describe the relevant optical and mechanical properties of these components. Moreover, because the objective of this section is restricted to showing how processing is dictated by these required characteristics, the discussions of optical and mechanical properties are not rigorous. Readers who are interested in a more complete analysis are referred to review articles which have treated either optical properties (Olshansky, 1979) or mechanical properties (Tariyal and Kalish, 1978).

The first part of this section defines optical fiber waveguides and describes their optical properties, including numerical aperture, attenuation, and pulse broadening. In this discussion, heavy use will be made of ray models to describe propagation; although solution of the wave equation is needed to fully describe propagation, the ray approach provides a conceptual model that can be used to understand the salient features.

The second part of the section treats the mechanical characteristics of fibers, i.e., strength and fatigue. As in the first part, the discussion will be limited to those aspects of the topic that affect the processing of optical fibers.

A. Optical Characteristics

An optical fiber, shown schematically in Fig. 1, consists of a cylindrical core of refractive index n_1, surrounded by an annular cladding of refractive index $n_2 < n_1$. As shown in Fig. 1, a ray of light in the core striking the core–cladding interface at an angle θ greater than the critical angle, $\sin^{-1}(n_2/n_1)$, will be totally reflected at that interface, thereby propagating down the fiber. This ray can be considered as the superposition of two rays: a ray propagating along the fiber axis and a standing wave perpendicular to that axis. Moreover, since the amplitude of the standing wave must go to zero outside the core, this wave is quantized such that one cycle of the corresponding ray consists of an integral number of wavelengths. (An alternative explanation of this condition is that succes-

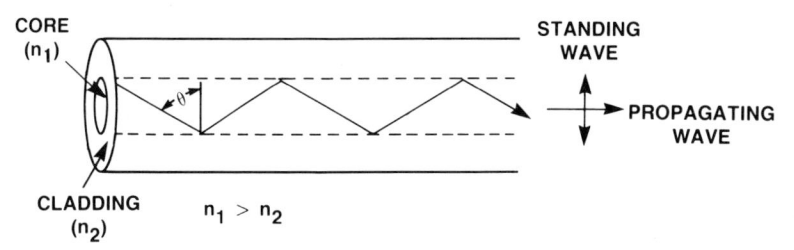

FIG. 1. Schematic of optical fiber.

sive rays all interfere constructively.) This condition implies that there are only a discrete number of incident angles greater than the critical angle which will yield a propagating ray. These angles are given by

$$\theta = \cos^{-1}(4an_1/m\lambda), \qquad (1)$$

where a is the core radius, λ the wavelength, and m an integer. Each of these angles can be associated with a mode of the fiber, and the number of modes in a given fiber varies as

$$m \simeq 2\pi^2 a^2(n_1^2 - n_2^2)/\lambda^2. \qquad (2)$$

Conventional multimode fibers, with $a = 25$ μm, $n_1 \simeq 1.47$, and $n_2 \simeq 1.46$, can support several hundred modes.

1. Numerical Aperture

The numerical aperture (NA) of a fiber is related to the angle of the cone of light it will accept and propagate. (Conversely, it is also the angle of the cone of light emitted by a fiber at its exit end.) Specifically, it is the sine of the half-angle of this cone, shown as ϕ in Fig. 2. Considering the definition of the critical angle and the fact that the ray of light will be refracted at the core/air interface, it is straightforward to show that

$$\mathrm{NA} = (n_1^2 - n_2^2)^{1/2} \simeq [2n_1(n_1 - n_2)]^{1/2} \qquad (3)$$

for the small differences $n_1 - n_2$ generally found in conventional optical fibers.

Since the numerical aperture defines the acceptance angle of the fiber in a manner analogous to the way in which f numbers characterize a lens,

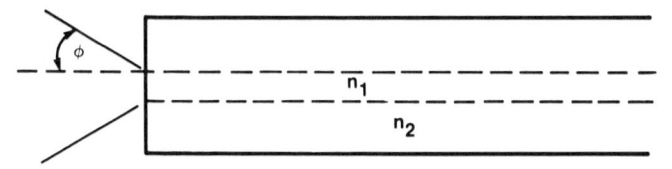

FIG. 2. Acceptance angle of optical fiber.

this quantity is related to the ability of the fiber to gather light from nonlaser sources. For example, the ability to couple a light-emitting diode to an optical fiber is proportional to the numerical aperture of the fiber.

2. Attenuation

Attenuation is a measure of the loss of signal that occurs as light propagates down a fiber. It is generally measured in decibels per kilometer (dB/km), i.e.,

$$\text{loss} \quad (\text{dB/km}) = 10 \log_{10}(I_{\text{in}}/I_{\text{out}}), \qquad (4)$$

where I_{in} and I_{out} are the intensities of the input and output light signals, respectively. The loss described in this manner, which is useful for system designers, is directly related to absorption coefficient A, which is a more common material constant defined by

$$I_{\text{in}}/I_{\text{out}} = \exp(At), \qquad (5)$$

where t is the thickness. Thus

$$\alpha \quad (\text{dB/km}) = 4.34 \times 10^5 A \quad (\text{cm}^{-1}). \qquad (6)$$

Since detection depends on the strength of the signal, attenuation is one factor that limits the distance over which useful optical signals can be transmitted.

Attenuation may be separated into intrinsic and extrinsic factors. The intrinsic factors that affect attenuation and give rise to a transmission window at wavelengths around 1 μm are the tail of the fundamental UV absorption and Rayleigh scattering on the short-wavelength side, and IR absorption due to vibrational modes on the long-wavelength side.

Fundamental absorption peaks corresponding to electronic transitions in the UV portion of the spectrum have tails which extend to the 1-μm wavelength region. These absorption edges, called Urbach tails, are due to locally varying electric microfields within the glass (Pinnow et al., 1973). Experimentally, it has been found that this loss varies exponentially with photon energy; for example, for GeO_2–SiO_2 glasses (Miya et al., 1979),

$$\alpha_{\text{UV}} = C_{\text{UV}} \exp(\nu/\nu_{\text{UV}}), \qquad (7)$$

where C_{UV} are ν_{UV} are material-sensitive constants and ν is the photon frequency.

Rayleigh scattering is caused by inherent fluctuations in density and composition which are present in all glassy materials. These fluctuations are distinct from the cords and striae that can be observed in poorly mixed glasses. The density fluctuations are defects that are frozen into the glass at its transition temperature T_g and, since the defect density is mainly dependent on temperature, are greater in materials with higher T_g's. Compositional fluctuations, which produce scattering because of the dependence of refractive index on composition, increase with the complexity of the glass considered. Losses caused by density fluctuations α_d (Pinnow et al., 1973) and compositional fluctuations α_c (Maurer, 1973) are given by

$$\alpha_d \simeq (8\pi^3/3\lambda^4)n^8p^2\beta T_g, \qquad \alpha_c \simeq (16\pi^3 n/3\lambda^4)(\partial n/\partial c)\overline{\Delta c^2}\,\delta V, \qquad (8)$$

where p is the photoelastic constant, β the compressibility, $\overline{\Delta c^2}$ the mean-square concentration fluctuation, and δV the volume over which the fluctuation occurs. Note that total Rayleigh scattering is proportional to λ^{-4}.

The third intrinsic loss mechanism, absorption caused by vibrational modes, dominates the infrared portion of the spectrum and increases with decreasing photon energy. This mechanism generally is only important at wavelengths greater than 1.4 μm. Absorption caused by this process is also exponentially related to photon energy (Olshansky, 1979):

$$\alpha_{IR} = C_{IR}\exp(-\nu/\nu_{IR}) \qquad (9)$$

where the constants C_{IR} and ν_{IR} depend on the glass composition. In silica-based glasses, the constants change slightly depending on the level of doping; in practical cases, the most important of these changes is caused by boron, which shifts the infrared absorption to shorter wavelengths, thereby narrowing the transmission window.

By combining Eqs. (7)–(9), it is possible to derive the fundamental transmission window that exists in glasses. This is done for a silica–germania glass, a common waveguide constituent, in Fig. 3. Note that attenuation is minimum at wavelengths $\simeq 1.5$ μm; this is one reason why optical communication systems have evolved from ones operating at $\simeq 0.8$ μm when they were introduced to ones operating at longer wavelengths.

Although levels of intrinsic absorption can be quite low, as shown in Fig. 3, attenuation in actual optical fiber waveguides can be dominated by extrinsic factors, and it is these factors which must be controlled during processing. One of these factors is impurity absorption. In the wavelength region of interest for optical communications, impurities which have the most significant effect on attenuation are transition metals and water or hydroxyl ions.

When present in conventional glasses, transition metals can cause coloration because of their incomplete d levels. In optical waveguides, this level of absorption can result in unacceptably high attenuation. For example, 1 ppb of Fe^{2+} adds 20 dB/km of attenuation at 1.0 μm. Table I (Gossink, 1977) shows the effect of a variety of transition-metal impurities on attenuation at 0.8 μm. Although the exact extent of loss caused by

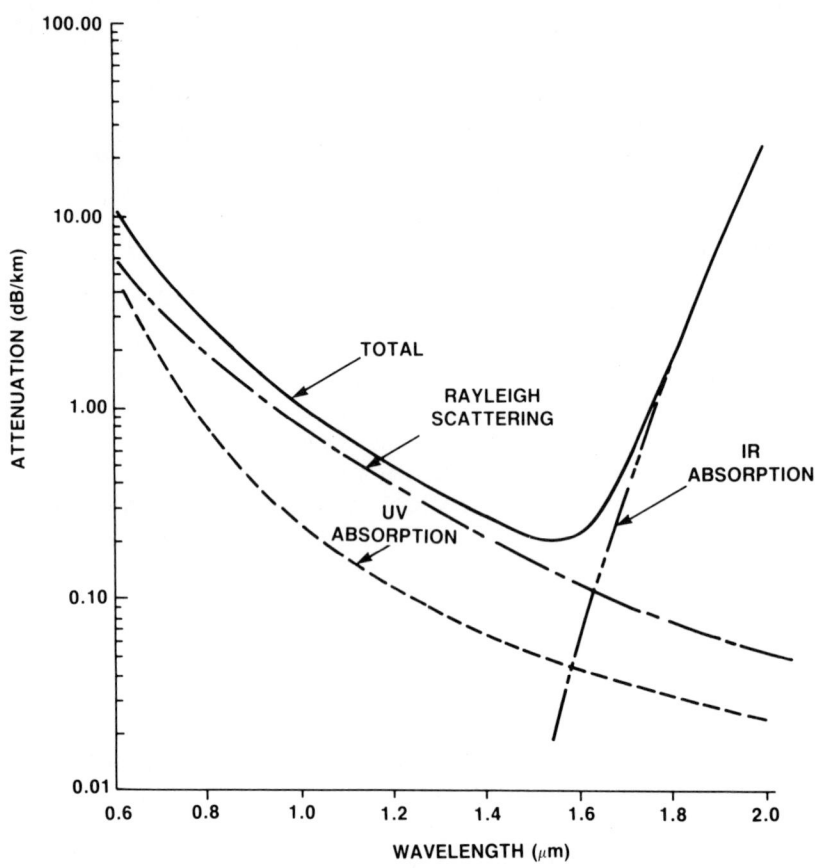

FIG. 3. Intrinsic transmission window in silica-based glasses.

TABLE I

ATTENUATION AT 800 nm CAUSED BY TRANSITION METALS (dB/km/ppm)[a]

	Soda–lime–silica	Sodium (T1) borosilicate	Fused silica
V	—	39	1050
Cr	20	50	610
Mn	80	11	20
Fe	100	10	50
Co	—	9	20
Ni	150	130	30
Cu	620	500	10

[a] From Gossink (1977). North-Holland Publishing Company, Amsterdam, 1977.

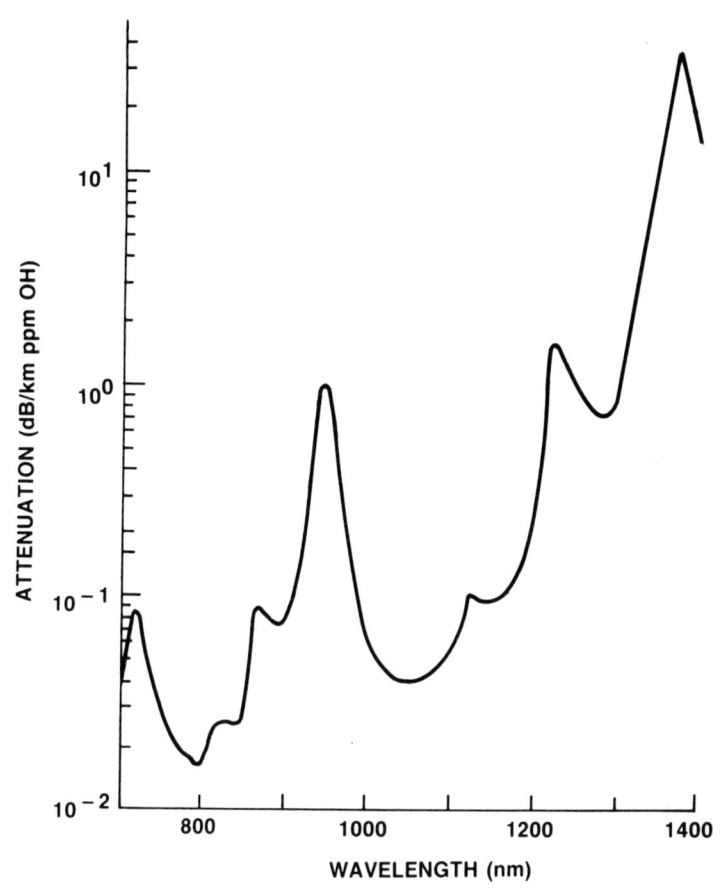

FIG. 4. Absorption due to hydroxyl impurities in silica-based glasses. [From M. K. Barnoski and S. D. Personick, Measurements in fiber optics, *Proceedings of the IEEE*. Copyright © 1978 IEEE.]

transition metals depends on the wavelength considered, Table I shows that the concentration of these impurities must be maintained at exceedingly low levels.

Water can also cause unacceptably high levels of attenuation in optical fiber waveguides. In this case, absorption is caused by overtones of the fundamental OH stretching vibration at 2.73 μm, and combinations of this band with the fundamental stretching frequencies of the Si–O network. Figure 4 (Keck *et al.*, 1973) shows how absorption due to hydroxyl impurities varies with wavelength. As shown, the strength of the absorption generally increases as the wavelength increases to the fundamental at 2.73 μm. This is why, whereas hydroxyl ion impurities of about 10 ppm are acceptable in waveguides geared for use at wavelengths of 800–900 nm (intrinsic fiber absorption ~3 dB/km), much lower levels of ~10–100 ppb are necessary when the waveguide is to be used in the 1.3-μm region (intrinsic fiber absorption ~0.5 dB/km).

The second extrinsic factor that can increase attenuation in practical optical fibers is waveguide scattering. This phenomena, which causes a wavelength-independent loss, can arise from waveguide imperfections and microbending.

Waveguide imperfections (Marcuse and Derosier, 1969) include defects at the core–cladding interface, variations in core diameter along the length of the fiber, and radial fluctuations in refractive index. Each of these deficiencies can cause light to be scattered into unguided modes which then escape from the fiber. Microbending (Gloge, 1975) is caused by external forces which produce lateral deformations and, thereby, mode coupling and loss. As in the case of waveguide imperfections, the result is a wavelength-independent increase in attenuation for defects that are generally larger than the wavelength of light being guided.

In summary, the intrinsic absorption window in silica-based glasses provides the possibility of low attenuation at wavelengths between 0.8 and 1.6 μm. Extrinsic factors that are process dependent produce additional absorption and scattering losses which can limit performance in practical fibers.

3. Pulse Broadening

Just as excessive attenuation can limit the distance over which useful signals can be transmitted using optical fibers, pulse broadening (which controls the density of information or bandwidth that can be transmitted) can also limit this distance. The reason for this is shown in Fig. 5. If a narrow pulse of light is introduced into a fiber, it will be broadened by a certain extent as it travels down the fiber; the amount of broadening depends on the fiber characteristics as well as the length of the fiber traveled. If one considers introducing information at a high rate in the form of closely spaced light pulses, then this broadening will eventually cause these pulses to overlap, and information will be lost. Pulse broadening in optical fibers can be ascribed to modal dispersion, material dispersion, and waveguide dispersion.

Modal dispersion arises from the fact that each of the different rays propagating in a fiber have a different path length in a given length of fiber. This is shown schematically in Fig. 6. For a step-index fiber; i.e., one in which the core has a uniform refraction index, the low-order modes (those traveling along the fiber axis) would arrive before the higher-order modes, represented by the zigzag ray, if they were both introduced into the fiber

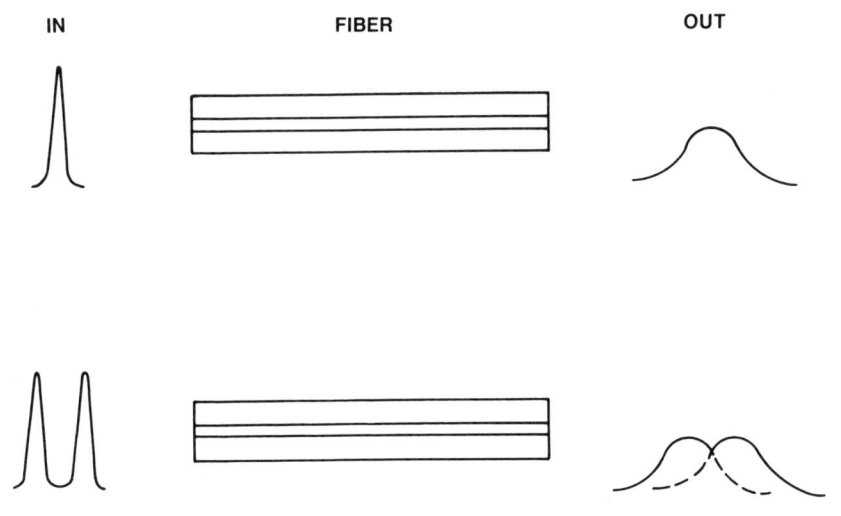

FIG. 5. Pulse broadening in optical fibers.

at the same instant. The net result when all possible modes are considered is that the pulse emerging from the fiber would be broadened when compared to the pulse introduced into the fiber. This broadening, 40 nsec/km (Gambling *et al.*, 1976), limits the bandwidth of step-index fiber to ~30 MHz km.

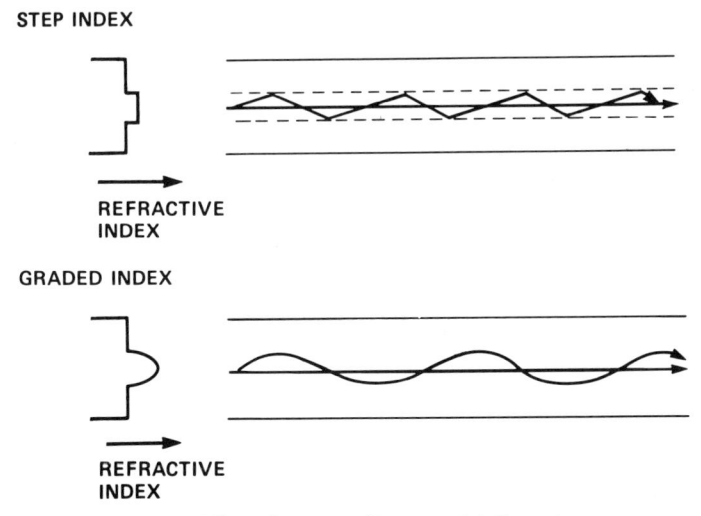

FIG. 6. Effect of core profile on modal dispersion.

The first means devised to increase fiber bandwidth by decreasing modal dispersion was the graded-index fiber in which the refractive index of the core varies in a near-parabolic fashion from the core center to the cladding (Gloge and Marcatili, 1973). Although this type of fiber will still have low-order and high-order modes with different path lengths (Fig. 6), the high-order modes, represented as a sinusoidal ray, will spend much of their time near the core–cladding interface. Since the refractive index in that region is relatively low, and since the velocity of light is inversely proportional to refractive index, the higher-order mode will travel faster while traversing a longer path length. The net result is that, in a well-designed graded-index fiber, all of the modes will have about the same transit time down the fiber in spite of the fact that they have different path lengths. In such fibers, pulse broadening can be reduced to 0.2 nsec/km (Campbell, 1977), resulting in bandwidths $\simeq 5$ GHz km.

The profile of a graded-index fiber, i.e., the variation of refractive index in the core, is given by

$$n(r) = n_0[1 - 2\Delta(r/a)^g]^{1/2}, \tag{10}$$

where n_0 is the index at the core center, $\Delta = n_0 - n(a)$, a is the core radius, and the profile parameter $g \simeq 2$. In order to obtain high bandwidths in graded-index fibers, it is necessary that the actual profile parameter be exceedingly close to the prescribed value, which depends on wavelength and glass composition. As shown in Fig. 7 (Gloge, 1979), small departures from the optimum parameter cause dramatic decreases in bandwidth. In addition, perturbations in the profile, such as ripples on the intended near-parabolic index variation, can also produce a large degradation in fiber bandwidth (Marcuse, 1979).

Since modal dispersion is caused by different modes having different transit times, the most effective means to eliminate the problem is to use a fiber which can only propagate a single mode. As can be shown through Eq. (2), this can be done by reducing the diameter of the core from 50 μm, for conventional multimode fibers, to ~ 10 μm, and by decreasing the difference in refractive index between the core and cladding from ~ 0.014 to 0.001. In these single-mode fibers, enormously high bandwidths (greater than 1000 GHz km) are possible, and bandwidths ~ 25 GHz km are quite practical (Cohen *et al.*, 1982).

In addition to modal dispersion, material dispersion and waveguide dispersion can also lead to pulse broadening in fibers. Pulse broadening per unit length caused by material dispersion τ_m arises from the facts that optical sources have a finite spectral width and that fiber characteristics are wavelength dependent; τ_m is given by (Olshansky, 1979)

$$\tau_m = (L/c)\lambda(d^2n_0/d\lambda^2)\,\delta\lambda, \tag{11}$$

9. OPTICAL FIBER WAVEGUIDES

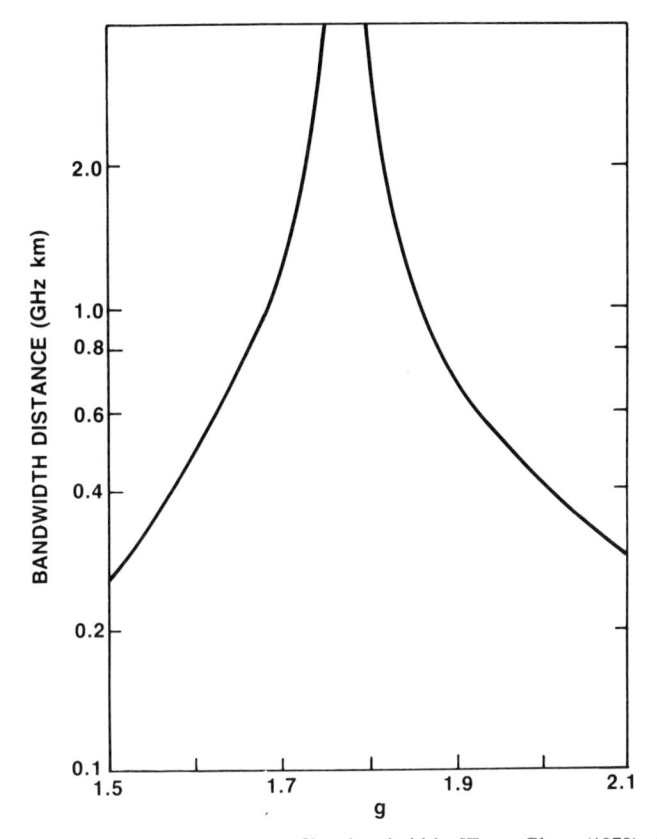

FIG. 7. Effect of profile parameter on fiber bandwidth. [From Gloge (1979). Copyright The Institute of Physics.]

where L is the fiber length, c the speed of light, and $\delta\lambda$ the spectral width of the source. At $\lambda = 0.9$ μm, τ_m is $\sim$1.5 nsec/km for light-emitting diode sources ($\delta\lambda = 20$ nm) and $\sim$0.8 nsec/km for diode laser sources ($\delta\lambda = 2$ nm). However, for silica-based fibers, $d^2n_0/d\lambda^2$ decreases to zero at $\lambda \simeq 1.3$ μm (Payne and Gambling, 1975). The elimination of material dispersion resulting from operating at this wavelength is, in addition to the reduced attentuation, a second driving force moving optical communication systems from their original operating region of 0.8–0.9 μm to longer wavelengths.

Waveguide dispersion, which arises from the dependence of light propagation on the size of the fiber core in relation to the wavelength of light being transmitted, is generally <0.015 nsec/km (Cohen and Lin, 1977) and therefore can be neglected in multimode fibers. In single-mode fibers, waveguide dispersion is controlled by the core diameter and NA and can be balanced against material dispersion in order to achieve zero total dispersion at a selected operating wavelength (Cohen *et al.*, 1979); this results in enormously high bandwidths at these wavelengths.

Therefore pulse broadening, in addition to attenuation, can limit the distance over which useful signals can be transmitted using optical fibers. In multimode fibers, processing must address the precise control required in the graded index needed to minimize modal dispersion and thereby achieve the high bandwidths that are possible. In single-mode fibers, control is required in the core diameter and refractive index difference between the core and the cladding in order to balance the effects of material and modal dispersion at the selected operating wavelengths; in this way, essentially limitless (>100 GHz km) bandwidths can be achieved.

B. Mechanical Characteristics

In addition to using processing techniques which ensure that optical properties meet performance requirements, manufacturers must also be able to provide optical fiber waveguides that are mechanically reliable. Of specific concern are strength and fatigue. First, the fiber must be strong enough to resist instantaneous failure when subject to the intermittent stresses expected in cabling the fiber and on cable deployment. Second, the fiber must have sufficient reliability to resist delayed or fatigue failure when exposed to the static stresses that may be present during its anticipated service life.

The purpose of this section is to briefly review these mechanical characteristics of optical fiber waveguides, highlighting those features important in fabrication. Those readers interested in more detail are directed to reviews on the mechanical characteristics of fibers (e.g., Tariyal and Kalish, 1978) or to a more general treatment on the strength of glasses which has been covered in a previous volume in this series (Uhlmann and Kreidl, 1980).

1. Strength

Whereas the theoretical tensile strength of glass is extremely high, measured values for typical samples are generally orders of magnitude lower. This discrepancy, illustrated in Table II, is attributed to the fact that glass is not a perfectly elastic solid. Being brittle, glasses do not deform plastically to relieve localized stress concentrations (Irwin and Wells, 1965). Because of this, inhomogeneities and flaws can control the strength of glass by providing a means by which a moderate applied stress can produce very high stresses in localized regions. When these localized stresses approach the theoretical bond strength, sequential bond failure can occur, leading to crack elongation and fracture.

TABLE II

Representative Strengths (GPa) of Glass

Ordinary glass	0.03
High-quality glass	0.10
High-quality optical fiber:	
Long lengths	1.0
Short lengths	6.0
Theoretical	20.0

A typical model of a flaw which can limit the strength of optical fiber waveguides is a straight surface crack with an elliptical cross section. In this case, the ratio of the local stress σ_1 at the crack tip to the average applied stress σ is

$$\sigma_1/\sigma = 2(\delta/\rho)^{1/2}, \qquad (12)$$

where δ is the crack length and ρ the radius of curvature at the crack tip. Assuming that ρ will be relatively constant, the relation shows that the strength of a given length of fiber will be determined by the single largest flaw within that fiber. For optical fiber waveguides, it has been estimated that a 1-μm flaw will cause failure if the applied load exceeds 0.58 GPa, or 84,000 psi (Miller, 1979).

Therefore a fiber manufacturer must take precautions to minimize the number and severity of strength-controlling flaws. Moreover, since such flaws are distributed randomly in both size and location on the fiber and, since there is no known means of locating these flaws nondestructively, the manufacturer must ensure that the fiber he produces has a specified minimum strength by stressing all of the fiber to this level, i.e., proof-testing.

2. Fatigue

As described above, fiber strength concerns that stress which, when applied to the fiber, causes immediate failure. However, the problem of guaranteeing the long-term mechanical reliability of optical fibers is complicated by the fact that strength-controlling flaws can grow in service, leading to delayed spontaneous failure. This process, by which brittle solids such as optical fiber waveguides fail after an extended time under stress, is called fatigue (Freiman, 1980).

Two conditions must be present in order for fatigue to occur. First, there must be a tensile stress in the vicinity of the flaw. Although optical fiber cables are designed to minimize residual stresses on fibers, a recent study (Tanaka *et al.*, 1982) has shown that fiber in even a well-designed cable can experience stresses that approach 20,000 psi, even when the cable is straight.

The second condition required for fatigue to occur is that the environment around the flaw must be chemically active. Water vapor, which can only attack silicon–oxygen bonds when they are strained, is the primary corrosive agent in all practical cases; optical fiber fatigue has been observed in as little as 2% relative humidity (Kalish and Tariyal, 1978).

Therefore, the expected mechanical lifetime of an optical fiber waveguide in service depends on its fatigue behavior. This, in turn, depends on the severity of the initial flaws as well as the rate at which these flaws grow in service. Thus both strength and fatigue depend on the size of the largest flaw existing within the length of fiber in question and, in order to ensure adequate mechanical reliability, fiber processing must ensure that such flaws are minimized.

III. Fiber Processing

As described in the preceding sections, the characteristics required of optical fiber waveguides make severe demands on the methods by which they can be processed. In summary, there are four characteristics which require specific attention:

(1) *Purity* In order to prepare fibers having near-intrinsic levels of attenuation, it is necessary to maintain exceedingly low concentrations of impurities in the fiber core. This is particularly true for transition metals and hydroxyl ions. Transition metals generally must be kept at parts per billion concentrations. Permissable hydroxyl ion concentrations depend on the intended operating wavelength, and vary from about 100 ppm at 0.8-μm operation to 100 ppb at 1.3-μm operation.

(2) *Dimension control* Control of fiber and core size and shape is important for several reasons. For example, excessive variations in core diameter can cause unacceptably high waveguide losses. Perhaps more importantly, variations in geometry will have a dramatic impact on the ability to splice and connect optical fibers; whereas fibers with elliptical cores can be made to have low attenuation, they will be impossible to splice to a circular-core fiber without suffering a loss penalty.

(3) *Compositional control* Precise control of the composition of the core and how it varies radially is important for three reasons. First, the composition of the core with respect to the cladding determines the numerical aperture of the fiber, and variations in NA from fiber to fiber make it impossible to obtain low-loss splices. Second, in order to obtain high bandwidths in graded-index fibers it is necessary to provide a core having

a precise variation in refractive index from the core center to the cladding; deviations in this variation lead to a dramatic decrease in fiber bandwidth. Finally, in single-mode fibers, balancing of material dispersion and waveguide dispersion at the intended operating wavelength requires that the numerical aperture, and, therefore, the core composition, be adequately controlled.

(4) *Low flaw density* In order to obtain fibers having sufficient mechanical integrity, it is necessary that the number of strength-limiting flaws be minimized. This raises two concerns. First, the process must not permit inclusions in the fiber cladding which, while not affecting attenuation, will reduce fiber strength. Second, the process must not introduce flaws on the fiber surface through abrasion.

In response to these requirements, several techniques have been developed by which optical fiber waveguides can be processed effectively. These methods fall into one of two general categories: processes which use vapor-phase techniques to prepare the fiber precursor, and processes which utilize liquid and solid-state glass fabrication methods for this purpose. These techniques will be described in the following two sections. Vapor-phase techniques, including the various manifestations of this approach, will be described first since it is these methods which were the first to successfully prepare fibers having usably low attenuations. In addition, these methods are still being used to prepare the highest quality fiber. Non-vapor-phase techniques will then be described, with attention given to the relative advantages and disadvantages of these methods compared to those based on vapor-phase reactions.

A. Vapor-Phase Techniques

There have been a variety of techniques developed which are based on the reaction of mixtures of silicon tetrachloride ($SiCl_4$) and other halides to form a doped silica glass. All of these techniques consist generally of two steps. First, vapor-phase techniques are used to fabricate a rod of glass, called a preform, which has the same relative radial variation in refractive index as desired in the final fiber. Second, the preform, which can be several centimeters in diameter, is drawn to a fiber which is generally 125 μm in diameter.

1. Preform Fabrication

The general approach to preform fabrication in all of the vapor-phase methods to be discussed is to use liquid halides as the raw material source and to bubble a carrier gas through these liquids so that it becomes saturated with the corresponding halide vapor. This vapor is then transported to the reaction zone. There are two specific advantages to this approach, and these advantages were critical in the successful fabrication of fibers which were less absorbing by more than a factor of 100 than the best glasses at the time. First, the halides which are needed, such as silicon tetrachloride, were common constituents in the semiconductor processing industry and were therefore available in a highly purified form. Perhaps more importantly, use of the carrier gas approach to entrain the halide vapor provides an additional distillation step, thereby further purifying the reactants. This is particularly effective for transition metals since, as shown in Fig. 8, the vapor pressures of transition-metal halides are generally several orders of magnitude lower than those of the reactant halides such as $SiCl_4$. Thus vapor-phase techniques provided the means by which glasses containing no more than parts per billion levels of transition-metal impurities could be fabricated. As discussed previously, this was precisely what was required to fabricate fibers with usable levels of attenuation.

Although vapor-phase methods have the tremendous advantage of being able to prepare highly pure glasses, their very nature limits the variety of glass compositions that can be used. Specifically, a potential index-modifying dopant must be available as a gas or, preferably, as a

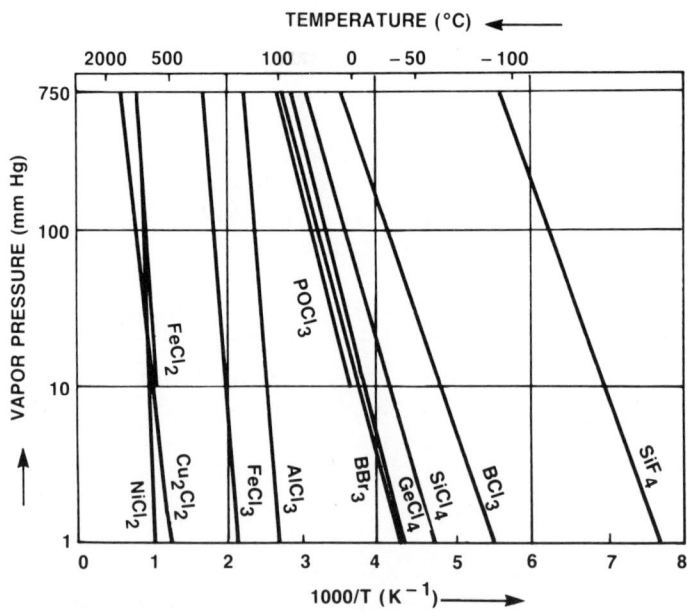

FIG. 8. Vapor pressure of reactant and impurity halides. [From Gossink (1977). North-Holland Publishing Company, Amsterdam, 1977.]

liquid with a vapor pressure near room temperature comparable to that of $SiCl_4$. This restriction has limited the practical dopants used in these methods to those shown in Table III.

In the following subsections, the four major approaches to vapor-phase preform fabrication techniques will be discussed. In these sections, each approach will be described; particular emphasis will be given to the source of energy used to drive the required glass-forming reaction, and the means by which compositional control is achieved.

a. Radial Flame Hydrolysis. The radial flame hydrolysis (RFH) technique, also known as outside vapor deposition (OVD), is based on high-temperature hydrolysis of the constituent halides by passing the vapor through a gas–O_2 or H_2–O_2 torch. [See Blankenship and Deneka (1982) for a recent review of this method.] It was the first method by which a fiber having the milestone attenuation of 20 dB/km was prepared (Kapron et al., 1970). The controlling chemical reaction is

$$SiCl_4(GeCl_4) + 2H_2O \longrightarrow SiO_2(GeO_2) + 4HCl. \tag{13}$$

In this process, a porous preform is formed around a target rod and is then sintered to full density.

The first step in the process is to carefully meter and mix the chemicals to be used. This is generally accomplished by passing carrier gases through the appropriate halides by using mass flow controllers and thermocouple detectors to control and monitor the amount of carrier gas and the ratio of halide to carrier gas. A typical delivery system, which can be used in all of the vapor-phase techniques, is shown in Fig. 9.

The vapor-phase reactants, which are mixed in the proportion required to provide the desired glass composition, are introduced into a deposition burner which generally uses fuel gas and oxygen. The burner is designed as a set of annular concentric orifices, with the center orifice being used for the halide reactants, the adjacent orifice for a shielding gas

TABLE III

COMMON INDEX-MODIFYING DOPANTS USED IN PREFORMS MADE BY VAPOR-PHASE TECHNIQUES

Dopant	Source	Effect on SiO_2 refractive index
Ge	$GeCl_4(l)$	Increase
P	$POCl_3(l)$	Increase
B	$BCl_3(g)$, $BBr_3(l)$	Decrease
F	$SiF_4(g)$, $CCl_2F_2(g)$	Decrease

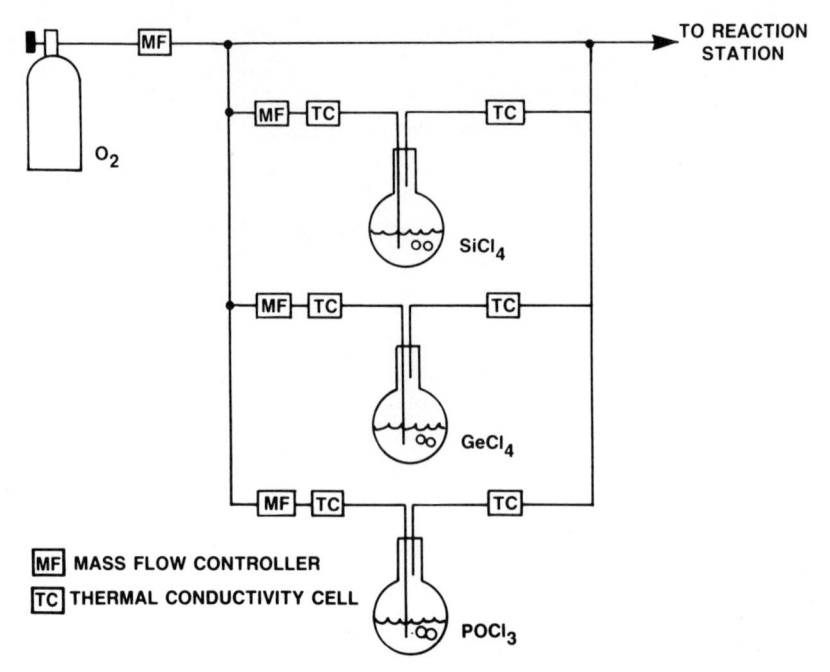

FIG. 9. Reactant delivery system for vapor-phase techniques.

which prevents premature reaction, and the outer orifice for the gas–O_2 mixture (Blankenship and Deneka, 1982). In this configuration, the halides react with H_2O, which is the combustion product of the flame, at the high flame temperatures producing small (~0.1-μm) spheres of doped silica glass having a composition which depends on the initial ratio of halides. These glass spheres, collectively called soot, are directed radially toward a target rod which is generally made of Al_2O_3 or graphite. During deposition, the target rod is both rotated and translated with respect to the burner as shown in Fig. 10a. The rotation ensures that the material is deposited in a radially uniform manner, while translation provides an axially uniform deposit. More important, the use of translational motion, through which the deposit is built up by layers, allows a preform for a graded-index fiber to be fabricated by changing the composition layer by layer in a manner prescribed in order to maximize bandwidth. Conventionally this is done by decreasing the GeO_2 concentration so that the refractive index decreases in a precise manner from the inner to the outer layers since, in this approach, core layers are deposited first, followed by layers corresponding to the cladding. About 1000 layers can be deposited by means of the RFH technique, so precise duplication of the desired refractive index profile is possible.

9. OPTICAL FIBER WAVEGUIDES

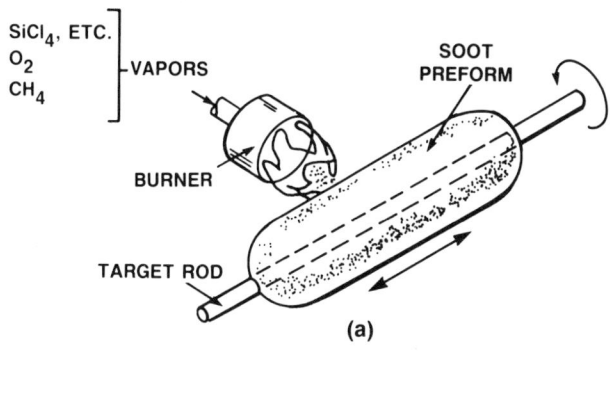

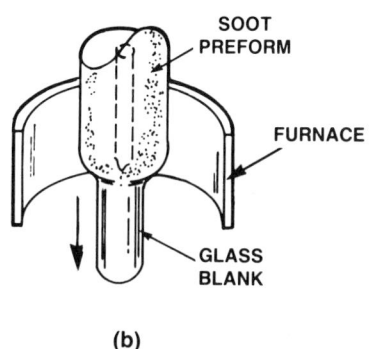

FIG. 10. Preform fabrication by the radial flame hydrolysis technique. [From M. G. Blankenship and C. W. Deneka, The outside vapor deposition method of fabricating optical waveguide fibers, *IEEE Journal of Quantum Electronics*. Copyright © 1982 IEEE.]

The overall efficiency of the deposition process is roughly 50%, with rates of deposition corresponding to as much as 2 g/min for core layers and 6 g/min for the cladding. In general, 200 core layers are used for graded-index fibers, although as many as 1000 have been reported (Blankenship *et al.*, 1982). The largest preform reported using this method has been 1.8 kg, corresponding to almost 70 km of 125-μm-diam fiber (Blankenship *et al.*, 1982). However, preforms made in production yield between 10 and 15 km of fiber.

Once deposition is complete the target rod is removed carefully in order to avoid disturbing the inner core layers. Then the porous preform, which is roughly 20% dense (Schultz, 1979a), is sintered to 100% density as shown in Fig. 10b in a manner analogous to zone refining at temperatures between 1400 and 1700°C (Rigterink, 1976). Sintering appears to be by a viscous flow mechanism (Yan *et al.*, 1980) and is often performed in an atmosphere of He since this gas, owing to its high diffusivity, promotes easy densification. This procedure, which can also eliminate the central hole left by the target rod, yields a solid glass preform tube or rod that is ready for drawing.

Because of the nature of the RFH technique, two problems must be solved in order to fabricate the highest-quality optical fiber waveguides. The first is due to the fact that the hydrolysis reaction produces a soot with hydroxyl ion impurity concentrations between 50 and 200 ppm (Schultz, 1979b). Although this impurity level might be tolerated in fibers to be used at short wavelengths, long-wavelength operation requires a significantly lower hydroxyl concentration in order to reduce attenuation. Reduced hydroxyl concentration has been obtained in practice by performing the sintering step in a controlled He–Cl$_2$ atmosphere. In this case, the chlorine can dehydrate the porous preform before consolidation by means of the reaction

$$2(\equiv\text{Si}-\text{OH}) + \text{Cl}_2 \longrightarrow (\equiv\text{Si}-\text{O}-\text{Si}\equiv) + 2\text{HCl} + \tfrac{1}{2}\text{O}_2. \qquad (14)$$

In this equation ($\equiv$Si—OH) is a siloxyl group bound to the network by three bridging oxygen ions. Hydroxyl concentrations of less than 50 ppb have been obtained with this approach in the laboratory (Blankenship and Deneka, 1982).

The second problem encountered in the RFH process is the tendency of preforms to fracture as a result of the thermal stress caused by the different thermomechanical properties of the deposited layers. This is a particular problem in this process because of the likelihood of flaws on the inner surface caused by the removal of the target rod. A consequence of this effect is a limitation in numerical aperture dictated by the limited compositional difference that can be tolerated between the core and the cladding. In this case, theoretical analyses (Scherer, 1979a,b) led to an effective solution of the problem based on compositional changes which helped to balance the stresses in the preform (Schultz, 1979b) and on the development of techniques to eliminate the central hold during sintering.

b. Internal Thermal Oxidation. The internal thermal oxidation (ITO) process, which is more generally known as modified chemical vapor deposition (MCVD) or inside vapor deposition (IVD), is based on the oxidation of the halide reactants within a confined substrate tube. [See Nagel, *et al.* (1982) and Partus and Saifi (1980) for general reviews of this process.] In this case, the controlling chemical reaction is

$$\text{SiCl}_4(\text{GeCl}_4) + \text{O}_2 \longrightarrow \text{SiO}_2(\text{GeO}_2) + 2\text{Cl}_2. \qquad (15)$$

The ITO technique (MacChesney *et al.*, 1974) derives from the chemical vapor deposition (CVD) process. However, in order to increase deposition rates, reactant concentrations and temperatures are increased such that the reaction occurs in the gas phase (homogeneously) instead of on the inside tube wall (heterogeneously) as in true CVD.

In ITO, reactants are metered and mixed by using techniques similar to those described for the RFH process. However, in this case, the vapors are introduced into a fused-quartz tube. This tube, which is typically 19 mm i.d. × 25 mm o.d., is held in a glass-working lathe with precisely aligned chucks (Fig. 11). Since the tube is rotated during processing, the reactants must be introduced by means of a gas-tight rotating coupler. The reaction is effected by means of an H_2–O_2 flame, which generally operates between 1300 and 1600°C and traverses the length of the tube on a fire carriage. The reaction product or soot deposits on the tube walls, downstream from the flame; in this way, in its traversal the flame also sinters the soot into a dense, glassy layer (Fig. 12). Therefore, in the ITO process, glass for the preform is deposited layer by layer in a manner similar to RFH, but, in contrast to that process, the layers are sintered individually rather than after all deposition is complete. Moreover, because of the geometry of the process, cladding layers are deposited first and then the core layers. Once deposition is complete, generally with 50–100 layers, the temperature of the flame is increased so that the tube collapses symmetrically to a preform rod.

The chemical processes that take place in the reaction zone are quite complex and have not been fully defined. Most results indicate that oxidation of $SiCl_4$ will go to completion at the temperatures required to sinter the layer (Powers, 1978; French *et al.*, 1978), with an intermediate oxychloride phase being formed at about 1000°C (Wood *et al.*, 1978). How-

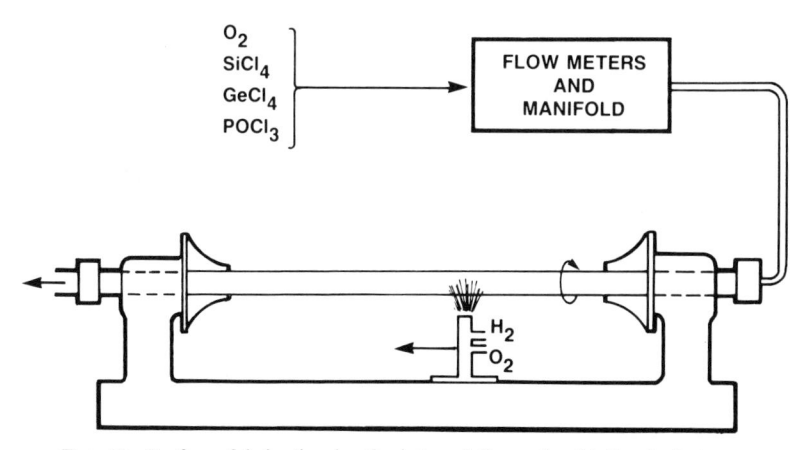

FIG. 11. Preform fabrication by the internal thermal oxidation technique.

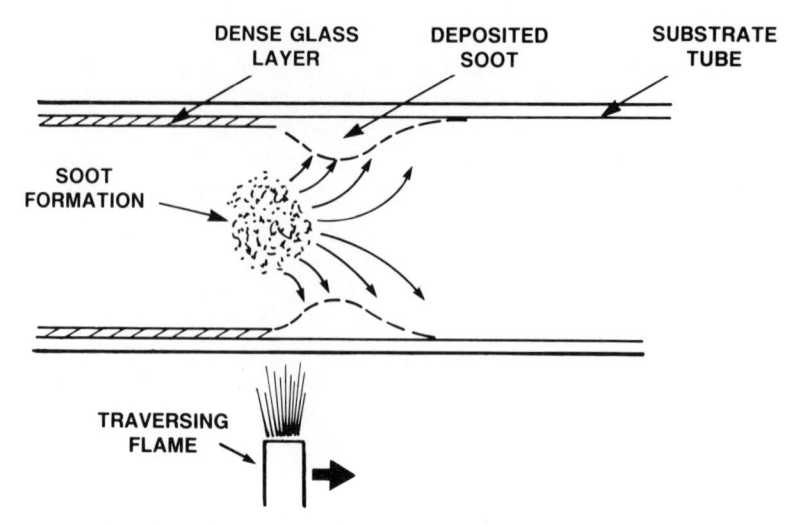

FIG. 12. Soot formation and deposition in the internal thermal oxidation technique.

ever, the situation for GeCl$_4$ is not so clear. For example, it has been found (Kleinert et al., 1980) that oxidation of this species is such a sensitive function of the partial pressure of oxygen and chlorine that, in the presence of SiCl$_4$, conversion to GeO$_2$ can be less than 50%. Moreover, formation of GeO$_2$ is found to occur only in a limited temperature regime (1100–1800°C) because of the high Cl$_2$ pressures generated by conversion of SiCl$_4$ at high temperatures (Wood et al., 1981). Because of these factors, the flame temperature and reactant concentrations must be very well controlled in the ITO process in order to control the composition of the deposited glassy layers.

In contrast to the RFH process, the chemistry of the ITO process assists in the removal of hydroxyl ion impurities. In the case of ITO, the controlling reaction generates copious amounts of Cl$_2$, which, as shown earlier, is an effective drying agent for the deposited soot (Wood and Shirk, 1981). For this reason, up to 4 ppm of H$_2$O in the reactant O$_2$ or 8 ppm of SiHCl$_3$ in the SiCl$_4$ causes only 1 ppb of hydroxyl ion impurities in the fiber. Moreover, purification techniques have been developed which can maintain the hydrogen-containing impurity concentrations in typical reactants at or below these levels (Barns et al., 1980).

Once the soot is formed in the gas phase, it is deposited onto the inner tube wall by a mechanism called thermophoresis (Simpkins et al., 1979; Walker et al., 1979). In this mechanism, particles which are suspended in a temperature gradient tend to travel down that gradient owing to differences in the momenta of impacting gas particles. Mathematical

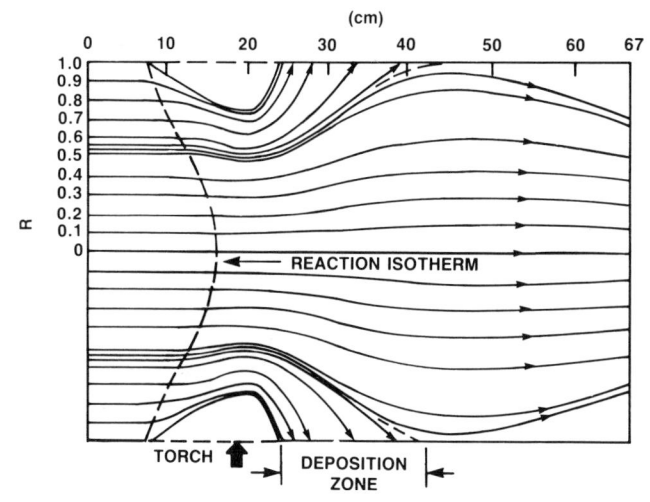

FIG. 13. Typical particle trajectories in the internal thermal oxidation process. [From S. R. Nagel, J. B. MacChesney, and K. L. Walker, An overview of the modified chemical vapor deposition (MCVD) process and performance, *IEEE Journal of Quantum Electronics.* Copyright © 1982 IEEE.]

analysis (Walker *et al.*, 1979; Weinberg, 1982) shows that the deposition efficiency ε is given as

$$\varepsilon \simeq 0.8[1 - (T_e/T_r)], \qquad (16)$$

where T_e and T_r (in K) are the temperature at which the gas and wall equilibrate downstream and the reaction temperature, respectively. According to this model, deposition efficiency is improved by minimizing this temperature ratio. In practice, this can be accomplished by downstream cooling.

Further analysis of thermophoresis, using actual temperature distributions found in the ITO process, shows the particle trajectories to be complex as illustrated in Fig. 13 (Walker *et al.*, 1980b). Moreover, these analyses indicate that there are two operating regimes in the ITO process. On the one hand, at nominal reactant flow rates and for normal size tubes, the chemical reaction goes to completion and the process will be limited by the transport of particles to the tube wall (i.e., a deposition-limited regime) as described in Eq. (16). However, as reactant flow rates and/or tube diameter increase, it is possible that the gases in the center of the tube do not reach the temperature required for their reaction; in those cases in which particle formation is limited by incomplete reaction, but all particles that do form are deposited, the process is said to be in a reaction-limited regime. For optimal process control, it is preferable to operate in the deposition-limited regime.

Once the particles are deposited on the tube wall, they will be sintered to full density by passage of the flame on the fire carriage. As in the case of RFH, the sintering is by viscous flow and care must be taken to avoid bubbles, which can arise from either incomplete consolidation or excessive temperatures (Walker et al., 1980a; Yan et al., 1980).

Once deposition is complete, the tube is collapsed to a solid rod by increasing the temperature of the traversing torch to about 1900°C while continuing to rotate the tube. The driving forces for collapse are the surface tension of the glass and the difference in pressure between the outside of the tube, resulting from the gases emanating from the torch, and the inside of the tube.

The stability of the collapse process plays an important role in determining the geometrical perfection of the core and outer diameter of the preform produced. In order to control the collapse, it is first necessary to have started with substrate tubes having minimal variations in wall thickness and good circularity (Nagel et al., 1982). Another important factor in obtaining preforms with circular symmetry is to maintain a small, positive pressure between the inside of the tube and the atmosphere during collapse (Tasker et al., 1978). This can be done by flame-sealing the downstream end of the preform prior to collapse and using a static gas pressure or by using a flowing carrier gas in conjunction with a pressure-controlling valve at the downstream end during collapse (Modone et al., 1982).

One problem which can arise during collapse is the selective volatilization of GeO_2 from the innermost core layers. This effect, which causes a dip in refractive index at the core center, can lead to a substantial degradation in the bandwidth of graded-index fibers (Kitayama et al., 1982). Although some volatilization occurs during sintering in the RFH process, the problem is more severe during the collapse phase of the ITO process because of the higher temperatures used in the latter. One way to minimize the problem is to flow a small amount of $GeCl_4$ along with the carrier gas during collapse in order to compensate for losses due to volatilization (Akamatsu et al., 1977). However, great care must be taken in using this procedure since overcompensation, which yields an index rise at the center corresponding to a "core within a core," can be more damaging to bandwidth than no compensation.

Although the chemical processes used in the ITO process assist in removal of hydroxyl ions, additional steps must be taken to reduce this impurity to the lowest possible level. For example, since no chlorine is present during collapse, hydrogen-containing impurities in the gases used to maintain the internal pressure in the tube can be absorbed by the deposited glass (Pearson, 1980). In this regard, it has been shown that the use of a small concentration of Cl_2 in this gas significantly reduces impurity absorption by hydroxyl ions (Nagel et al., 1982). As mentioned previously, another method of reducing hydroxyl impurities is to remove hydrogen-bearing impurities from both the oxygen and halide reactants (Barns et al., 1980). In addition, since substrate tubes can contain 100–200 ppm water, measures must be taken to avoid diffusion into the core during fabrication (Yoshida et al., 1981). Typical precautions include the deposition of a thick cladding layer (up to 1.5 mm in the preform) and limiting the temperatures used for deposition and collapse. Finally, more esoteric techniques based on the use of isotopic exchange of OD for OH in both the substrate tube and the preform have been used to minimize optical absorption by hydroxyl ions (Modone and Roba, 1981; Stone and Lemaire, 1982); such methods make use of the fact that deuterium substitution shifts the vibrational absorption to longer wavelengths, away from the transmission window.

Using the ITO process, preforms yielding up to 40 km of fiber have been reported (Nagel et al., 1982), although this certainly is not typical. Deposition rates in the standard process, typically 0.35–0.5 g/min, are less than those in the RFH process. In comparing the two, however, it must be recognized that a significant amount of the cladding is obtained from the substrate tube using ITO, whereas all of the cladding is generally deposited using RFH. Some effort has gone into increasing the deposition rate of the ITO process, specifically by increasing deposition efficiency. The most successful technique has been a plasma-assisted ITO process (Fleming and O'Connor, 1981) in which an atmospheric O_2 rf plasma centered in the substrate tube is used for deposition. The temperature of the plasma is more than adequate to achieve complete homogeneous reaction; more importantly, by making the substrate tube sufficiently large (46 mm i.d. × 50 mm o.d.), the fireball can be spaced roughly 1 cm from the tube wall, which provides a tremendous thermophoretic driving force for deposition. By using this technique, which is particularly suitable for the light doping levels and minimal number of compositional changes required for single-mode fibers, 3.5 g/min of material can be deposited (Fleming and Raju, 1981).

c. Axial Flame Hydrolysis. The axial flame hydrolysis (AFH) technique, which is also known as vapor-phase axial deposition (VAD), is similar to the RFH technique in that it also involves high-temperature hydrolysis in a torch. However, in contrast with that technique, in the AFH method the soot is deposited along the axis of the preform instead of along its radius. Because of this, the method of forming the desired refractive index gradient is quite different. [See Inada (1982) for a recent review of this process.]

Preparing a preform using the AFH method involves four steps: porous preform fabrication, dehydration and sintering, preform elongation, and jacketing. As shown in Fig. 14, the first two steps can be performed in tandem.

In the AFH process, the porous preform is made by depositing soot from a hydrogen–oxygen torch onto the end of a seed rod which is rotated and pulled upward so that the growth interface remains at a constant position relative to the torch (Fig. 14). In this way, a porous preform about 5 cm in diameter is grown in an axial direction and the process can be continuous. A graded-index preform can be formed by using the AFH technique because of the fact that, for a given ratio of reactant halides, the concentration of GeO_2 in the deposited particles of soot is a strong function of the temperature of the substrate (Kawachi *et al.*, 1980). Thus, by controlling the temperature distribution across the end face of the growing preform, it is possible to control the radial variation in composition and,

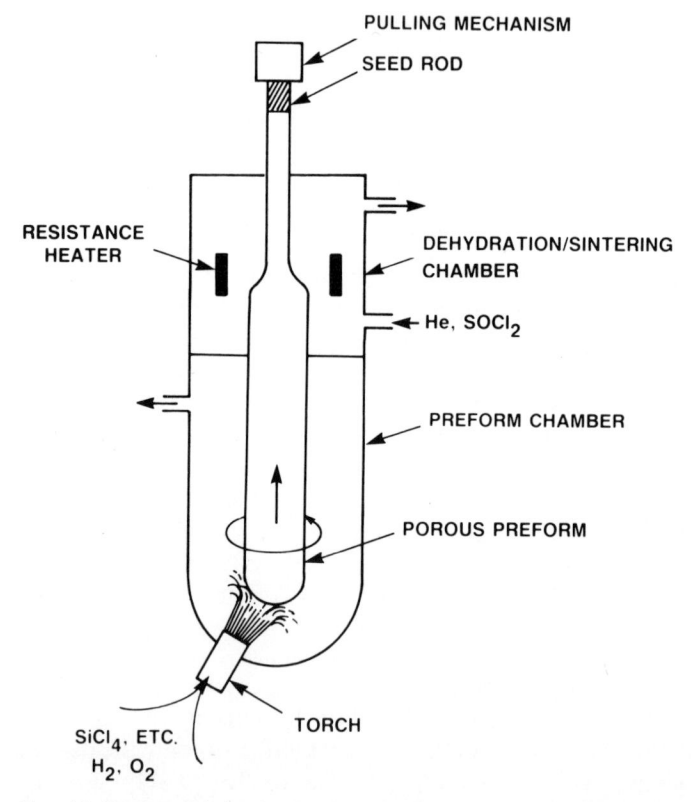

FIG. 14. Preform fabrication by the axial flame hydrolysis technique.

thereby, the refractive index profile. In order to do this, it is necessary to control the H_2 and O_2 flow rates, the position of the end face with respect to the burner, and the diameter and shape of the growing porous preform (Imoto and Sumi, 1981). By using laser monitors to control the position and diameter of the porous preform (Watkins, 1982), infrared pyrometers to monitor the temperature distribution on the end face (Chida et al., 1981), and various means to decrease fluctuations in the burner flame and pressure in the preform chamber (Imoto and Sumi, 1981), it is possible to make graded-index fibers with bandwidths of up to 6.5 GHz km (Nakahara et al., 1980). One reason for these high bandwidths is that the process gives no refractive index depression at the core center as there is in preforms made by the ITO and, to a lesser extent, the RFH processes. Further, the profile is not approximated by discrete layers as it is in those techniques.

The second step in the AFH technique is a combined dehydration–sintering operation to form a fully dense, transparent rod. In contrast with the RFH technique, axial growth of the preform using the AFH method makes it possible to perform this operation in tandem with the porous preform fabrication step, as illustrated in Fig. 14. Both dehydration and sintering are performed at temperatures between 1200 and 1500°C, as in the RFH process. Dehydration is performed at 1200°C using either $SOCl_2$ (Sudo et al., 1978) or Cl_2 (Moriyama et al., 1980) in the He gas which assists densification. By using these dehydration aids, and by avoiding contamination in subsequent processing steps, it is possible to reduce the hydroxyl ion concentration to less than 1 ppb (Chida et al., 1982).

After densification, the preform is about 2.5 cm in diameter and 30 cm long (Inada, 1982). The next step in the process is to elongate the preform such that its diameter is reduced to about 1 cm. This is generally performed on a glass-working lathe by using a traversing hydrogen–oxygen torch while applying tension between the head stock and tail stock. Remote diameter measurement equipment can be used to provide feedback control over this process.

The final step of this process is to jacket the preform rod by collapsing a thick-walled silica tube onto the elongated preform so that the diameter is increased to about 2.5–3.0 cm. This procedure, which adds cladding to the preform in order to obtain the correct ratio of core to cladding, can also be performed on a glass-working lathe in a manner similar to the collapse process in the ITO technique.

The deposition efficiency of the AFH process is typically 60%, compared to 50% for the RFH and ITO methods. Deposition rates, around 0.5 g/min, are closer to those obtained in the ITO process than to the 1.5–2 g/min obtained with RFH. Although deposition rates of up to 2 g/min can be achieved by using AFH, this generally degrades the achievable bandwidth and deposition efficiency (Inada, 1982).

One advantage of the AFH technique is that, because the preform is grown axially in a semicontinuous manner and the jacketing process can increase preform diameter, exceptionally large preforms can be made. For example, preforms yielding up to 220 (Chida *et al.*, 1982a) and 300 km (Kyoto *et al.*, 1983) of multimode fiber have been reported. However, the use of jacketing is not necessary and sometimes has been specifically avoided (Kawachi *et al.*, 1982). The reason for this is that the silica tube used in this step often contains inclusions which can limit the strength of the fiber. Thus, in applications such as submarine cables where high strength is required, preforms synthesized entirely by soot deposition may be preferred.

d. Internal Plasma Oxidation. The internal plasma oxidation (IPO) technique, also called plasma chemical vapor deposition (PCVD), is the only vapor-phase technique currently used for preform fabrication in which the glass is produced via a heterogeneous reaction; i.e., it is deposited directly on the substrate wall without being created in the gas phase. Although it bears some similarity to the ITO process, using the same reactants and delivery system, the driving force for the reaction is a nonisothermal plasma produced by a microwave cavity, rather than the elevated temperatures produced by a hydrogen–oxygen flame in the ITO method. In the IPO technique (Geittner *et al.*, 1976; Kuppers *et al.*, 1976), deposition also occurs inside a fused-silica substrate tube (Fig. 15). In contrast to the ITO technique, oxidation of the halide reactants is effected by a nonisothermal plasma produced by a microwave cavity (2.45 GHz) that travels down the length of the tube. (A nonisothermal plasma is one in which the electron temperature is much greater than the ion temperature.) The plasma, which occurs when the internal tube pressure is maintained at subambient conditions (1–100 torr), creates reactive species in the gas phase which deposit directly as a glassy layer on the tube wall. Thus, since no soot is formed, no sintering is necessary. As in the ITO process, the deposit is built up in the prescribed fashion, layer by layer, by changing the reactant composition.

Although the plasma deposition reaction will occur at room temperature, in actual practice a stationary furnace is used to maintain the entire substrate tube at a temperature between 800 and 1200°C (Kuppers *et al.*, 1976). The reason for this is that at lower temperatures substantial levels of chlorine are incorporated in the deposited film, which tends to cause the deposited glass to crack and peel off the substrate (Geittner *et al.*, 1976). However, by maintaining the tube temperature at roughly 1000°C,

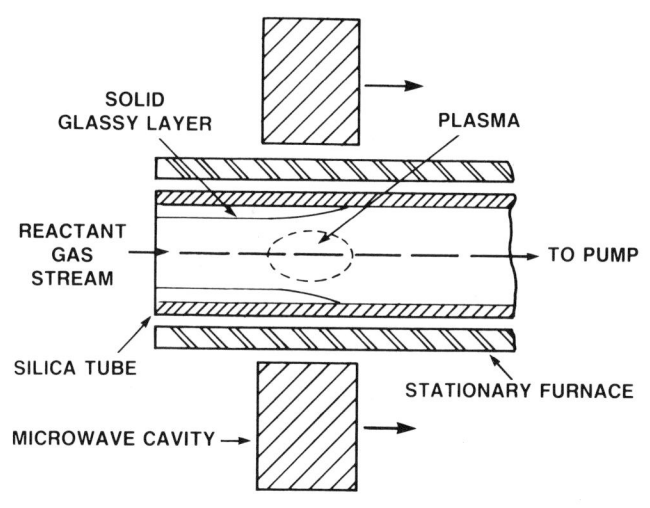

FIG. 15. Preform fabrication by the internal plasma oxidation technique. [From Beales and Day (1980).]

the chlorine content is reduced to 0.1% and the film can maintain its integrity. In any case, the temperature is sufficiently low that the tube need not be rotated during deposition.

One peculiarity of the IPO process is that, owing to differences in reaction kinetics and diffusion rates of the reactive species to the tube wall, different dopants can be deposited at different average distances downstream from the microwave cavity (van Ass *et al.*, 1976b). For the practical case of a moving microwave cavity, this difference implies that the composition of the individual layers will vary radially. For example, since the deposition zone for GeO_2 appears to be more elongated than that for SiO_2, each layer deposited will vary from SiO_2-rich near the top to GeO_2-rich near the bottom. In order to minimize this effect, which could easily hamper the achievement of a high bandwidth, the microwave cavity is traversed at high speeds (about 8 cm/sec) and deposition occurs while the cavity is moving both with and against the gas flow. In this way, up to 2000 layers can be deposited in a reasonable time, each thin enough that the diffusion which occurs during subsequent processing steps can smooth out the profile.

It should be noted that the IPO process is fundamentally different from the plasma-assisted ITO process described in Section III.A.1.b. In that technique an isothermal plasma produced by an rf resonator within a tube at ambient pressure was used to create soot in the gas phase and drive it toward the cool tube wall; in the IPO technique, reaction occurs heterogeneously directly on the tube wall.

Once deposition is complete, the tube is collapsed by methods identical to the ITO technique. Thus, along with that process, the IPO technique suffers from GeO_2 volatilization at the core center, leading to an index depression.

Good process control has been demonstrated with the IPO method (Peelen *et al.*, 1978). However, the spectral attenuation curves obtained from fibers made with this method show substantial levels of hydroxyl ion impurities (about 20 ppm). It is not clear if this is an incidental effect or if it is due to the fact that the chlorine produced in the reaction is a less efficient scavenger of water under the conditions of the IPO process than it is under those of the ITO process. Recent evidence indicates that the raw materials used must be free of all hydrogen-bearing species in order to obtain fibers with low levels of hydroxyl ion contamination (Koel, 1983).

Because of its nature, deposition efficiency in the IPO process is almost 100%, although the deposition rate is generally only 0.3–0.5 g/min. Increasing this rate has proved to be difficult owing to the tendency to obtain porous deposits when the partial pressure of the reactants is increased (Gossink, 1977). However, recently rates of greater than 1 g/min have been reported for the IPO method (Peelen and Koenings, 1981).

2. *Fiber Drawing*

The fabrication of a preform is only one step in making an optical fiber waveguide. In this section the other step, drawing of the preform into a fiber that can be tested and cabled, is treated. The purpose of this procedure is to reduce the diameter of the preform, generally 10–20 mm, to the required fiber diameter, generally 125 μm, without altering the prescribed refractive index profile. In addition to diameter control, which is a direct function of the conditions used in this process, fiber drawing can also affect strength and attenuation.

The apparatus used for fiber drawing, called a draw tower, is shown schematically in Fig. 16. At the top of the structure, the preform is fed at a set rate into a source of heat which is generally at a temperature between 1900 and 2100°C. The fiber is then drawn downward from the viscous molten glass which forms at the end of the preform. The rate of drawing, which can vary between 0.5 and 10 m/sec, is such that the masses of glass entering and leaving the heat source are equal. This condition assists in controlling fiber diameter, although, as shown later, periodic disturbances can occur. Drawing was initially performed by winding the fiber on a takeup drum. However, now it is more common to use a capstan or pinch wheels, and spool the fiber separately. While drawing, the fiber diameter is measured by using a noncontact device, and an electrical signal propor-

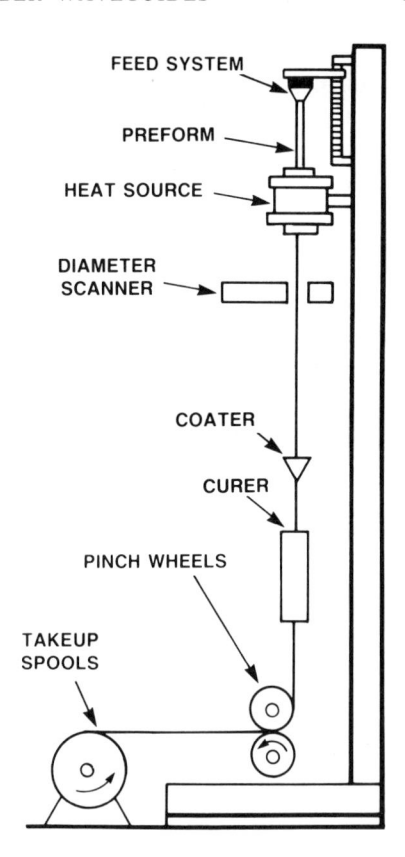

FIG. 16. Optical fiber draw tower.

tional to the difference between the measured and set diameter is used to provide a feedback signal to the drawing mechanism in order to control fiber diameter. Finally, in order to preserve the high pristine strength of the freshly drawn fiber, it is coated on-line. Polymer coatings, which are generally used, must be fully cured before the fiber reaches the drawing mechanism since direct contact with the fiber can induce strength-degrading flaws.

In the following subsections, three subsystems of the draw tower, i.e., the heat source, diameter control, and fiber coating, will be described in detail. In these sections, particular attention will be paid to how these features can affect and be used to control fiber diameter, strength, and attenuation.

a. Heat Sources. The obvious purpose of the heat source is to raise the preform to the temperature necessary to draw the fiber. Generally, viscosities between 10^3 and 10^5 P are needed. For the high-silica glasses characteristic of preforms prepared by vapor-phase techniques, this corresponds to temperatures around 2000°C. The particular temperature used is determined by two competing effects. On the one hand, it has been found that high temperatures and low draw tensions (<5 g) are critical to obtaining high fiber strengths (DiMarcello and Hart, 1978; DiMarcello et al., 1979); this is presumably due to the healing of defects on the surface of the preform rod by viscous flow, pyrolysis, or dissolution. On the other hand, high draw temperatures have been found to be responsible for increases in attenuation in certain preform compositions (Yoshida et al., 1977a, 1978). In this case, the high draw temperatures appear to cause defects in the glass, called drawing-induced defects, which have absorption peaks at short wavelengths (0.5–0.7 μm) with tails extending into the transmission windows of interest. For the common germania–phosphorus pentoxide–silica compositions, draw tensions between 15 and 30 g have been found to minimize loss (Yoshida et al., 1977b).

In addition to providing the required operating temperature, the heat source used in optical fiber drawing should be able to provide a radially symmetric and compact hot zone, a clean environment for the preform and fiber, and a low thermal mass in order to allow fast response times and easy temperature control. To meet these objectives, sources based on resistance heating, rf heating, CO_2 laser heating, and oxy-hydrogen torches have been devised.

The most popular resistance-heated furnace used to draw optical fiber is based on a graphite element (Payne and Gambling, 1976). Although the graphite element must be protected from the atmosphere by using inert-gas curtains, its low thermal mass can attain high temperatures while being compatible with SiO_x vapors that can be volatilized from the preform. Generally, these furnaces utilize a tubular element which is shaped to provide the desired hot zone. The furnace itself is designed to permit the preform to enter the top and the fiber to exit the bottom without direct contact. However, conditions around the element must be inert in order to avoid oxidation.

Detailed studies of graphite resistance furnaces (Nakahara et al., 1978) have shown the effect these heat sources can have on strength and fiber diameter control. To avoid fiber strength degradation, the element must not contaminate the fiber or preform with carbon particles (Versluis and Peelen, 1979). Also, control of fiber diameter, to be discussed in detail in the following subsection, is hampered by temperature fluctuations which can upset the long-term mass balance of the drawing process and yield instabilities. Rapid temperature fluctuations are a function of both element design and inert-gas flow in graphite resistance furnaces.

The second form of heat source used to draw optical fibers is the rf induction furnace, which utilizes zirconia susceptors (Runk, 1977). In such furnaces, yttria-stabilized zirconia rings, heated to temperatures above 1000°C by coupling the rf power to an inserted carbon rod, are sufficiently conductive to absorb the rf power directly. For a zirconia susceptor wall thickness of a few millimeters, rf power at 1–10 MHz can be used to heat the furnace to over 2000°C.

One advantage of the zirconia furnace is its long element life compared to graphite resistance furnaces. Moreover, high-strength fiber can be achieved (DiMarcello and Hart, 1978), although care must be taken to replace the susceptor rings before they degrade mechanically and contaminate the preform and fiber with particulates. Another advantage of this furnace is that, since it can operate in air and does not require a protective gas curtain, the turbulence that degrades fiber diameter control in the graphite resistance furnace can be reduced. However, owing to the high draw temperature, convective turbulence can still be a problem which must be controlled in order to reduce high-frequency diameter fluctuations (Smithgall and Myers, 1980).

The third type of heat source, based on heating with a CO_2 laser (Jaeger, 1976), has been demonstrated in the laboratory but has not yet proven to be suitable for production draw systems. In this approach, the 10.6-μm radiation emitted by the laser is strongly absorbed by the preform, causing it to reach the required draw temperature. A key problem in using laser heating is to distribute the radiant energy in a radially uniform manner around the neckdown region of the preform. Two methods were initially proposed (Jaeger, 1976): rotating the preform and using a stationary beam; and using a rotating lens to move the beam around the preform. However, both approaches have drawbacks, including limitations on preform size, the high laser powers needed when lenses are used, and the criticality of optics and preform alignment. This situation was improved by the introduction of galvanometer-based beam scanning (Oehrle, 1979). In this method, two galvanometer-driven mirrors are sinusoidally oscillated 90° out of phase to produce an annular beam which is reflected onto the preform by using a conical reflector. Using this approach, preforms greater than 8 mm in diameter can be drawn—but, while greater than that possible using rotating lenses, this is still much less than the diameters that can be drawn by using furnace heating. However, one advantage of this method is that, because of the ability to maintain exceptionally clean environments, fibers with very low flaw densities and therefore high strengths are possible (Schonhorn et al., 1976; Paek et al., 1980).

The final heat source that has been used to draw optical fiber is the oxy-hydrogen burner (Wang and Zupko, 1980). The key problem in using oxy-hydrogen flames for fiber drawing is to reduce the turbulence, which can adversely affect fiber diameter. This was accomplished by designing a surface mixing torch, i.e., one in which mixing of the H_2 and O_2 only occurs at the nozzle exits immediately before ignition. This technique has provided fibers with good diameter control and high strength, but, as in the case of laser heating, it has not been found suitable for production.

b. Diameter Control. As described in Subsection II.A.2, diameter variations can increase fiber attenuation because of waveguide losses. However, a more serious consequence of diameter variations is the difficulties they create in attempting to couple fibers. Since many splicing and connecting techniques are based on alignment using an external reference plane, diameter differences lead to a misalignment of the fiber cores. Even for those techniques which use self-centering, diameter variations will cause a proportionate variation in core diameter which will also produce a high coupling loss.

Several theoretical treatments of fiber drawing serve to illustrate how instabilities in the fiber drawing process can occur (Geyling, 1976). These instabilities, which arise because the preform is drawn to a fiber without lateral restraints as in drawing from a die, are tensile, leading to a continuous decrease in diameter until fracture; capillary, in which surface tension causes a tendency to "beading"; or drawing resonance (Geyling and Homsy, 1980), which causes periodic changes in diameter.

In practice, diameter variations in fibers are of two types, high frequency and low frequency (Nakahara *et al.*, 1978). High-frequency variations; i.e., those which occur over fiber lengths less than about 1 m, cannot be controlled by feedback since this distance is comparable to the distance between the furnace and the point at which the diameter is measured. As discussed earlier, turbulence in the drawing furnace is one source of high-frequency diameter fluctuations. In this case, furnace and element designs aimed at controlling gas movement are important in minimizing these variations. Bubbles in the preform rod, either introduced during vapor-phase processing or as remnants from the original substrate tube, can also cause high-frequency fluctuations in diameter. Here control depends on identifying and eliminating the source of the defects.

Low-frequency diameter variations can be caused by changes in draw temperature and variations in preform diameter (Nakahara *et al.*, 1978). Temperature changes of less than 10°C can cause a short-term 1-μm change in fiber diameter. In cases in which the feed and draw rates are constant, a change in preform diameter will induce a proportionally equal change in fiber diameter.

Diameter variations which occur over fiber lengths in excess of 1 m can in principle be controlled by using appropriate measuring devices and feedback loops on the draw speed. Although there are a variety of methods to monitor the diameter of fiber without direct contact as it is being drawn, the two most widely used are based on laser scanning and light scattering. The laser technique, available as a commercial device, is based on scanning the beam across a field perpendicular to the fiber and onto a detector. Intensity variations caused by the beam first intercepting and then leaving the fiber are used by a clock circuit to compute fiber diameter. The light scattering technique (Watkins, 1974) is based on the interference pattern produced when the fiber scatters light from a laser in the forward direction, i.e., in the direction of the beam. Since the diameter is proportional to the reciprocal of the period of the interference fringes, detection of the light as a function of scattering angle can be used to obtain fiber diameter (Smithgall, 1977). Once the diameter is measured, the offset signal can be used to control the draw speed so as to minimize low-frequency diameter variations. Designs of such circuitry have been discussed for both the laser scanning (Hoshikawa *et al.*, 1977) and forward scattering (Smithgall, 1979) techniques.

c. Fiber Coating. The coating applied to an optical fiber as it is drawn should serve two functions, each of which makes opposite demands on the coating characteristics. On the one hand, the coating must mechanically protect the fiber from abrasion damage which degrades fiber strength; i.e., the coating should be hard and tough. On the other hand, the coating must be able to shield the fiber from microbending losses; i.e., the coating should provide a compliant cushion for the fiber. Although single coatings can be used as a compromise, the best approach to this problem is a dual coating, i.e., a soft inner layer and a hard outer shell (Gloge, 1975). In addition to these requirements, the coating and means of application must satisfy constraints imposed by the fabrication techniques as well as practical considerations of use (Montierth, 1977; Schonhorn *et al.*, 1979; Miller, 1979). In the case of the former, for example, the coating should be able to be applied concentrically and cured at high draw speeds. Practical considerations imply that the coating should be strippable and should not degrade in time.

Organic coatings are the most common material used today to coat optical fibers. Three types have been used: UV-cured, thermally cured, and hot-melt thermoplastics. The UV-cured materials are generally epoxy acrylates or urethane acrylates (Schonhorn *et al.*, 1979; Paek *et al.*, 1980; Krause, 1980). A great advantage of this type of coating is that the resin, while having a low viscosity, has no solvent to remove during curing. This allows for easy application without bubble entrapment. Curing is performed by commercially available UV lamps in elliptical housings which allow the light to be focused on the fiber as it is being drawn, thereby providing short curing times. Using this approach, it is important to avoid forming defects in the glass from the UV radiation since such defects can be a significant source of attenuation (Blyler *et al.*, 1980; Stone, 1980; Levy, 1981). However, by using well-centered coatings with the correct amount of photoinitiator, these loss mechanisms can be avoided (Stone and Eichenbaum, 1980).

Silicone resins are the most common thermally cured coating material (Newns *et al.*, 1977; Tasker *et al.*, 1978; France *et al.*, 1979). The precursor resins have a low surface tension and are rapidly cured by infrared radiation, so they are also applicable to high-speed coating (Kimura *et al.*, 1980). However, they generally have a low modulus in the cured state. While this is desired for ensuring low microbending losses, such coatings do not have adequate abrasion resistance. Therefore they are usually used with an overcoat of extruded nylon and sometimes a harder inner coat in order to optimize their characteristics (Mochizuki *et al.*, 1979; Yaminishi *et al.*, 1979; Ishihara *et al.*, 1979; Yoshizawa *et al.*, 1981).

Hot-melt thermoplastic elastomers are much less commonly used in this application than the other two classes of coating materials. A hot-melt adhesive of ethylene–vinyl acetate (EVA) has been used to obtain fibers with a strength distribution equivalent to that obtained with silicone-coated fiber (Miller *et al.*, 1978). In this case, curing is performed by cooling the coated fiber sufficiently before it reaches the drawing mechanism.

In order to avoid fiber damage during the application of the coating, it is important that the coating materials be free of particulates and that contact with the bare fiber be avoided. Other requirements of the coating process are that it should provide bubble-free centered coatings of relatively constant diameter. Constant-diameter coatings are an advantage in handling the fiber during subsequent cabling operations. Coating concentricity and freedom from bubbles are important in order to avoid producing fibers containing sections that are only minimally coated and thereby prone to abrasion-related damage. In addition, noncentered coatings can lead to increased microbending losses (Miller, 1979).

Based on these requirements, die application has evolved as the technique of choice in coating optical fibers. Using this method it is possible to avoid fiber contact with the applicator. Moreover, the technique has been shown to be fundamentally more stable than other methods (Homsy and Geyling, 1977).

As shown in Fig. 17, die coating involves passing the fiber through a reservoir filled with coating resin and exiting the fiber through a shaped tip. The ability of this process to produce centered coatings without

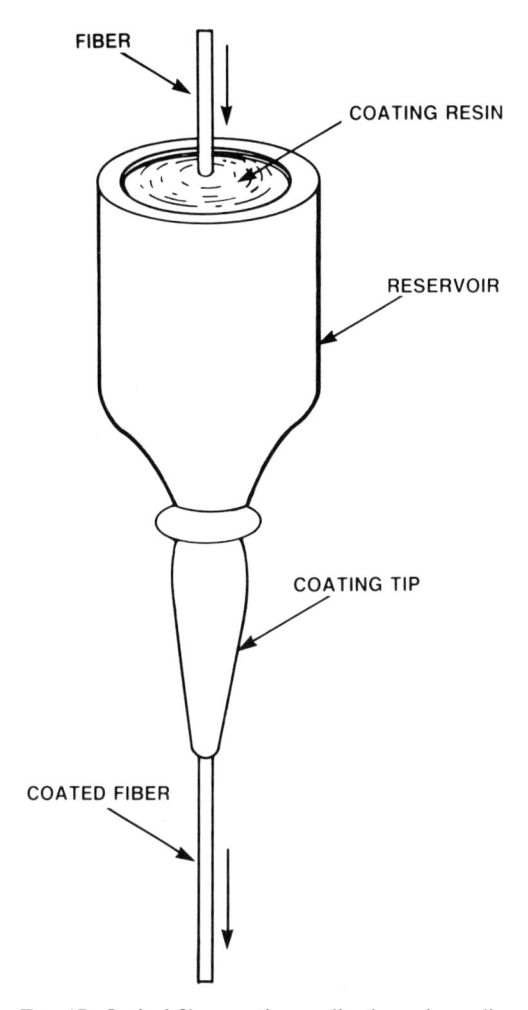

FIG. 17. Optical fiber coating application using a die.

abrading the fiber is based on the fact that, with a properly shaped conical tip, hydrodynamic forces created by the moving fiber and the converging flow field in the tip can act to center the fiber (Torza, 1976; Paek and Schroeder, 1979a; Eichenbaum, 1980a). Both flexible tips and hard tips have been used. In the case of flexible tips (Hart and Albarino, 1977), the centering forces act on the tip itself. The hard-tip applicators (Newns *et al.*, 1977; France *et al.*, 1979) have been based on split, tapered crucible applicators in which the taper angle, about 2°, is chosen to maximize the centralizing forces that act, in this case, on the fiber.

The achievement of centered coatings applied without fiber damage is greatly assisted by the use of devices which allow the concentricity of the coating to be monitored as it is being applied. These devices are based on either backward (Marcuse and Presby, 1977) or forward (Eichenbaum, 1980b) scattering patterns produced when the coated fiber is illuminated transversely. In the case of forward scattering, it is found that the scattering pattern produced is either symmetric or asymmetric about the fiber axis depending on whether or not the coating is centered in that direction (Fig. 18). Thus, by using two orthogonal beams, it is possible to evaluate the concentricity of the coating on-line and, in fact, use appropriate detectors, signal, and feedback circuits to obtain concentricity control (Smithgall and Frazee, 1981).

At the present time, the coating and curing of optical fibers are key factors which limit the speeds at which fibers can be drawn (Paek and Schroeder, 1981). There are several reasons for this. Obviously, curing of

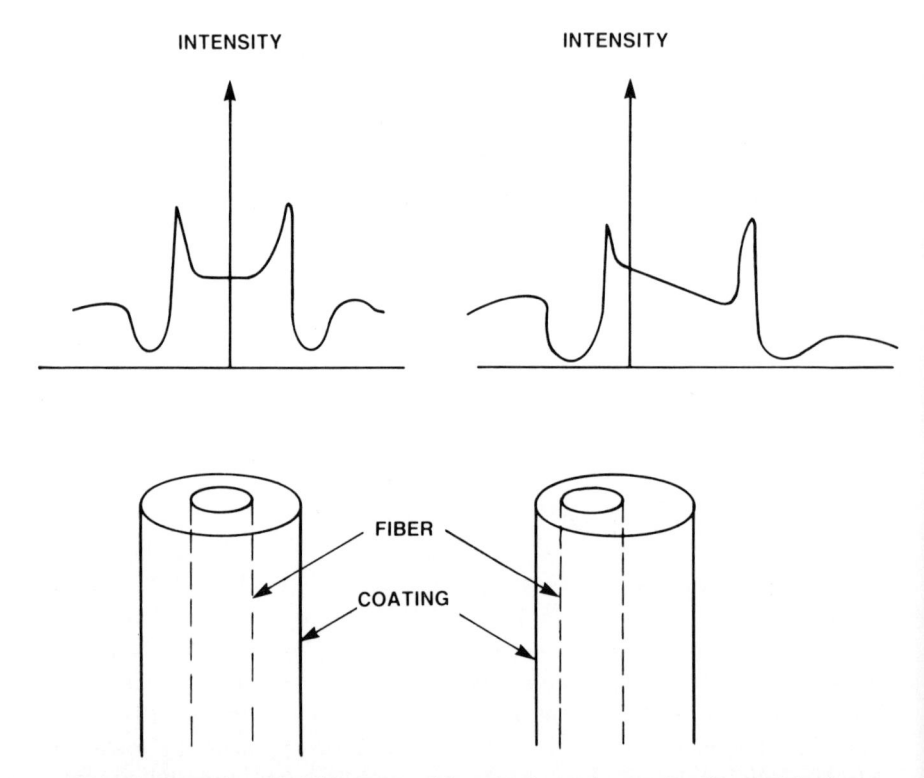

FIG. 18. Detection of coating concentricity using forward scattering. [From Eichenbaum (1980b). Reprinted with permission from *The Bell System Technical Journal,* Copyright 1980, AT&T.]

the coating and maintaining concentricity are more difficult as drawing speeds increase. In addition, in order to avoid degradation of the coating resin, it is necessary that the fiber temperature be below about 300°C before it contacts the resin (Paek and Schroeder, 1979b). Thus as draw speeds increase, either the distance between the furnace and coating reservoir must be increased or special means must be used to cool the fiber. The increasing shear stress at the fiber–resin interface is another limitation on draw speed. In this case, it has been found that this shear stress can increase draw tension to almost 100 g at a 4 m/sec draw rate from the 30-g optimum tension at 1 m/sec (Paek and Schroeder, 1981). Another adverse effect of this increased shear stress is that it can cause an unacceptable change in resin viscosity (Wagatsuma et al., 1982). Finally, as draw speeds increase the depth of the meniscus that forms at the point where the fiber enters the reservoir can be increased to a point where little effective coating occurs. One approach to this problem is to use a pressurized coating die (Chida et al., 1982b) which delivers coating to the fiber under an externally applied pressure so as to control the miniscus.

This concludes the discussion of vapor-phase techniques used for processing optical fibers. Although these methods were the first to produce low-loss fibers, and are still the techniques of choice for high-quality fiber waveguides, other methods based on bulk glass fabrication are still being explored because of their potential for reducing production costs. These techniques are the subject of the next subsection.

B. Non-Vapor-Phase Techniques

Although vapor-phase techniques have been developed which are very successful in preparing high quality optical fibers, these processes have several disadvantages. For example, they use expensive raw materials, they are not suited to batch-type processing, and they are restricted in the degree of flexibility in composition that can be obtained. The compositional restrictions arise from the need to have reactants at an elevated vapor pressure at reasonable temperatures as well as from the need to have the fabricated glasses thermally compatible with fused silica. The ability to utilize the broad spectrum of chemistries that is characteristic of conventional glasses would be an advantage in optical fibers for two reasons (Rigterink, 1976): compositions with low glass transition temperatures could be used in order to reduce intrinsic scattering by density fluctuations (Tynes et al., 1979), and fibers with high numerical aperatures could be prepared since compositional adjustments could be made to produce large differences in refractive index between the core and cladding without generating excessive thermal stresses because of concomitant differences in expansion coefficients and setting temperatures.

Because of these potential advantages, a significant amount of research has gone into alternative methods of fabricating optical fiber waveguides. Three of these will be discussed: conventional glassmaking, phase-separated glasses, and sol–gel techniques. However, before describing the techniques themselves, methods used to select an appropriate glass composition will be described. After fabrication of the glasses is reviewed, fiber drawing will be discussed.

1. Glass Fabrication

a. Choice of Glass Composition. Several criteria have been proposed to guide the choice of a candidate glass system for optical fiber waveguides (Maurer, 1973):

(1) The glass should have a low melting temperature to minimize contamination from the furnace during melting.
(2) Volatile constituents should be avoided to minimize unintentional composition variations.
(3) Glasses with low refractive indices are preferred in order to reduce material dispersion and, thereby, the optimum operating wavelength.
(4) The core and cladding glasses should provide an optimum difference in refractive index, yet be thermally and chemically compatible in order to avoid high stresses, reboiling, and precipitation at the core–cladding interface.
(5) The components of the core glass should have comparable molar refractivities in order to minimize scattering by compositional fluctuations.
(6) The glass system should have at least one mobile component which controls refractive index so that graded-index profiles can be formed during fiber drawing.

Based on these criteria, several glass systems have been selected for extensive investigation including soda–lime silicates (Imagawa and Ogino, 1977), sodium borosilicates (Koizumi *et al.*, 1974; Beales *et al.*, 1977), alkali germanosilicates (van Ass *et al.*, 1976a) and alkali–alkaline earth phosphates (Spierings *et al.*, 1981). By far the most widely studied of these has been the sodium borosilicate system. This is because of its low firing temperature (Beales *et al.*, 1974), its ability to be melted in fused-silica crucibles (Aulich *et al.*, 1978), and its low inherent scattering loss (Tynes *et al.*, 1979).

b. Direct Melting. Clearly, the most straightforward way to prepare glasses for optical fiber waveguides is to melt them directly. However, extreme care must be used throughout the process in order to obtain homogeneous glasses with very low impurity levels.

Obtaining appropriate glasses starts with using high purity raw materials. Although some suppliers can provide oxides and carbonates with low levels of transition-metal impurities, separate purification may be required. In that case the particular method used, including solvent extraction, ion exchange, mercury cathode electrolysis, or low-temperature sublimation, depends on the raw material to be purified (Mitchell, 1982).

After the appropriate raw materials are obtained, they must be mixed and transferred to the furnace. In order to avoid contamination during these processes, these steps should be performed in a high-quality clean room (Pearson, 1974; Takahashi et al., 1974; Gossink, 1977). Also, mixing can be done in plastic containers so that abraded contaminants can be pyrolyzed during melting. A more elegant solution is to use an enclosed mixture preparation system which can feed the crucible to the furnace directly (Takahashi and Kawashima, 1977).

In melting the glasses, it is important to obtain homogeneous, bubble-free materials without adding additional contaminants which can result from airborne dust, dissolution of the crucible, and vapor-phase transport from the furnace (Beales et al., 1974). Thus, in designing an appropriate furnace, it is important that all refractory materials in contact with the atmosphere surrounding the melt should be of the highest purity available and that granular refractories used for thermal insulation should be completely enclosed in an impervious, nonfriable container (Scott and Rawson, 1973). In addition, it is helpful to be able to continuously purge the furnace to sweep vapor or particulate contaminants away from the melt.

No single crucible material has been found to be optimum. Iron impurities are a problem in Al_2O_3 crucibles (Beales et al., 1974), although this can be reduced by baking (Scott and Rawson, 1973). Impurities can also be a problem in Pt crucibles (Newns et al., 1973) and, more important, the melt can be contaminated by colloidal Pt (Imagawa and Ogino, 1977). In this case, annealing the crucible (Shibata and Takahashi, 1977) or firing under reducing conditions (Scott and Rawson, 1973) can reduce this problem. Finally, although fused-silica crucibles can be obtained with appropriately high purities, dissolution can produce striae in the glass at temperatures above 1250°C. One effective way to minimize many crucible-related problems is to use direct rf heating of the batch (after preheating to temperatures around 1000°C) and a water-cooled crucible (Beales et al., 1974; Gossink, 1977; Smith and Denton, 1980).

The conditions used during melting are also important in controlling the effect of contaminants on absorption in the glass. For example, since absorption due to iron is much lower if it is oxidized, while the opposite is true for copper, melt atmospheres should be tailored to the relative amounts of these contaminants (Gossink, 1977).

Homogenization is generally performed by bubbling a gas through the melt which, by use of low-water-content gases, can also help dry the molten glass (Gloge, 1979). Bubbling is often preferred to stirring because of its mechanical simplicity and its ability to quickly equilibrate the melt with the surrounding atmosphere in order to control the oxidation state of contaminant ions. Problems in fining a melt that has been homogenized by bubbling are greatly reduced by using either oxygen or helium (Scott and Rawson, 1973).

Once the glass is melted, homogenized, and fined, it must be extracted from the crucible. Three approaches have been used (Gossink, 1977). The most homogeneous samples are obtained by allowing the glass to cool in the crucible and then machining the piece required for fiber making. However, this raises the possibility of contamination. The second method is to cast the required shape, but this generally results in glasses with poor homogeneity. The final technique used to extract the glass is to draw rods from the melt; as will be described in Subsection III.B.2, these rods are the most generally used to draw these glasses into fibers.

Although the progress made in reducing attenuation in fibers made from direct-melted glass has been slower than for those made from vapor-phase techniques, significant gains have been achieved. For example, attenuation at 850 nm in fibers made from soda–lime silicate glasses have been reduced from 25–30 dB/km (Beales *et al.*, 1974; Pearson, 1974) to 4.2 dB/km (Takahashi and Kawashima, 1977). Attenuation at the same wavelength in similarly prepared sodium borosilicate fibers has gone from 15–25 dB/km (Beales *et al.*, 1974) to under 4 dB/km (Beales *et al.*, 1980).

In spite of these achievements, fibers made from directly melted glasses have not yet made inroads into markets served by fibers made from preforms prepared by vapor-phase techniques. The principal reason for this is that the effort required to purify the raw materials and maintain this purity throughout fabrication has thus far prevented this technique from becoming the low-cost alternative to vapor-phase processing that it was projected to be. Consequently, alternative approaches are still being sought.

c. Phase-Separated Glasses. In 1976, a technique was introduced which was aimed at greatly reducing the cost of preparing preforms suitable for direct drawing into fibers (Macedo *et al.*, 1976). The key to this approach was that it utilized a natural process, phase separation, to purify the glass as it is being prepared.

The process starts with reagent grade powders which are melted to form a specific sodium–potassium borosilicate composition (Kilroy and Moynihan, 1978). In particular, the composition was one that could be phase separated through appropriate heat treatment into two continuous, interconnected phases. One of these phases, containing almost all of the alkali and boron, is soluble and can be leached in acid so as to leave a microporous, high-silica rod. Moreover, it has been found that impurities present in the raw materials are preferentially incorporated in the alkali borate phase and so are also removed by leaching. Because of this, the remaining glass is sufficiently pure to produce absorption losses under 10 dB/km in spite of the fact that reagent grade materials are used in their fabrication (Simmons et al., 1979).

Preforms are made from the microporous rod by doping it with a solution of a salt of an ion which, when incorporated in the high-silica matrix, can increase its refractive index. By doping in this fashion and then removing the dopant from the surface by immersing the rod in a solvent with either a low or high solubility for the dopant salt, step-index or graded-index profiles are obtained, respectively (Simmons et al., 1979). After the dopant is distributed in the prescribed manner, the rod is carefully dried and consolidated by thermal treatment to produce a preform.

The phase separation process has been principally applied to sodium borosilicate glasses in which $CsNO_3$ dissolved in hot water is the dopant solution (Simmons et al., 1979). The advantage of this approach is that the $CsNO_3$, which can be precipitated in the pores by cooling the saturated rod, decomposes to Cs_2O at temperatures well below those required for consolidation, thereby avoiding entrapment of decomposition products. In an alternative approach, GeO_2 is added to the base sodium borosilicate composition (de Panafieu et al., 1980). In this case, the leaching process described above is sufficient to produce a refractive index gradient; this is because of the slight solubility of germanium in the leaching solution, which creates an almost parabolic variation in germanium concentration from the axis to the circumference of the leached rod.

Both of these compositional approaches have yielded fibers with losses below 10 dB/km. Moreover, fibers have been made with index variations showing rms deviations from the desired profile of less than 2% which provide bandwidths of about 300 MHz km (Simmons et al., 1979). However, in spite of these successes, and in spite of the fact that more complicated doping procedures have been devised which theoretically should provide even higher bandwidths, commercialization of this process has been very slow. This is perhaps due to the fact that, although reasonable quality preforms can be made with this method, uncontrolled factors, as yet unidentified, have thus far prevented the process from being reproducible (Mohr, *et al.*, 1977).

d. Sol–Gel Glasses. Recently, a very different method of preform fabrication was reported based on the sol–gel technique of preparing glasses without melting (Susa *et al.*, 1982; Puyané *et al.*, 1982). This method, derived from a chemical technique of obtaining intimate mixtures of silicates (Roy, 1956), has been used to prepare complex glasses (Dislich, 1971) as well as binaries such as SiO_2–GeO_2 (Goerlich *et al.*, 1976) which are common constituents of optical fiber waveguides.

The basic sol–gel technique, which has recently been reviewed (Mukherjee, 1980; Sakka, 1982), consists of three steps. The first is to mix the appropriate metal alkoxides and, possibly, water-soluble salts and ethanol solutions to obtain the desired cation ratio. The second step in the process is to hydrolyze this mixture with a water-based acid solution which can be diluted with alcohol; this causes the mixture to polymerize and form a gelled mass through the reaction

$$M(OR)_n + nH_2O \longrightarrow M(OH)_n + nROH, \tag{17}$$

where M is the metal and R is an alkyl group such as $(C_2H_5)^-$. In this part of the process, conditions can be varied to obtain bulk, powder, or fibrous materials. In the final step of the process, the gel is heated slowly to obtain the oxide glass, i.e.,

$$M(OH)_n \longrightarrow MO_{n/2} + \tfrac{1}{2}nH_2O. \tag{18}$$

In addition to removing the water, heating also serves to eliminate any residual organics and to sinter the resultant porous glassy oxide in order to obtain bulk glasses that are essentially indistinguishable in density, refractive index, and thermal expansion from conventionally melted glasses of the same composition.

Sol–gel methods have several advantages in glass fabrication that are particularly relevant for optical fiber waveguides. As mentioned previously, the technique can be applied to a variety of compositions; moreover, since the method involves lower temperatures than those used in conventional glass melting, compositions which cannot now be used because of crystallization or volatilization might be possible. Another advantage is that sol–gel processing can lead to very high purity glasses. This is because the liquid alkoxide raw materials can be effectively purified by using distillation or recrystallization, and because the low process temperatures serve to minimize contamination. A final advantage is that since the reactants are mixed in the liquid state, the homogeneity of the materials fabricated tends to be exceptionally high.

Based on these advantages, it is no wonder that the sol–gel technique has been applied to optical fibers. In both published reports of fiber fabrication using sol–gels the approaches used were similar to that described above. In one case (Puyané et al., 1982), sol–gel techniques were used to fabricate a SiO_2–GeO_2 core rod as well as a cladding tube. Fiber drawn from the rod in the tube (see Subsection III.B.2) exhibited a loss minimum of 22 dB/km at 830 nm. The relatively high losses were ascribed to impurities in the germanium alkoxide used and to residual hydroxyl ion contamination (~30 ppm) in the glass. More impressive results were obtained in the other published report (Susa et al., 1982). In this case, the sol–gel technique was used to prepare a pure SiO_2 core rod which was then drawn to a fiber with a cladding tube made by depositing B_2O_3–SiO_2 glass on the inside of a silica tube using vapor-phase techniques. This fiber had a loss minimum of 6 dB/km at 0.85 μm, which was achieved in part by using Cl_2 at temperatures above 600°C to dry the porous rod before sintering. The result compares favorably to those for fibers made from conventionally melted glasses despite the much shorter time that sol–gel techniques have been applied to this problem.

Although fibers made thus far using sol–gel techniques have had simple step refractive index profiles, a method has been suggested whereby graded-index preforms could be made (Harmer et al., 1982). As illustrated in Fig. 19, an appropriate solution is applied to the inside of a vertically suspended glass tube, leaving a thin layer. Because it is so thin, this layer can be gelled, dried, and sintered in a short time (~30 min), after which the next layer, with a different composition, is applied as in the ITO technique.

2. Fiber Drawing

Since the objectives in drawing fiber from glasses made by non-vapor-phase techniques are similar to those discussed previously for vapor-phase-prepared preforms, it is not surprising that many of the techniques involved in diameter control and maintenance of strength are the same for the two approaches. However, in the case of glasses prepared by non-vapor-phase techniques, special preform or glass configurations demand different methods of actually drawing the fiber, and it is these methods that will be stressed in this section.

One of the first methods suggested to draw optical fiber, called the rod-in-tube approach, is appropriate when the preform glass is available as a core rod and a cladding tube (Kapany, 1967). Thus, for example, this method can be used for glasses made by phase separation techniques in which the cladding layer fabricated directly by exosolution of the index-modifying dopant is not thick enough to provide fiber with the desired

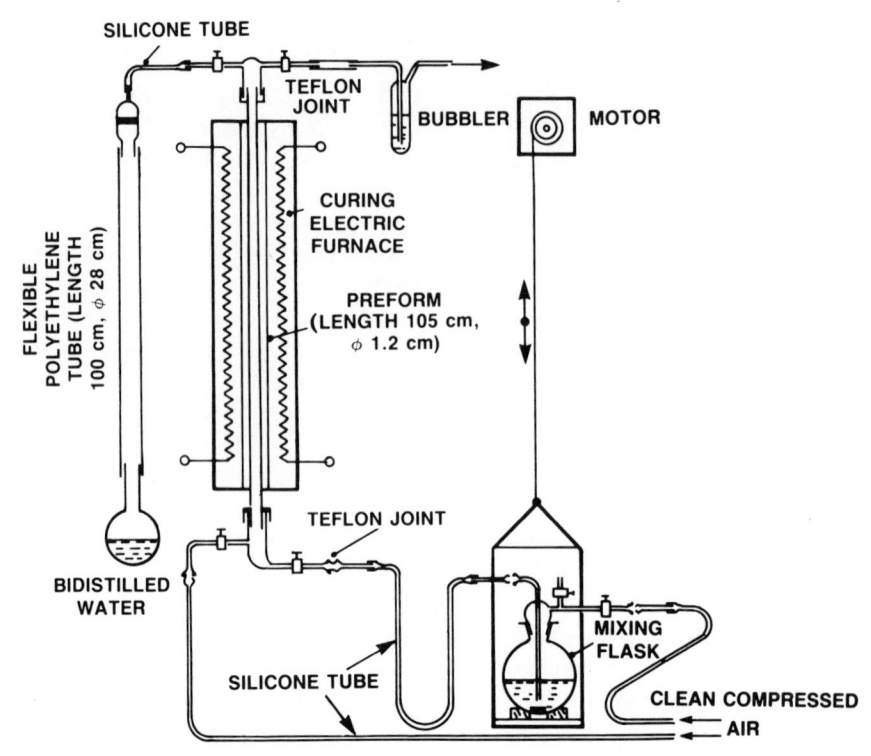

FIG. 19. Proposed technique for graded-index preforms using sol–gels. [From Harmer et al. (1982). Copyright Information Gatekeepers, Inc.]

core:cladding diameter ratio. In that case, rod-in-tube drawing can provide additional cladding.

In practice, rod-in-tube drawing is performed in one of two ways. The most straightforward approach is to draw the core rod and cladding tube simultaneously by using equipment similar to that described in Subsection III.A.2. However, with this approach it is often found that gas bubbles are trapped at the core–clad interface, which can cause a reduction in fiber strength and, if the cladding on the rod is too thin, an increase in attenuation due to Mie scattering (Pearson, 1974; French et al., 1975). One way to circumvent this problem is to collapse the cladding tube onto the core rod prior to drawing the fiber. By performing this step carefully on a glassworking lathe using techniques similar to the collapse step in the ITO process, such defects can be avoided.

Although rod-in-tube methods can be applied to some of the preform-making techniques described in the preceding section, they are not appro-

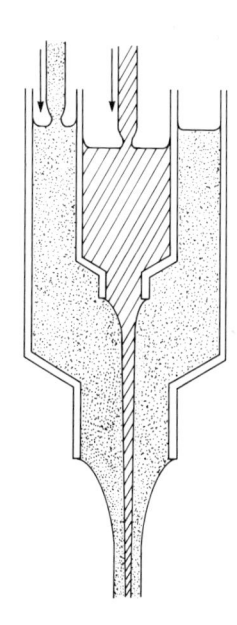

FIG. 20. Fiber drawing from concentric crucibles.

priate for glasses made by direct melting since, in that method, the core and cladding glasses are usually both in the form of rods. In this case, the concentric- or double-crucible method is the one most widely used to provide cladded optical fiber waveguides.

The concentric-crucible method of fiber drawing, which has been reviewed (Aulich *et al.*, 1978; Beales *et al.*, 1980), is illustrated in Fig. 20. As shown, rods of glass of the core and cladding composition are introduced at a controlled rate into the center and annual regions, respectively, of a pair of concentric crucibles which are positioned in a furnace. The feed rate is adjusted such that the amount of core and cladding glass entering the crucible equals the corresponding amounts exiting the crucible as fiber through the bottom orifice.

As currently practiced, the furnace is adjusted to provide glass viscosities on the order of 10^3 P; this corresponds to a temperature of around 900°C for borosilicate glasses. Feed rates are limited to about 2 cm/min (Newns, 1976) since higher speeds tend to introduce bubbles in the melt which are incorporated in the drawn fiber (Aulich *et al.*, 1978). Because of this limitation, draw rates are restricted to roughly 0.3 m/sec (Schultz, 1979b). Platinum has been found to be the most effective crucible material for this application, but it does have drawbacks. The problem of Pt purity, discussed in the section on direct melting of glasses, can be overcome by long-term melting of a sacrificial charge of glass (Newns, 1976). A more subtle difficulty is the generation of O_2 bubbles at the core–cladding interface due to electrolytically induced reboiling (Day et al., 1974). In this case, eliminating oxygen from the atmosphere around the crucible has been found to be effective in suppressing this effect (Newns, 1976).

The key to the popularity of the concentric-crucible method for drawing fibers from bulk melted glasses is that by allowing diffusion between the core and cladding glasses to occur in a controlled way, graded-index fibers can also be prepared (Koizumi et al., 1974). The way this is generally done is to allow a sufficient distance between the bottoms of the orifices of the inner and outer crucibles that the desired amount of diffusion can occur before the fiber exits the furnace (van Ass et al., 1976a).

The diffusion equations which govern this process are well known (Koizumi et al., 1974; Ishikawa et al., 1977; Yamazaki and Yoshiyagawa, 1977). As summarized by Beales and Day (1980), the controlling equation is

$$(\partial^2 N/\partial r^2) + (1/r)(\partial N/\partial r) - (1/D)(\partial N/\partial t) = 0, \quad (19)$$

where N is the concentration of the critical ion after time t, r the radial distance from the center of the core, and D the diffusion coefficient. If D is constant with N, then the equation can be solved to yield

$$\frac{N}{N_0} = \int_0^\infty \exp\left(-\frac{Dt}{A^2} u^2\right) J_0(uR) J_1(u) \, du, \quad (20)$$

where N_0 is the initial concentration and $R = r/A$ and A is the radius of the orifice of the inner crucible.

In order to obtain a near-parabolic variation in concentration, it has been shown (Dyott and Brain, 1974) that the factor Dt/A^2 should be about 0.08. Based on this, it is apparent that the resultant refractive index profile is controlled by four parameters (Beales et al., 1980): the diffusion length, the rate of glass flow, the temperature, and the compositions of the core and cladding glasses.

In selecting the particular glass composition, it is desirable to find one in which there is a particularly mobile ion which has a large effect on refractive index in order to maximize the numerical aperture of the resultant fiber. Moreover, the core and cladding compositions should then be selected to provide similar thermal properties in spite of the different concentrations of this mobile ion; this will minimize thermal stresses.

The vast majority of the research in this area has been done on the sodium borosilicate systems, mainly for the reasons described in Subsection III.B.1.a, although a variety of mobile ions have been used. Initial efforts in this system were based on one of two approaches. One utilized diffusion of Na^+ alone since sodium is known to increase the refractive index of the base glass (Newns, 1976; Newns et al., 1977). The other initial approach involved the interdiffusion of Tl$^+$ from the core glass and Na$^+$ from the cladding glass. The use of alkali ions as the diffusing species in concentric-crucible fibers has several drawbacks. On the one hand, sodium diffusion alone is slow and practical numerical apertures are limited to 0.15 (Beales and Day, 1980); on the other hand, using Tl$^+$ interdiffusion to alleviate these difficulties introduces the problem of toxicity. Because of these problems, alkaline-earth ions are now commonly used as the diffusing species in sodium borosilicate glasses (Beales et al., 1980; Vacha et al., 1980). Use of these ions can provide fiber with a high numerical aperture and, although they exhibit low diffusion coefficients, long diffusion times can be used to provide a suitable refractive index profile.

While most results reported on this process have used sodium borosilicates, alkali–lime germanosilicates can also be used to provide graded-index fibers (van Ass et al., 1976b; Spierings et al., 1980). In this case, Na$^+$ in the core exchanges for K$^+$ in the cladding. Although this system can provide a higher numerical aperture than the same ions in sodium borosilicate glasses, it suffers from the disadvantage that high concentrations of germanium are required.

The results obtained on graded-index fibers drawn from concentric crucibles have been very impressive, particularly for the sodium borosilicate materials. For example, attenuation, which had been about 20 dB/km at 800 nm (Koizumi et al., 1974), has been reduced to 3.4 dB/km at 840 nm and <4 dB/km over the wavelength range from 780 to 920 nm (Beales et al., 1977). Similar progress has been made in reducing dispersion; in this case the lowest reported value, 0.5 nsec/km (Ishikawa et al., 1977), begins to approach that obtained reproducibly in vapor-phase processes (~0.3 nsec/km).

In spite of these achievements, there are significant difficulties with the double-crucible process which must be solved before this method can become a commercially viable competitor of vapor-phase processes. First, the presence of boron in most of the compositions and the difficulty in removing hydroxyl ion impurities from bulk melted glasses yield high attenuation losses at wavelengths greater than 1.2 μm (Schultz, 1979b). This problem is still being addressed (Beales et al., 1982). In the case of dispersion, deviations from a strictly parabolic profile at the core–cladding interface, which limit the attainable bandwidth, are an inherent feature of this process, which depends on diffusion to form the profile (Beales and Day, 1980). However, it has been suggested that this problem may be circumvented by using a triple concentric crucible (Ishikawa et *al.*, 1977; Yamazaki and Yoshiyagawa, 1977). A more subtle disadvantage of fibers drawn from concentric crucibles may be that, since they require high diffusion coefficients in order to obtain a suitable profile shape, they probably will be inherently less chemically durable than fibers made by vapor-phase techniques. This could cause problems in long-term degradation of the fiber end faces used in coupling as well as in an enhanced rate of mechanical fatigue.

One of the potential applications that has been suggested for these fibers is in short-haul optical fiber systems. The reason for this is that in such systems higher fiber attenuation and lower bandwidth can be tolerated, particularly if compensated by a higher numerical aperture which decreases coupling losses. However, before such applications become a reality for the concentric-crucible fibers, the economics of their fabrication, currently limited by the cost of purification of raw materials and the low draw speeds, must be improved.

IV. Summary

In this chapter, an attempt has been made to show how materials scientists and engineers have responded to the stringent demands placed on optical fiber waveguides by the requirements of commercially viable optical communication systems. This response has resulted in optical fibers with attenuation at near-intrinsic levels and bandwidths which were thought to be impossible only ten years ago. These achievements are so significant that optical fiber waveguides, which were once the limiting elements in optical communication systems, are now the driving force behind research to improve the characteristics of other elements of the system. However, in spite of this progress, much remains to be done. For example, attempts are being made to reduce the cost of fibers made by current vapor-phase techniques by improving manufacturing efficiencies and utilizing less expensive dopants. In addition, the search continues for alternative, non-vapor-phase techniques which will be viable competition for the standard processes in at least a segment of the market. Finally, there is a significant amount of interest in novel glass systems. These glasses are based on anions other than oxygen; having infrared absorption edges at longer wavelength than those which characterize oxide glasses in general and silica-based glasses in particular, they may have intrinsic attenuation limits which are orders of magnitude below those of current fiber waveguides. Although the particular directions that these research paths will take are uncertain, it is clear that the story of optical fiber waveguide fabrication is far from complete.

References

Akamatsu, T., Okamura, K., and Ueda, Y. (1977). *Appl. Phys. Lett.* **31**, 515–517.
Aulich, H. A., Grabmaier, J. B., Eisenrith, K. H., and Kinshofer, G. (1978). *Siemens Forsch. Entwicklungs Ber.* **7**, 298–304.
Barnoski, M. K., and Personick, S. D. (1978). *Proc. IEEE* **66**, 429–441.
Barns, R. L., Chandross, E. A., and Mellier-Smith, C. M. (1980). *IEE Conf. Publ.* **190**, 26–28.
Beales, K. J., and Day, C. R. (1980). *Phys. Chem. Glasses* **21**, 5–21.
Beales, K. J., Midwinter, J. E., Newns, G. R., and Day, C. R. (1974). *Post Off. Electr. Eng. J.* **67**, 80–87.
Beales, K. J., Day, C. R., Duncan, W. J., and Newns, G. R. (1977). *Electron. Lett.* **13**, 755–756.
Beales, K. J., Day, C. R., Duncan, W. J., Dunn, A. G., Dunn, P. L., and Newns, G. R. (1980). *Phys. Chem. Glasses* **21**, 25–29.
Beales, K. J., France, P. W., and Partington, S. (1982). *Am. Ceram. Soc. Bull.* **61**, 1228–1231.
Blankenship, M. G., and Deneka, C. W. (1982). *IEEE J. Quantum Electron.* **QE-18**, 1418–1423.
Blankenship, M. G., Morrow, A. J., and Silverman, L. A. (1982). *Opt. Fiber Commun., OFC '82, 1982,* p. 18–19.
Blyler, L. L., Jr., DiMarcello, F. V., Simpson, J. R., Sigety, E. A., Hart, A. C., Jr., and Foertmeyer, V. A. (1980). *J. Non-Cryst. Solids* **38-39**, 165–170.
Campbell, L. L. (1977). *Fiber Integr. Opt.* **1**, 21–37.
Chida, K., Sudo, S., Nakahara, M., and Inagaki, N. (1981). *Conf. Dig., Eur. Conf. Opt. Commun., 7th, 1981,* Paper 6.3, pp. 1–4.
Chida, K., Okazaki, H., and Nakahara, M. (1982a). *Electron. Lett.* **18**, 330–331.
Chida, K., Sakaguchi, S., Wagatsuma, M., and Kimura, T. (1982b). *Electron. Lett.* **18**, 713–715.
Chida, K., Hanawa, F., and Nakahara, M. (1982c). *IEEE J. Quantum Electron.* **QE-18**, 1883–1889.
Cohen, L. G., and Lin, C. (1977). *Appl. Opt.* **16**, 3136–3139.
Cohen, L. G., Lin, C., and French, W. G. (1979). *Electron. Lett.* **15**, 334–335.
Cohen, L. G., Mammel, W. L., and Lumish, S. (1982). *IEEE J. Quantum Electron.* **QE-18**, 49–53.
Day, C. R., Midwinter, J. E., Newns, G. R., Uffen, R. W. J., and Worthington, R. (1974). *Electron. Lett.* **10**, 450–451.
de Panafieu, A., Nemaud, Y., Baylac, C., Turpin, M., Faure, M., and Gauthier, F. (1980). *Phys. Chem. Glasses* **21**, 22–24.
DiMarcello, F. V., and Hart, A. C., Jr. (1978). *Electron. Lett.* **14**, 578–579.
DiMarcello, F. V., Hart, A. C., Jr., Williams, J. C., and Kurkjian, C. R. (1979). *In* "Fiber Optics: Advances in Research and Development" (B. Bendow and S. S. Mitra, eds.), pp. 125–135. Plenum, New York.
Dislich, H. (1971). *Angew. Chem., Int. Ed. Engl.* **10**, 363–370.
Dyott, R. B., and Brain, M. C. (1974). *Electron. Lett.* **10**, 131–132.
Eichenbaum, B. R. (1980a). *Ind. Eng. Chem. Prod. Res. Dev.* **19**, 132–135.
Eichenbaum, B. R. (1980b). *Bell Syst. Tech. J.* **59**, 313–332.
Fleming, J. W., and O'Connor, P. B. (1981). *Adv. Ceram.* **2**, 21–26.
Fleming, J. W., and Raju, V. R. (1981). *Electron. Lett.* **17**, 867–868.
France, P. W., Dunn, P. L., and Reeve, M. H. (1979). *Fiber Int. Opt.* **2**, 267–286.

Freiman, S. W. (1980). "Glass: Science and Technology" (D. R. Uhlmann and N. J. Kreidl, eds.), Vol. 5, pp. 21–78. Academic Press, New York.
French, W. G., MacChesney, J. B., and Pearson, A. D. (1975). *Annu. Rev. Mater. Sci.* **5**, 373–394.
French, W. G., Pace, L. J., and Foertmeyer, V. A. (1978). *J. Phys. Chem.* **82**, 2191–2194.
Gambling, W. A., Payne, D. N., Hammond, C. R., and Norman, S. R. (1976). *Proc. IEE* **123**, 570–576.
Geittner, P., Kuppers, D., and Lydtin, H. (1976). *Appl. Phys. Lett.* **28**, 645–646.
Geyling, F. T. (1976). *Bell Syst. Tech. J.* **55**, 1011–1056.
Geyling, F. T., and Homsy, G. M. (1980). *Glass Technol.* **21**, 95–102.
Gloge, D. (1975). *Bell Syst. Tech. J.* **54**, 245–262.
Gloge, D. (1979). *Rep. Prog. Phys.* **42**, 1777–1824.
Gloge, D., and Marcatili, E. A. J. (1973). *Bell Syst. Tech. J.* **52**, 1563–1578.
Goerlich, E., Kuciel, E., Niesulowska, C., Sieminska, G., and Stoch, A. (1976). *Rocz. Chem.* **50**, 1673–1679.
Gossink, R. G. (1977). *J. Non-Cryst. Solids* **26**, 112–157.
Harmer, A. L., Puyané, R., and Gonzalez-Oliver, C. (1982). *Int. Fiber Opt. Commun.*, November/December, pp. 40–44.
Hart, A. C., Jr., and Albarino, R. V. (1977). *Tech. Dig.—Opt. Fiber Transm., 2nd, 1977*, Paper TuB2, pp. 1–4.
Homsy, G. M., and Geyling, F. T. (1977). *AIChE J.* **23**, 587–590.
Hoshikawa, M., Yoshida, M., Suzuki, S., Yoshimura, K., Yamanishi, T., Yoneji, S., Yokoda, H., Takimoto, H., and Matsuno, K. (1977). *Sumitomo Electr. Tech. Rev.* **17**, 77–88.
Imagawa, H., and Ogino, N. (1977). *Int. Conf. Integr. Opt. Opt. Fiber Commun., 1977*, pp. 613–615.
Imoto, K., and Sumi, M. (1981). *Electron. Lett.* **17**, 525–526.
Inada, K. (1982). *IEEE J. Quantum Electron.* **QE-18**, 1414–1431.
Irwin, G. R., and Wells, A. A. (1965). *Metall. Rev.* **10**, 223–270.
Ishihara, K., Tokuda, M., and Seikai, S. (1979). *Rev. Electr. Commun. Lab.* **27**, 949–959.
Ishikawa, R., Seki, M., Kaede, K., Koizumi, K., and Yamazaki, T. (1977). *Int. Conf. Integr. Opt. Opt. Fiber Commun., 1977*, pp. 301–304.
Jaeger, R. E. (1976). *Am. Ceram. Soc. Bull.* **55**, 270–273.
Kalish, D., and Tariyal, B. K. (1978). *J. Am. Ceram. Soc.* **61**, 518–523.
Kapany, N. S. (1967). "Fiber Optics," pp. 111–117. Academic Press, New York.
Kapron, F. P., Keck, D. B., and Maurer, R. D. (1970). *Appl. Phys. Lett.* **17**, 423–425.
Kawachi, M., Sudo, S., Shibata, N., and Edahiro, T. (1980). *Jpn. J. Appl. Phys.* **19**, L69–L71.
Kawachi, M., Sudo, S., Shibata, N., and Edahiro, T. (1982). *Electron. Lett.* **18**, 328–330.
Keck, D. B., Maurer, R. D., and Schultz, P. C. (1973). *Appl. Phys. Lett.* **22**, 307–309.
Kilroy, W. P., and Moynihan, C. T. (1978). *Am. Ceram. Soc. Bull.* **57**, 1034–1039.
Kimura, T., Sakaguchi, S., Namikawa, H., and Yoshida, K. (1980). *IEE Conf. Publ.* **190**, 57–60.
Kitayama, K. I., and Seikai, S., and Morishita, K. (1982). *IEEE J. Quantum Electron.* **QE-18**, 838–843.
Kleinert, P., Schmidt, D., Kirchof, J., and Funke, A. (1980). *Krist. Tech.* **15**, 85–90.
Koizumi, K., Ikeda, Y., Kitano, I., Furukawa, M., and Sumimoto, T. (1974). *Appl. Opt.* **13**, 255–260.
Koel, G. J. (1983). *Ann. Telecommun.* **38**, 36–46.
Krause, J. T. (1980). *J. Non-Cryst. Solids* **38-39**, 497–502.

Kuppers, D., Koenings, J., and Wilson, H. (1976). *J. Electrochem. Soc.* **123**, 1079–1083.
Kyoto, M., Satoh, H., Watanabe, M., Nishimura, M., and Yano, K. (1983). *Tech. Dig.— Top. Meet. Opt. Fiber Commun., 6th, 1983,* Paper WC2, pp. 80–82.
Levy, N. (1981). *Appl. Opt.* **20**, 460–464.
MacChesney, J. B., O'Connor, P. B., and Presby, H. M. (1974). *Proc. IEEE* **62**, 1278–1279.
Macedo, P. B., Simmons, J. H., Olson, T., Mohr, R. K., Samanta, M., Gupta, P. K., and Litovitz, T. A. (1976). *Eur. Conf. Opt. Fibre Commun., 2nd, 1976,* pp. 37–39.
Marcuse, D. (1979). *Appl. Opt.* **18**, 4003–4005.
Marcuse, D., and Derosier, R. M. (1969). *Bell Syst. Tech. J.* **48**, 3217–3232.
Marcuse, D., and Presby, H. M. (1977). *Appl. Opt.* **16**, 2383–2390.
Maurer, R. D. (1973). *Proc. IEEE* **61**, 452–462.
Miller, R. A. (1979). In "Fiber Optics: Advances in Research and Development" (B. Bendow and S. S. Mitra, eds.), pp. 77–103. Plenum, New York.
Miller, T. J., Hart, A. C., Jr., Vroom, W. I., Jr., and Bowden, M. J. (1978). *Electron. Lett.* **14**, 603–605.
Mitchell, J. W. (1982). *Pure Appl. Chem.* **54**, 819–834.
Miya, V., Terunuma, Y., Hosaka, T., and Miyashita, T. (1979). *Electron. Lett.* **15**, 106–108.
Mochizuki, S., Ishihara, K., and Nakatani, N. (1979). *Rev. Electr. Commun. Lab.* **27**, 210–216.
Modone, E., and Roba, G. (1981). *Electron. Lett.* **17**, 815–817.
Modone, E., Parisi, G., and Roba, G. (1982). *Electron. Lett.* **18**, 721–722.
Mohr, R. K., Macedo, P. B., and Litovitz, T. A. (1977). Final Report, Contract N00019-76-C-0674, AD A049168.
Montierth, M. R. (1977). *J. Electron. Mater.* **6**, 349–372.
Moriyama, T., Fukuda, O., Sanada, K., Inada, K., Edahiro, T., and Chida, K. (1980). *Electron. Lett.* **16**, 698–699.
Mukherjee, S. P. (1980). *J. Non-Cryst. Solids* **42**, 477–488.
Nagel, S. R., MacChesney, J. B., and Walker, K. L. (1982). *IEEE J. Quantum Electron.* **QE-18**, 459–476.
Nakahara, M., Sakaguchi, S., and Miyashita, T. (1978). *Rev. Electr. Commun. Lab.* **26**, 476–483.
Nakahara, M., Sudo, S., Inagaki, N. Yoshida, K., Shibuya, S., Kokura, K., and Kuroha, T. (1980). *Electron. Lett.* **16**, 391–392.
Newns, G. R. (1976). *Eur. Conf. Opt. Fibre Commun., 2nd, 1976,* pp. 21–26.
Newns, G. R. Pantelis, P., Wilson, J. L., Uffen, R. W., and Worthington, R. (1973). *Optoelectron (London)* **5**, 289–296.
Newns, G. R., Beales, K. J., and Day, C. R. (1977). *Tech. Dig.—Int. Conf. Integr. Opt. Opt. Fiber Commun., 1977,* pp. 609–612.
Oehrle, R. C. (1979). *Appl. Opt.* **18**, 496–500.
Olshansky, R. (1979). *Rev. Mod. Phys.* **51**, 341–367.
Paek, U. C., and Schroeder, C. M. (1979a). *Fiber Integr. Opt.* **2**, 287–298.
Paek, U. C., and Schroeder, C. M. (1979b). *J. Appl. Phys.* **50**, 6144–6148.
Paek, U. C., and Schroeder, C. M. (1981). *Appl. Opt.* **20**, 4028–4034.
Paek, U. C., Spainhour, C. D., Schroeder, C. M., and Kurkjian, C. R (1980). *Am. Ceram. Soc. Bull.* **59**, 630–634.
Partus, F. P., and Saifi, M. A. (1980). *West. Electr. Eng.* **24**, 39–47.
Payne, D. N., and Gambling, W. A. (1975). *Electron. Lett.* **11**, 176–178.
Payne, D. N., and Gambling, W. A. (1976). *Am. Ceram. Soc. Bull.* **55**, 195–197.
Pearson, A. D. (1974). *Int. Congr. Glass, 10th, 1974,* pp. 6/31–6/39.
Pearson, A. D. (1980). *IEE Conf. Publ.* **190**, 22–25.

Peelen, J. G. J., and Koenings, J. M. J. (1981). *Eur. Conf. Opt. Commun., 7th, 1981,* pp. 1.1–1.4.
Peelen, J. G. J., Versluis, J. W., and Vervaart, A. P. (1978). *Eur. Conf. Opt. Commun., 4th, 1978,* pp. 67–71.
Pinnow, D. A., Rich, T. C., Ostermayer, F. W., Jr., and DiDomenico, M., Jr. (1973). *Appl. Phys. Lett.* **22,** 527–529.
Powers, D. L. (1978). *J. Am. Ceram. Soc.* **61,** 295–297.
Puyané, R., Harmer, A. L., and Gonzalez-Oliver, C. J. R. (1982). *Eur. Conf. Opt. Commun., 8th, 1982,* pp. 623–628.
Rigterink, M. D. (1976). *Am. Ceram. Soc. Bull.* **55,** 775–780.
Roy, R. (1956). *J. Am. Ceram. Soc.* **39,** 145–146.
Runk, R. B. (1977). *Tech. Dig.—Opt. Fiber Transm., 2nd, 1977,* Paper TuA1, pp. 1–4.
Sakka, S. (1982). "Treatise on Materials Science and Technology," Vol. 22 (M. Tomozawa and R. H. Doremus, eds.), pp. 129–167. Academic Press, New York.
Scherer, G. W. (1979a). *J. Non-Cryst. Solids* **34,** 223–238.
Scherer, G. W. (1979b). *J. Non-Cryst. Solids* **34,** 239–256.
Schonhorn, H., Kurkjian, C. R., Jaeger, R. E., Vazirani, H. N., Albarino, R. V., and DiMarcello, F. V. (1976). *Appl. Phys. Lett.* **29,** 712–714.
Schonhorn, H., Torza, S., Albarino, R. V., Vazirani, H. N., and Wang, T. T. (1979). *J. Appl. Polym. Sci.* **23,** 75–84.
Schultz, P. C. (1979a). *In* "Fiber Optics: Advances in Research and Development" (B. Bendow and S. S. Mitra, eds.), pp. 3–31. Plenum, New York.
Schultz, P. C. (1979b). *Appl. Opt.* **18,** 3684–3693.
Scott, B., and Rawson, H. (1973). *Glass Technol.* **14,** 115–124.
Shibata, S., and Takahashi, S. (1977). *J. Non-Cryst. Solids* **23,** 111–122.
Simmons, J. H., Mohr, R. K., Tran, D. C., Macedo, P. B., and Litovitz T. A. (1979). *Appl. Opt.* **18,** 2732–2733.
Simpkins, P. G., Greenberg-Kosinski, S., and MacChesney, J. B. (1979). *J. Appl. Phys.* **50,** 5676–5681.
Smith, B. A., and Denton, B. (1980). *J. Mater. Sci.* **15,** 2515–2519.
Smithgall, D. H. (1977). *West. Electr. Eng.* **21,** 54–59.
Smithgall, D. H. (1979). *Bell Syst. Tech. J.* **58,** 1425–1435.
Smithgall, D. H., and Frazee, R. E. (1981). *Bell Syst. Tech. J.* **60,** 2065–2080.
Smithgall, D. H., and Myers, D. L. (1980). *West. Electr. Eng.* **24,** 49–61.
Spierings, G. A. C. M., Jochem, C. M. G., Meeuwsen, T. P. M., Meyer, F., and Severin, P. J. W. (1980). *Phys. Chem. Glasses* **21,** 30–33.
Spierings, G. A. C. M., Jochem, C. M. G., and Meeuwsen, T. P. M. (1981). *Glass Technol.* **22,** 243–246.
Stone, F. T. (1980). *Opt. Lett.* **5,** 507–509.
Stone, F. T., and Eichenbaum, B. R. (1980). *J. Non-Cryst. Solids* **38-39,** 189–194.
Stone, J., and Lemaire, P. J. (1982). *Electron. Lett.* **18,** 78–80.
Sudo, S., Kawachi, M., Edahiro, T., Izawa, T., Shioda, T., and Gotoh, H. (1978). *Electron. Lett.* **14,** 534–535.
Susa, K., Matsuyama, I., Satoh, S., and Suganuma, T. (1982). *Electron. Lett.* **18,** 499–500.
Takahaski, S., and Kawashima, T. (1977). *Int. Conf. Integr. Opt. Opt. Fiber Commun., 1977,* pp. 621–624.
Takahaski, S., Miyashita, T., Edahiro, T., Horiguchi, M., and Masuno, K. (1974). *Int. Congr. Glass, 10th, 1974,* pp. 6/24–6/29.
Tanaka, S., Kameo, Y., Ichikawa, O., Hoshikawa, M., Kurauchi, N., Katsuyama, Y., and Mitsunaga, Y. (1982). *Sumitomo Electr. Tech. Rev.* **21,** 47–51.

Tariyal, B. K., and Kalish, D. (1978). *Fract. Mech. Ceram.* **3,** 161–175.
Tasker, G. W., French, W. G., Simpson, J. R., Kaiser, P., and Presby, H. M. (1978). *Appl. Opt.* **17,** 1836–1842.
Torza, S. (1976). *J. Appl. Phys.* **47,** 4017–4020.
Tynes, A., Pearson, A. D., and Northover, W. R. (1979). *J. Am. Ceram. Soc.* **62,** 324–326.
Uhlmann, D. R., and Kreidl, N. J. (eds.) (1980). "Glass: Science and Technology," Vol. 5. Academic Press, New York.
Vacha, L., Granberg, M., Marcolla, P., and Lindborg, U. (1980). *J. Non-Cryst. Solids* **38-39,** 797–802.
van Ass, H. M. J. M., Gossink, R. G., and Severin, P. J. W. (1976a). *Electron. Lett.* **12,** 369–370.
van Ass, H. M. J. M., Geittner, P., Gossink, R. G., Kuppers, D., and Severin, P. J. W. (1976b). *Philips Tech. Rev.* **36,** 182–189.
Versluis, J. W., and Peelen, J. G. J. (1979). *Philips Telecommun. Rev.* **37,** 215–230.
Wagatsuma, M., Kimura, T., Shuto, Y., and Yamakawa, S. (1982). *Electron. Lett.* **18,** 731–732.
Walker, K. L., Homsy, G. M., and Geyling, F. T. (1979). *J. Colloid Interface Sci.* **69,** 138–147.
Walker, K. L., Harvey, J. W., Geyling, F. T., and Nagel, S. R. (1980a). *J. Am. Ceram. Soc.* **63,** 96–102.
Walker, K. L., Geyling, F. T., and Nagel, S. R. (1980b). *J. Am. Ceram. Soc.* **63,** 552–558.
Wang, T. T., and Zupko, H. M. (1980). *Fiber Integr. Opt.* **3,** 73–87.
Watkins, L. S. (1974). *J. Opt. Soc. Am.* **64,** 767–772.
Watkins, L. S. (1982). *Proc. IEEE* **70,** 626–634.
Weinberg, M. C. (1982). *J. Am. Ceram. Soc.* **65,** 81–87.
Wood, D. L., and Shirk, J. S. (1981). *J. Am. Ceram. Soc.* **64,** 325–327.
Wood, D. L., MacChesney, J. B., and Luongo, J. P. (1978). *J. Mater. Sci.* **13,** 1761–1768.
Wood, D. L., Walker, K. L., Simpson, J. R., MacChesney, J. B., Nash, D. L., and Angueira, P. (1981). *Eur. Conf. Opt. Commun., 7th, 1981,* Paper 1.2, pp. 1–4.
Yamanishi, T., Yoshimura, K., and Suzuki, S. (1979). *Electron. Lett.* **16,** 100–101.
Yamazaki, T., and Yoshiyagawa, M. (1977). *Int. Conf. Integr. Opt. Opt. Fiber Commun., 1977,* pp. 617–620.
Yan, M. F., MacChesney, J. B., Nagel, S. R., and Rhodes, W. W. (1980). *J. Mater. Sci.* **15,** 1371–1378.
Yoshida, K., Furui, Y., Sentsui, S., and Kuroha, T. (1977a). *Electron. Lett.* **13,** 608–610.
Yoshida, K., Sentsui, S., Shii, H., and Kuroha, T. (1977b). *Tech. Dig.—Int. Conf. Integr. Opt. Opt. Fiber Commun., 1977,* pp. 327–330.
Yoshida, K., Sentsui, S., Shii, G., and Kuroha, T. (1978). *Trans. IECE Jpn.* **E61,** 181–184.
Yoshida, K., Furui, Y., Sentsui, S., and Kuroha, T. (1981). *Opt. Quantum Electron.* **13,** 85–89.
Yoshizawa, N., Yabuta, T., Kojima, N., and Negishi, Y. (1981). *Appl. Opt.* **20,** 3146–3151.

Materials Index*

A

Ag, in glass coatings, 256–258, 266, 267
AgIn$_2$S$_8$, in glass coatings, 261
Ag soaps as glass coatings, 266
Al
  in glass coatings, 263, 265
  in roller composition, 64
  in solder glasses, 180
Al(O–iso-C$_3$H$_7$)$_3$, gel synthesis from, 220
Al(O–iso-C$_4$H$_9$)$_3$, gel synthesis from, 220
Al$_2$O$_3$
  in compositions for flat glass, 57, 58
  in glass, 3, 13
  in glass coatings, 263, 273
    characteristics, 270
  in glass container formulas, 110, 111
  in glass melts, 5
  transparent films from, 213
Al$_2$O$_3$–SiO$_2$, glasses from gels of, 213
Aluminum phosphate, glass coatings from, 273–274
Ar/SF$_6$, as windowpane filling, 272
As, glass decolorizing by, 16
As$_2$O$_3$
  effect on foaming, 27
  as fining agent, 20, 21, 24
Au, in glass coatings, 256–258, 266, 267

B

B
  in glass coatings, 263
  as optical fiber dopant, 301
  in solder glasses, 180
Ba, in solder glass, 180
BaSO$_4$, in glass container formulas, 111
B$_4$C, as glass filler, 174
Be, in glass coatings, 263
BeO, in glass coatings, 263
BEt$_3$, glass coatings from, 265
Bi, in glass coatings, 266
B$_2$O$_3$
  in gels, 219
  in glass shell composition, 163
  loss from glass melts, 39
B$_2$O$_3$–SiO$_2$, in sol–gel glasses for optical fibers, 329
B(OCH$_3$)$_3$, gel synthesis from, 220

C

C
  in chemistry of glass melting, 15
  in glass coatings, 263
  in roller composition, 64
CaCO$_3$, dissolution in molten Na$_2$CO$_3$, 11
CaO
  in compositions for flat glass, 57, 58
  in glass, 3, 13
  in glass container formulas, 110, 111
Ca(O–C$_2$H$_5$)$_2$, gel synthesis from, 220
CaSO$_4$ · 2H$_2$O, in glass container formulas, 111
Cd, in glass coatings, 261
CdSnO$_3$, in glass coatings, 261
Cd$_2$SnO$_4$, synthesis of, 276
CdS$_x$Se$_{1-x}$, in glass coatings, 261
Ce
  in glass coatings, 263
  glass decolorizing by, 16
CeO$_2$
  effect on foaming, 27
  in glass coatings, characteristics of, 270
Ce$_2$SnO$_4$, in glass coatings, 261
CH$_3$–Si(OR)$_3$, in optical coating preparation, 279

*Formulas given only. For given or trivial name, see Subject Index.

MATERIALS INDEX

Co
  effect on optical fiber waveguide attenuation, 290
  in glass coatings, 255, 260, 266
  physical decolorizing of glass melts by, 16
  in solder glasses, 180
CO_2
  bubbles in glass melts, 21
  conversion to O_2 in glass melts, 20
  effect on reboil in glass melts, 27, 28
  in glass melts, 19–20
  solubility role in glass melting, 13, 26
Cobalt oxide
  in glass coatings, 273
    characteristics, 271
Cr
  in colored glass, 15
  effect on optical fiber waveguide attenuation, 290
  in glass coatings, 255, 256, 263
  in roller composition, 64
  in solder glasses, 180
Cr_2O_3
  in glass coatings, 103
    characteristics, 271
  in glass container formulas, 111
$CsNO_3$, as dopant for optical fiber glasses, 327
Cu
  in colored glass, 15
  effect on optical fiber waveguide attenuation, 290
  in glass coatings, 256–258, 260, 261, 266, 267
  in roller composition, 64
$CuInSe_2$, in glass coatings, 261
CuO, in glass coatings, characteristics of, 271

D

N,N-Dimethyl-thiourea, in glass coating process, 261

F

F, as optical fiber dopant, 301
Fe
  in colored glasses, 14–16
  effect on optical fiber waveguide attenuation, 290
  in glass coatings, 255, 260, 266
  in solder glasses, 180
Fe_2O_3
  in compositions for flat glass, 57
  in glass coatings, 263, 273
    characteristics, 271
  in glass container formulas, 111

G

Gd, in glass coatings, 263
Ge
  in glass coatings, 263
  as optical fiber dopant, 301
  requirement in concentric crucible method for optical fibers, 333
$GeCl_4$, in optical fiber waveguide processing, 308
$Ge(O-C_2H_5)_4$, gel synthesis from, 220
GeO_2, in optical fiber waveguide formation, 313, 314
GeO_2-SiO_2 glasses, attenuation in, 288, 289

H

Hf, in solder glass, 180
HfO_2, in glass coatings, characteristics of, 270
H_2O, solubility role in glass melting, 13

I

In, in glass coatings, 256, 261, 266
In_2O_3
  in glass coatings, 263, 273
    characteristics, 270

K

KNO_3, decomposition in glass melting, 15
K_2O
  in glass, 3
    flat glass compositions, 57
    glass container formulas, 110, 111
  in glass shell composition, 163
K_2O-SiO_2 system, bubble formation in, 27

MATERIALS INDEX

L

La$_2$O$_3$, in glass coatings, characteristics of, 270
Lead borate melts, volatilization from, 38–39
Li, in solder glass, 180
Li$_2$O, in glass shell composition, 163
Li$_2$O · Al$_2$O$_3$ · 2SiO$_2$ (β-Eucryptite)
    as glass filler, 178
    CTE of, 174
Li$_2$O–SiO$_2$ system, bubble formation in, 27
Lithium silicate, iron in, 14

M

Mg, in solder glasses, 180
MgF$_2$, in glass coatings, 258
MgO
    brick, in glass furnace construction, 123
    in flat glass formulas, 57
    in glass, 3
    in glass container formulas, 110, 111
    in sheet glass, 4
2MgO · 2Al$_2$O$_3$ · 5SiO$_2$ (Cordierite), as glass filler, 174, 178
Mn
    in colored glass, 15, 16
    effect on optical fiber waveguide attenuation, 290
    in roller composition, 64
MnO$_2$, effect on foaming, 27
Mo
    in glass coatings, 263
    in roller composition, 64

N

N$_2$: Ar ratios, in glass melts, 20
Na$_2$Ca(CO$_3$)$_2$, formation in glass melts, 11
NaCl
    effects on segregation during melting of, 11
    in glass container formulas, 111
Na$_2$CO$_3$
    in chemistry of glass melting, 9–12
    melts of, vitreous silica rod behavior in, 10
    silica reaction with in batch heating, 8–9
Na$_2$CO$_3$ + CaCO$_3$, vitreous silica rod behavior in, 10
Na$_2$CO$_3$–CaCO$_3$–SiO$_2$ system
    chemistry of melting of, 10, 11, 12
    decolorizing in, 15–16
Na$_2$CO$_3$–SiO$_2$, chemistry of melting of, 11
Na$_2$HAsO$_4$, as leaching agent, 278
NaNO$_3$, effects on segregation during melting of, 11
Na$_2$O
    in glass, 3, 13–14
    flat glass, 57
    in glass container formulas, 110, 111
    in glass melts, 11
    in glass shell composition, 163
    loss from glass melts, 39
Na$_2$O–B$_2$O$_3$ melts, volatilization from, 39
Na$_2$O–CaO–Al$_2$O$_3$, flow in melts of, 31
Na$_2$O–CaO–SiO$_2$ glass melt
    foaming in, 27
    inhomogeneity of, 28
    Knudsen effusion experiments on, 40
    volatilization from, 39
Na$_2$O–K$_2$O–CaO–MgO–Al$_2$O$_3$–SiO$_2$ glasses, viscosity–temperature relations for, 4
Na$_2$O–SiO$_2$ glass melts
    bubble formation in, 27
    inhomogeneity of, 28
    sulfate solubility in, 27
    volatilization from, 39
Na$_2$SiO$_3$, formation in glass melts, 10
Na$_2$Si$_2$O$_5$–SiO$_2$, eutectic point of, 10–11
Na$_2$SO$_4$
    effects on segregation during melting of, 11
    in glass container formulas, 111
Nd, in colored glasses, 16
Ni
    in colored glasses, 16
    effect on optical fiber waveguide attenuation, 290
    in glass coatings, 255, 256, 260, 263, 266, 267
    in roller composition, 64
Nickel oxide
    in glass coatings, 103, 273
    characteristics, 271
NO$_x$, as pollutants from glass manufacture, 134

MATERIALS INDEX

O

$O_2 + CO_2$, bubbles, behavior in water, 23
$O_2-H_2O-N_2$ atmospheres, effect on foaming in glass melts, 27
Os, in glass coatings, 263

P

P
  as optical fiber dopant, 301
  in roller composition, 64
Pb
  in glass coatings, 261
  in solder glass, 180
PbO
  in glass coatings, 273
    characteristics of, 270
  in solder glasses, 182
    effect on chemical durability, 192
$PbO-ZnO-B_2O_3$
  in solder glass composition, 172, 179–180
  devitrification in, 175
  ternary phase diagram, 179
Pb_3O_4, decomposition in glass melting, 15
$PbTiO_3$, as glass filler, 178
Pd, in glass coatings, 265
$PdCl_2$, surface sensitization by, 266
P_2O_5, in cerdip devices, 195
Potassium silicates
  iron in, 14
  silica sols from, 214
Pt, in glass coatings, 265

R

Re, in glass coatings, 263
Rhodium oxide, in glass coatings, characteristics, 271
Ru, in glass coatings, 265
Ruthenium oxide, in glass coatings, characteristics, 271

S

S
  in glass coatings, 261, 265
  in roller composition, 64
Sb_2O_3
  effect on foaming, 27
  as fining agent, 20, 24
Sb_2S_3, photosensitive glass coatings from, 274
Se, in glass coatings, 261, 265
Si
  in glass coatings, 263
  in roller composition, 64
  in solder glasses, 180
$SiCl_4$
  in optical fiber waveguide compositions, 299
  silica sols from, 214
Silicon carbide, as glass filler, 174
SiO_2
  in chemistry of glass melting, 10
  in glass, 3, 13
    flat glass, 57
  glass coatings of, 256, 258, 263, 264, 265, 269, 272, 273
    characteristics, 270
  in glass container formulas, 110
  glasses from gels of, 212–214
    drying, 222–223
    by hot-pressing, 240
    TTT diagrams, 243
  in glass shell composition, 163
  Na_2CO_3 reaction with in batch heating, 8
  synthesis from $Si(OCH_3)_4$, 273
$SiO_2-Al_2O_3$
  glasses from gels of, 212, 213
    devitrification behavior, 244
$SiO_2-Al_2O_3-CaO$, glasses from gels of, devitrification, 244
$SiO_2-Al_2O_3-K_2O$, devitrification of gel glasses from, 244
$SiO_2-Al_2O_3-Li_2O$, devitrification of gel glasses from, 244
$SiO_2-Al_2O_3-MgO$, glasses from gels of, devitrification behavior, 244
$SiO_2-Al_2O_3-MgO-P_2O_5-B_2O_3-CaO-BaO-As_2O_3$, glass layers from gels of, 211
$SiO_2-Al_2O_3-Na_2O$
  devitrification of gel glasses from, 244
  gel synthesis from, 220
  glasses from gels of, 211
$SiO_2-Al_2O_3-P_2O_5-Li_2O-MgO-Na_2O-TiO_2-ZrO_2$, glasses from gels of, 211

MATERIALS INDEX

SiO$_2$–Al$_2$O$_3$–ZnO–Li$_2$O–TiO$_2$–ZrO$_2$–BaO–MgO–CaO–K$_2$O, glasses from gels of, 211
SiO$_2$–B$_2$O$_3$, glass from gels of, 212, 213, 244
SiO$_2$–B$_2$O$_3$–Al$_2$O$_3$–Na$_2$O–BaO, glasses from gels of, 213
SiO$_2$–B$_2$O$_3$–Al$_2$O$_3$–Na$_2$O–K$_2$O, glasses from gels of, 211
SiO$_2$–B$_2$O$_3$–Na$_2$O
  gel synthesis from, 220
  as leached glass coating, 279
SiO$_2$–CaO, glasses from gels of, 214
SiO$_2$–CaO–Na$_2$O, glasses from gels of, 214
SiO$_2$–Fe$_2$O$_3$, glasses from gels of, 214
SiO$_2$–GeO$_2$, glasses from gels of, 214
SiO$_2$–K$_2$O, glasses from gels of, 212
SiO$_2$–La$_2$O$_3$
  glasses from gels of, 212
  devitrification behavior, 244
SiO$_2$–La$_2$O$_3$–Al$_2$O$_3$, glasses from gels of, 212
SiO$_2$–La$_2$O$_3$–ZrO$_2$
  glasses from gels of, 212
  devitrification behavior, 244
SiO$_2$–Na$_2$O, glasses from gels of, 214, 245
SiO$_2$–PbO–Na$_2$O, glasses from gels of, 211
SiO$_2$–P$_2$O$_5$, glass from gels of, 212, 214, 245
SiO$_2$–SnO$_2$, glasses from gels of, 213
SiO$_2$–SrO, glasses from gels of, 214
SiO$_2$–TiO$_2$, glasses from gels of, 212–214
SiO$_2$–Y$_2$O$_3$, glasses from gels of, 214
SiO$_2$–ZrO$_2$
  glasses from gels of, 212, 213
  devitrification behavior, 244
SiO$_2$–ZrO$_2$–Na$_2$O, glasses from gels of, 212
Si(OCH$_3$)$_4$, gel synthesis from, 220
Si(OC$_2$H$_5$)$_4$
  gel synthesis from, 220
  glass coatings from, 265
Si(OR)$_4$, silica sols from, 214, 215
Sn
  in chemistry of glass melting, 15
  in glass coatings, 256, 261, 264, 266
SnCl$_4$, glass coatings from, 261
SnO$_2$
  in glass coatings, 253, 257, 258, 260, 262–265
  characteristics, 270
  as glass filler, 174, 178

SO$_x$, as pollutants from glass manufacture, 134
Sodium silicate
  iron in, 14
  silica sols from, 214
SO$_2$ + O$_2$
  bubbles, in glass melts, 20–21
  solubility role in glass melting, 13
SO$_3$
  in compositions for flat glass, 57
  in glass container formulas, 110
  solubility role in glass melting, 13
Sodium metasilicate, see Na$_2$SiO$_3$

T

Ta$_2$O$_5$
  in glass coatings, 273
  characteristics, 270
Te, in glass coatings, 265
Teflon, glass coating by, 279
Teflon FEP, structure of, 280
Tetramethoxysilane, monolithic glasses from gels of, 228, 230
ThO$_2$, in glass coatings, characteristics, 270
Ti
  in glass coatings, 256, 257, 260, 261, 266
  in solder glasses, 180
Ti(O–C$_2$H$_5$)$_4$, gel synthesis from, 220
Ti(O–iso-C$_3$H$_7$)$_4$, gel synthesis from, 220
Ti(O–C$_4$H$_9$)$_4$, gel synthesis from, 220
Ti(O–C$_5$H$_7$)$_4$, gel synthesis from, 220
TiO$_2$
  glass coatings of, 103, 253, 255, 261, 263, 269, 272, 273
  characteristics, 270
  IROX glass, 271–272
  in glass container formulas, 111
TiO$_2$–SiO$_2$, glasses from gels of, 213
Titanium tetraisopropylate, glass coatings from, 261
TMS, see Tetramethoxysilane

U

Uranium oxide, in glass coatings, characteristics, 271

V

V
  effect on optical fiber waveguide attenuation, 290
  in glass coatings, 260, 263
Vanadium oxide in glass coatings, 103
  characteristics, 271

W

W, in glass coatings, 263

Y

Y(O–C$_2$H$_5$)$_3$, gel synthesis from, 220
Y$_2$O$_3$, in glass coatings, characteristics, 270

Z

Zn
  in glass coatings, 264, 266
  in solder glasses, 180
Zn$_x$Cd$_{1-x}$S, in glass coatings, 261
ZnEt$_2$, glass coatings from, 265
Zr, in solder glass, 180
ZrO$_2$
  in glass coatings, 273, 275
    characteristics, 270
  as glass filler, 178
  in glass melts, 5
Zr(O–*iso*-C$_3$H$_7$)$_4$, gel synthesis from, 220
Zr(O–C$_4$H$_9$)$_4$, gel synthesis from, 220
ZrSiO$_4$ (Zircon)
  as glass filler, 174, 178
  as paving material for glass furnaces, 121, 122
  as refractory heavy mineral, 112

Subject Index

A

Acetylacetonates, oxide coatings from, 259, 265
Adhesion, of glass and plastics, improvement by coatings, 281
Aerogels
  in gel glass preparation, 230
  monolithic type, 232
  weight losses from, 235
Aesthetic appeal, requirements for, in glass, 101
Aging in gel formation, 217–218
Airplane windows, coatings for, 259
Air pollution, container manufacture and, 133–134
Alkali–alkaline earth phosphates, in glasses for optical fibers, 324
Alkali borosilicate glasses, coatings for, 278
Alkali germanosilicates, in glasses for optical fibers, 324
Alloy 42, in cerdip leadframes, 173
Alloys, in glass coatings, 255
Alpha Beta alumina, use in glass furnace construction, 121, 122
Alpha particles
  measurements of, 202
  role in soft errors in microelectronic devices, 200–204
  solder glasses low in, 203–204
Alumina (Al_2O_3)
  in ceramic substrates, 173
  effect on foaming in glass melts, 28
  raw materials, specifications for glass, 111
Aluminum, molten, hydrogen dissolution in, 27
Aluminum alkoxide, monolithic gels from, 229
Amber glasses, reboil studies on, 27–28
Ammonium silicates, silica sols from, 214
Annealing lehr, for container glass, 130–131

Annealing of glass, 2
Anticorrosion coatings for glass, 253, 259, 280
Antifriction glass coatings, 253, 255, 261
Antireflection films for glass, 213, 258–259, 261
Argon, glass spheres filled with, 165
Arsenic, in glass, limits of, 7
Artifacts of glass, role in glass history, 107–108
Atomic absorption, use in glass oxide analysis, 128
Auto windshields, flat glass production for, 83
Axial flame hydrolysis technique, for optical fiber waveguides, 309–312

B

Backwalls, in furnaces for container manufacture, 120–121
Barytes, specifications for glass raw materials, 111
Bastick model for fining behavior, 17–18
Batch heating of glass, 8
  thermal conductivities, 9
Batch house, for glass container manufacture, 112–114
Batch mixing of glass, 7–8
Batch particle size, role in glass melting, 11
Beaded movie screens, glass sphere use on, 149
Bells, for tubing manufacture, 142, 143
Benzene, as outgassed product from solder glass, 196
Bicheroux rolling process for flat glass, 48
  description and history, 59–60
Biocompatible glass surfaces, 281
Bloom in float glass, 99
Blow and blow process for making glassware, 126

347

Blown cylinder process for flat glass, 48, 70
Blow pipe, discovery, 108
Blow suck air support tray, for double-roll process, 67
Body-tinted glass for radiation control, 102
  properties of, 104–105
Borosilicate glass
  fusion drawn sheet glass from, 83
  from gels, 211
  lead and nickel use in, 16
  melting of, 6
  for optical fibers, 324, 326, 332, 333
  silica scum formation in melts of, 38
  softening and working points of, 172
  sulfate as poor fining agent for, 24
Boudin continuous double-roll process for flat glass, 60–61
Breastwalls, in furnaces for container manufacture, 120
Bridgecovers, in furnaces for container manufacture, 120
Bridgewall, in furnaces, 116
Bromine, glass spheres filled with, 165
Brounshtein equation, 10
Bubbles
  foaming from, 24–28
    observation of individual bubbles, 20–21
    theoretical models for, 21–24
  in glass melts, 16–28
Burner blocks, in furnaces for container manufacture, 122
Butane, as outgassed product from solder glass, 196

C

Cab-O-Sil
  densification of, 211
  gels from, 219
Cadmium stannate, glass coatings, 257, 258
Capacitors, flat glass for, 47
Capillary forces, role in glass gel drying, 223–225
Carbonates
  gas formation from, in glass melting, 15
  natural, specifications for glass raw materials, 111
Carbon dioxide, as outgassed product from solder glass, 196
Carbon dioxide laser, as heating element for optical fiber waveguide processes, 317
Catalytically active coating of glass, 281
Catenary drop, from drawing glass tubing, 145
Caustic soda, as wetting agent for container raw materials, 113–114
Ceramics, from gels, 213
Cerdip packages
  assembly of, 183–185
    diagram, 184
  chip attachment in, 185
  coatings for, 204
  cross-sectional view of, 173
  electroplating of, effect on solder glass in, 192
  failure of, due to moisture entrapment, 194–195
  future perspectives of, 204–206
  glass properties and processing effections on, 186–194
  insulation resistance of, 186
  lead attachment in, 183, 185
  lead finish of, 185
  leadframe designs in, 192
  moisture level in sealed cavity of, 205
  moisture measurement methods for, 195–197
  outgassing of solder glasses, during sealing of, 194–200
  sealing of, 185
  soft error in, 200–204
    future trends in, 205
    lower temperature for, 204
  solder glass use in, 170, 181–186
    dry processing, 199–200
    strength of, 204
  stress testing of, 186
  temperature cycling of, 186
  torque testing on, 186
  wire bonding in, 185
Chance process, for double-rolled glass, 59
Checkers, for container furnaces, 124
Chemical decoloring in glass melting, 15–16
Chemical polymerization of glasses, 210
Chemical vapor deposition
  glasses obtained by, 212

method for glass coatings, 252, 254, 255, 262–265
Chlorination, of glass gels, 235
Chrome, glass color from, 112
Clean Air Act, 133
Closed-pore model of glass gel texture, 235–237
Coalescence, role in gas evolution from glass melts, 19
Coatings, for optical fiber waveguides, 319–323
Coatings on glass, 251–283
    chemical vapor deposition, method for, 252, 254, 255
    coloration and decoration by, 281
    for containers, 129–133, 261
    dip coating deposition of, 254, 267–277
    flat glass, 253, 255, 260, 263–264
    from gels, 212, 213
    lamps, 261
    leaching process for, 254, 255, 277–279
    liquid spray deposition method for, 254, 255
    physical vapor deposition method for, 254
    processes for making, 254–255
    reasons for, 253–254
    spray processes for, 252, 259–262
    tubing interiors, 264–265
    vacuum processes for, 252, 256–259
        plasma polymerization, 259
        sputttering method, 257–259
        thermal evaporation, 256–257
    wet reduction deposition of, 254, 255, 265–267
Cobalt, coating of radiation-control glass, 102
Cobalt oxide
    coating of glass, 103
    in solder glasses, 179
Colburn process for flat glass, 49, 70, 75–77
    advantages and disadvantages, 76
    compositions used, 57, 58
    control variables, 76
    equipment and operation, 75–76
    Glaverbel modification of, 77
Cold-end coating, of container glass, 131–133

Color of glass, iron effects on, 14–15
Communication fibers, polymer coatings for, 259
Computers, application to float process, 84, 85
Concentric-crucible method of fiber drawing, 331–332, 334
Container glass
    composition of, 3
    flow in glass melts of, 31
    melting of, 6
Container manufacture, 107–136
    coatings used in, 129–133
    composition formulas
        range, 110
        specific, 111
    forming machines for, 125–129
    furnaces for, 114–125
    glass analytical procedures in, 127–129
    history of, 108
    inspection techniques, 132
    packaging procedures, 133
    pollution control, 133–136
    pollutants, 134
    raw material for, 127–129
    in United States, 109
Containers, coatings for, 129–133, 261
Copper, coating of radiation-control glass, 102
Cordierite, as glass filler, 174, 178
Cosmic-ray background, soft error in cerdip devices, 204
Crown process, for flat glass, 47–48, 70
Crystalline coatings, for glass, 275
Cullet, specifications for glass raw materials, 111
Cup molds, first use of, 108
CVD, *see* Chemical vapor deposition
CV97 solder glass, thermal analysis of, 190
CV111 solder glass
    chemical durability of, 193
    emission spectrographic analysis, 180
    moisture sorption mechanism in, 199
    properties, 178, 181
    strength, 204
    thermal analysis, 190
Cylinder glass, homogenization by diffusion in melts of, 29

D

Danner process, for tubing and rod manufacture, 138–140
  disadvantages, 139–140
DCD, see Dip coating deposition
Decolorizing in glass melting, 15–16
Density of glass, checks, for containers, 128
Destabilization of sols, in gel formation, 214–219
Deuterium, glass spheres filled with, 165
Devitrification kinetics, of gel glasses, 240–244
Differential thermal analysis, segregation during melting, studies using, 11
Diffusion, glass-melt homogenization by, 28–30
Dip coating deposition, of glass coatings, 254, 267–277
  history of, 268, 269
Dip coating technique, for gel glasses, 213
Doghouses, in charging systems for container manufacture, 114, 119–120
Dolomite, specifications for glass raw materials, 111
Double-roll processes for flat glass, 48, 59–62
  continuous methods, 60–67
  intermittent method, 59–60
Downdraw process for flat glass, 49
  basic science of, 55–56
DRAM, see Dynamic random access memories
Drawing operations, for tubing and rod manufacture, 144–147
Drawn cylinder process, for flat glass, 70, 71
Drop-generator process
  for production of high-precision glass spheres, 154–167
    drop generator for, 156–157
    vertical tube furnace for, 157–158
Dynamic random access memories, soft error prevention in, 171

E

EDAC, see Error detection and correction
Egypt, early use of glass in, 109

Electrochromic coatings of glass, 281
Electro-float process
  for tinted glass, 104–105
  glass properties, 104–105
Electron microprobe analysis, of radioactive impurities, in solder glasses, 202
Emission spectroscopy, of radioactive impurities, in solder glasses, 202
Endport furnace, for container industry, 115
Energy requirements, in glassmaking, 5–7
Environmental Protection Agency, in pollution control for container manufacture, 133
EPA, see Environmental Protection Agency
Error detection and correction, of cerdip devices, 203
Ethylene-vinyl acetate, as coating for optical fiber waveguides, 320
β-Eucryptite
  as glass filler, 178
  CTE of, 174
EVA, see Ethylene-vinyl acetate
Eyeglasses, antireflection coatings for, 258

F

Federal Water Pollution Control Law, 135
Feldspathic sand, specifications for glass raw materials, 111
Ferrous–ferric equilibria of iron in glass melts, 14, 15
Fiber light guides, coatings for, 253
Fick's second law, 36
Fining, of glass melts, 16–17, 20, 24
Fisher's theory for bubble nucleation and growth, 25, 26
Flame process, for glass sphere production, 153–154
Flat drawn sheet processes, for flat glass, 70–83
Flat glass
  composition of, 3, 56–58
  development of, 46–58
  pattern of, 47–49
  flat drawn sheet processes for, 70–83
  lines and distortion during drawing process for, 54

manufacturing processes for, 45–106
  basic science of, 50–58
  modern techniques for, 49
  products made from, 46, 47
  for radiation control, 100–106
  specialty types of, 47
  stress and movement in drawing of, 50–54
Flint glass, colorless, 15
Float glass, 48, 49, 83–100
  chemical aspects of, 97–100
  coatings for, 260, 262, 263–264
  colored, 254
  compositions, 56–58
  development, 83–85
  diagram, 86
  equilibrium float variant, 92–93
  equilibrium thickness, 87–88
  essential features, 85–87
  flatness related to forces acting on ribbon, 90–91
  float ribbon formation in, 91–97
  gravity and surface tension forces in, 88–89
  high-output variants, 94–95
  market, 49
  optical distortion in glass, 91
  PPG modification, 100
  problems, 98–100
  production of specified width ribbon, 92
  theory, 87–91
  thin film production, 104–105
  tin properties, 97–98
  uses of, 46, 47
Flow, in glass melts, 30–36
Fluorine, losses of, from glass melts, 40
Fluoropolymers, as anticorrosion coatings for glass, 259
Foaming, in glass melts, 24–28
Forehearth
  with furnace for glass container manufacture, 116–117, 124–125
  role in continuous double-roll process, 62
Fourcault process for flat glass, 49, 70, 71–83
  compositions used in, 56–58
  control variables in, 73–75
  equipment and operation, 71–72
  quality control in, 74–75
Freon TE in glass sphere manufacture, 166

Froth flotation, of sand grains in glass melts, 11
Fuel capsules, glass sphere use in, 150
Fumed silica, gels from, 219
Furnaces
  for glass container manufacture, 114–125
  operation details, 117–119
  homogeneity in flows of, 34–35
Fusion downdraw process for flat glass, 81–83
  advantages of, 83
  equipment and operation of, 82
  glass quality from, 83

G

Gas chromatography, of gas composition in seed, 19
Gases
  composition, in seed, 19–20
  role in glass melting, 12, 15
Gas-filled glass spheres, 165
Gelation, precipitation compared to, 218
Gel glass, 209–249
  devitrification kinetics of, 240–244
    heterogeneous nucleation in, 243–244
    nucleation-growth conditions
    TTT diagrams, 240–241
    viscosity-time equivalence, 242–243
  dip-coating technique for, 213
  economics, 214
  gel drying process for, 221–231
    phenomenological approach, 221–223
    rheological aspects, 227–228
    structural approach, 223–228
    structure models, 233
  gel preparation for, 214–220
    from organometallic compounds, 214, 219–220
    by sol destabilization, 214–219
  gel textures in, 233
  gel-to-glass transformation in, 234–235
  historical aspects of, 210–214
  hot-pressing technique for, 212, 238–240
  optical fiber preforms from, 328–329
  sintering by viscous flow, 235–240
    closed-pore model, 235, 236–237
    open-pore model, 235, 237–238
  special features of, 244–245
Gel molding, 219
Gels, polymerization steps leading to, 216

Georgian wire, for wire mesh, 69–70
Ginstling equation, 10
Glass
  basic operations in making, 2
  coatings, 251–283
  color, iron effects, 14–15, 16
  composition of choice, 2–4
  from gels, *see* Gel glass
  methods for producing, 209
  new requirements for, 101
7740 Glass, coating, 278
Glass ceramic coatings, 275
Glass fibers
  failure studies, 26
  from gels, 212
    drying, 221
Glass fillers
  coefficient of thermal expansion, 174
  glass solders as, 172–174
Glass lubricants, 131–132
Glass melting and melts
  batch heating of, 8–9
  batch mixing in, 7–8
  bubbles in, 16–28
  chemistry, 9–12
  decolorizing, 15–16
  energy requirements, 5–7
  fining, 16–17
  flow, 30–36
  in furnaces, flow, 34–35
  gas role, 12
  homogenizing, 28–36
  ideal, 2
  inhomogeneity, sources of, 28
  oxidation control, 12–16
  principles, 1–44
  raw material choice for, 4–5
  reboil and foaming, 24–28
  supersaturation, 26
  thermal conductivity, 9
  volatilization from, 36–40
Glass oxides, analysis for, 128
Glass products, standard definition of, 109–110
Glass rods, for optical fibers, from gels, 214
Glass spheres
  commercial processes, 152–154
  deuterium–tritium fuel-filled, 150
  drop-generator process, 154–167

flame process, 153–154
gas-filled, 165
gelation problem, 166
from gels, 212, 213
hollow, from liquid drops, 158–160
  glass composition and surface control, 163–164
  glass formation and fining, 160–162
  mass production, 165–166
  measurements, 164–165
  range of sizes, 166
  Rayleigh technique, 150–152
uses of, 149
Glazing, of solder glass in cerdip packages, 182–183
Gob feeder, for container forming machines, 125
Gold, coating of radiation-control glass, 102–103
Gold plating, of cerdip packages, 192
Graphite resistance furnaces, for optical fiber waveguide processes, 316
Griffith flaws in containers, coating protection against, 253, 255, 261
Gypsum, specifications for glass raw materials, 111

H

Halogenides, oxides from, 259
Hard errors, in microelectronic devices, 200
Hazardous and toxic wastes, from glass industry, 135–136
Heat mirrors
  coated glasses as, 254, 255
    from float glass, 264
  principles, 256
  processes, 257, 258, 276–277
Helium, glass spheres filled with, 165
Hermetic sealing, solder glass use in, 170
  by diffusion, 28–30
  flow effects on, 30–36
  in furnaces, 34–35
Homogenization of glass melts, 28–36
  of pot and laboratory melts, 35–36
  surface tension role in, 32–34
Hot-end coating, of container glass, 129–131

SUBJECT INDEX

Hot-pressing techniques, for gel glasses, 211, 212, 238–240
Hot-stage microscopy, in studies of glass melting, 11
Hydrocarbons, as outgassed products from solder glass, 196
Hydrochloric acid, effect on solder glasses, 193
Hydrogels, in gel glass preparation, 221
Hydrogen
  dissolution in molten aluminum, 27
  glass spheres filled with, 165
Hydroxyl impurities, in silica-based glasses, absorption due to, 291
Hypercritical solvent evacuation, from glass gels, 230–231

I

IC, see Integrated circuit
ICF, see Inertial confinement fusion
Inclusion, infinitely deformable, in glass melts, 31–32
Indium-tin oxide, glass coatings, 257, 258, 261
  characteristics of, 277
  for heat mirrors, 276–277
Inertial confinement fusion
  glass sphere targets for, 150, 154, 164, 167, 213
  design specifications, 155
Infrared absorption, in moisture detection in cerdip devices, 196
Infrared spectroscopy, of gel synthesis, 220, 234
Inorganic salts, sphere production from, 166
Insulation resistance test, on cerdip packages, 186
Integrated circuit, solder glass use in, 170
Intermittent double-roll process, for flat glass, 59–60
Internal thermal oxidation process, for optical fiber waveguide, 304–309
Iron
  in glasses
    chemical behavior, 14, 16
    color from, 15, 112
Iron bar, mechanical stirring of glass melts by, effect on glass quality, 35

Iron oxide
  in body-tinted glass, 102
  coating of glass, 103
  in solder glasses, 179
IROX sun-protection glass, process for, 267, 271–272
IS machines, for container formation, 125–126
ITO, see Indium-tin oxide, Internal thermal oxidation

J

Jebsen–Marwedel model for fining behavior, 17–18

K

Kauzman paradox in glass technology, 14
KCl solder glass, properties of, 181
KClM solder glass
  emission spectrographic analysis of, 180
  properties of, 181
KC 400, properties, 181
KC 402 solder glass, properties, 181
KC 405 solder glass
  chemical durability of, 193
  properties, 181
Kidney, artificial, biocompatible glass in, 280
Knudsen effusion experiments, on soda-lime-silica glasses, 40
Kovar, in cerdip leadframes, 173
Kyanite, as refractory heavy mineral, 112

L

Laboratory melts, homogenizing of, 35
Laminar flow, effect on glass melt inhomogeneity, 30–31
Lamps, coatings for, 261
Large-scale glass manufacturers, optimum batch charging techniques in, 9
Laser, protection filters, coated optical glasses for, 274, 278–279
Leaching process for glass coatings, 254, 255, 277–279

Lead crystal, range of compositions in melts of, 28
Lead glasses
 chemical durability of, 191
 lead and nickel use in, 16
Lead silicate glass
 melting of, 6
 softening and working points, 172
Lea Ronal Aural 92, gold plating bath for cerdip packages, 192
Leaving plate, in continuous double-roll process, 66–67
Lehr, *see* Annealing lehr
Lens-type thermometer tubing, manufacture, 143
Lime, in glass, limits of, 7
Limestone, specifications for glass raw materials, 111
Liquid-drop process, for glass spheres, 156
Liquid spray deposition, method for glass coatings, 254, 255
LSD, *see* Liquid spray deposition
LS0110 (KC1, NCG-556) solder glass, properties, 178
LS0113 (KC1M, NCG560) solder glass
 chemical durability, 193
 properties, 178
LS0803 (KC400, NCG-564) solder glass, properties, 178
LS2001 (KC405) solder glass
 properties, 178
 scanning electron micrograph, 176
Ludox
 densification, 211
 destabilization, for gel preparation, 214, 216

M

Mackenzie–Shuttleworth model of glass–gel texture, 236–237, 238
Magnesium aluminum spinel as glass coating, 275–276
Mass spectrometry
 of gas composition in seed, 19
 in moisture detection, in cerdip packages, 195, 196
MEA, *see* Moisture evolution analysis
Melter crowns, in furnaces, for container manufacture, 121

Melters, for glass container manufacture, 119
Metal alkoxides, gel preparation from, 219–220
Metallic spheres, 166
Metal oxides, as glass coatings, 270
Metals, molten
 equilibria, 13
 gas dissolution in, 27
Methane, as outgassed product from solder glass, 196
Microbalance–mass spectrometry method for moisture detection in solder glasses, 196
Microscope cover glass, from flat glass, 47
Mie scattering, attenuation in optical fibers, 330
Mirrors, coatings, 265
Mixed drinks, glass sphere use, 149
Moisture evolution analysis, of solder glasses, 196–197
Moisture outgassing, of cerdip packages, 194–200
Moisture-sensor integrated circuits, in detection of moisture in cerdip devices, 195–196
Moisture stress, in glass gel drying, 225–227
Monolithic glasses, from gels, 212, 213–214
 drying, 221
 preparation, 228–231
Monomethylpolysiloxane, structure, 279
MRW, *see* Murray–Rodgers–Williams approximation
Multicomponent oxide coatings, of glass, 274
Murray–Rodgers–Williams approximation, for gel glass formation, 238–240

N

National Pollutant Discharge Elimination System, 135
Natural strain, of infinitely deformable inclusions in glass melts, 32
Neon, glass spheres filled with, 165
Neutron activation analysis, of radioactive impurities in solder glasses, 202
New Source Performance Standards, 134

SUBJECT INDEX

Newtonian viscous flow, role in densification of glass gels, 235
Nitrates, in gels, 219
Nitric acid, effect on solder glasses, 191, 193
Nitrogen
  as outgassed product from solder glass, 196
  solubility in reduced glasses, 26
NSPS, see New Source Performance Standards
Nucleation of bubbles, theory, 25–26

O

Open-pore model of glass gel texture, 235, 237–238
Optical coatings, for glass, 278–279
Optical color filters, from flat glass, 47
Optical fiber waveguides, 285–339
  attenuation, 288
  axial flame hydrolysis process, 309–312
  coating, 319–323
  diameter control, 318–319
  dimension control, 298–299
  fatigue, 297–298
  fiber characteristics, 286–298
  fiber drawing, 314–322, 329–334
  glass fabrication, 324–329
  internal plasma oxidation process, 312–314
  internal thermal oxidation process, 304–309
  low-flaw density requirements, 299
  mechanical characteristics, 296–298
  non-vapor-phase techniques, 323–334
  numerical aperture, 287, 288
  optical properties, 286–296
  from phase-separated glasses, 326–328
  preform fabrication, 299–314
  processing, 298–334
  pulse broadening, 292–296
  purity requirements, 298
  radial flame hydrolysis technique, 301–304
  schematic, 287
  from sol–gel glasses, 328–329
  strength, 296–297
  uses, 285

Optical interference coatings for glass, 269–272
Optical melts, of optical glass, 35
Optical waveguides
  coatings for, 264–265
  glass coatings as, 213
Organic adhesives, unsuitability for hermetic sealing, 170
Organometallic compounds, gel preparation from, 214, 219–220
Ostwald ripening, in sol destabilization, 215, 216
Outgassing of solder glasses, for cerdip packages, 194–200
Owens forming machine, for container manufacture, 125
Oxidation, control in glass melting, 12–16
Oxides, coating of glass, 103
Oxidic semiconductor coatings, of glass, 253
Oxygen
  bubbles, in glass melts, 28
  contamination of tin by, infloat process, 98
  ions, role in glass structure, 13
  as outgassed product from solder glass, 196
  solubility in oxidized glasses, 26
Oxy-hydrogen burner, for optical fiber waveguide processes, 317

P

Patterned rolled glass, 48
  use of, 47
Phenol, as outgassed product from solder glass, 197
Phosphate coatings, for glass, 273
Photochromic coatings, of glass, 281
Physical decolorizing, of glass melts, 16
Physical vapor deposition, of glass coatings, 254
Pilkington double-pass wired process, 68–69
Pilkington process for flatglass, 49, 60
  float process for, see Float glass
  flow process for continuous double rolling, 60, 61–62
Plasma polymerization, glass coating deposition by, 259

Plastic coatings for glass, 280
Plastics, glass sphere use as fillers in, 149
Plate glass
　market for, 49
　use of, 47
Platinum, early glass melts made in, 35
Platinum crucibles, for fiber drawing, 331–332
Poisson's ratio, definition of, 50–51
Polished plate process for flat glass, 70
Pollution control, in container manufacture industry, 133–136
Polymeric materials, sphere production from, 166
Polyimides, as coatings for cerdip devices, 204
Pond/Boudin process for wired rolled glass, 67–68
Ports in furnaces, for container manufacture, 122
Pot melts, homogenizing, 35–36
PPG Pennvernon flat glass process, 70, 77–80
　compositions used in, 57
　equipment and operation, 77–78
　glass quality, 80
　operational variables, 78–80
Precipitation, gelation compared to, 219
Press and blow process, for making glassware, 126, 127
PVD, see Physical vapor deposition
Pyrolytic methods, for tinted glass production, 103
　glass properties, 104–105

Q

Quartz, dissolution kinetics, 9

R

Radial flame hydrolysis, for optical fiber waveguides, 301–304
Radiation-control glass, 100–106
　body-tinted glass, 102
　coating by wet chemical processing, 102–103
　processes for production of, 102–105
　properties of, 105–106
　pyrolytic methods for, 103
　vacuum coating technique, 103
RAM, see Random access memory
Raman spectroscopy, of gel-to-glass transformation, 234
Random access memory, dynamic-type, soft error in, 200
Rare earths, use in glass compositions, 109
Raw materials, for glass melts, 4–5
Rayleigh technique, liquid sphere generation, 150–152
Reactive planar magnetron sputtering, glass coating deposition by, 257
Rear-view mirrors, coatings, 273
Reboil, in glass melts, 24–28
Redox pairs, mutual interaction of, in glass decolorizing, 16
Refiners, in furnaces for container manufacture, 121–122
Refining agents in glass, limits, 7
Refractories, for glass container manufacture, 119–125
Refractory oxides, in glass melts, 5
Regenerative furnaces, for container industry, 115
Regenerators, in furnaces, for container manufacture, 123–125
Reynolds number, of molten glasses, 30
RFH, see Radial flame hydrolysis
Rheological behavior, of glass gels, 227–228
Rod manufacture, see Tubing and rod manufacture
Rolled glass
　composition, 57
　procedures of, 58–70
Rolling machines, for double-roll process, 67
Rolls
　composition of, 64–65
　role in continuous double-roll process, 62–66
Roto-Tap machine, 127

S

Saltcake, specifications for glass raw materials, 111
Sample size, in batch mixing of glass, 7–8

SUBJECT INDEX

Sand
  in glass, limits, 7
  grains, dissolution in glass melting, 11
  high-silica, specifications for glass raw materials, 111
  particle size, role in glass melting, 12
Sandwich glass, homogenization by diffusion in melts, 29
SAXS, *see* Small-angle x-ray scattering
Scherer's model for glass gel texture, 237–238
Scriven's model for bubble growth, 22–23, 26
Seal boat, for cerdip packages, 185, 194
Seed
  gas composition in, 19–20
  removal from glass melts, 17–19
Semiconductor industry, solder glass use, 170, 178
SG–200 solder glass
  chemical durability, 193
  properties, 178, 181
  thermal analysis, 188
Shading coefficient, of radiation control glass, 102
Sheet glass
  market, 49
  MgO in, 4
  uses, 47
Sideport furnace, for container industry, 115
Sieman's regenerative-type furnace
  description, 114–116
  development, 109
Silane gas (SiH_4), reflective glass production by, 103
Silica scum formation, 38
Silica sols, destabilization, for gel preparation, 214–216
Silicate glass,
  as mixtures of pure oxides, 13–14
  structure, 12–13
Silicate melts, heats of mixing, 5
Silicate–phosphate glass film, synthesis, 275
Silicone resins, as coatings, for optical fiber waveguides, 320
Siliconizing, of medical ampoules, 253, 280

Sillimanite, as refractory heavy mineral, 112
Siloxanes, in glass sphere production, 167
Sink–float density test on container glass, 128
Sintering, glass gel densification as, 235–240
Slab glass, homogenization by diffusion in melts of, 29
Slot bushing downdraw process for flat glass, 81
Small-angle x-ray scattering
  of gel synthesis, 220, 233
  in studies of glass gels, 23
Soda, in glass, limits, 7
Soda ash, specifications for glass raw materials, 111
Soda-lime–silica glass
  for optical fibers, 324
  softening and working points, 172
Sodium vapor lamps, 261
Softening points, of various glasses, 172
Soft error
  in solder glasses in cerdip packages, 200–204
  mechanism, 201
Solar cells
  antireflective coatings from gels for, 213
  coatings for, 262
Solar control coatings, for glass, 254, 255, 267–268, 271–272
Solar control glass, 101, 102
  by float glass process, 85
Solarization of old glass windows, 16
Solder glass, 169–207
  alpha activity in, 202
  glasses low in, 203–204
  in cerdip processing, 170, 173, 181–186
  future prospects in, 204–205
  glass property effects on, 186–194
  chemical durability of, 171, 191–192
  coefficient of thermal expansion, 181, 187
  composition, 179–180
  density, 181
  description and uses of, 170
  devitrification in, 174–180
  effect on cerdip packages, 187–191

dry processing of, 199–200
emission spectrographic analyses of, 180
evolution of, 172–180
as glass fillers, 172–174
low seal temperature, 171
moisture measurement methods for, 195–197
moisture sorption mechanism in, 197–199
outgassing of, 194–200
  control, 171
properties, 181
radioactive impurities in, 202
raw glass manufacturing of, 182
requirements for, 171
scanning electron micrographs of, 176, 177
seal temperatures of, 181
softening and working points of, 172, 181
surface reduction of, 192
thermal analysis, 188–190
thermal expansion matching and mechanical durability of, 171
7583 Solder glass
  emission spectrographic analysis, 180
  properties, 178, 181
  thermal analysis, 188, 189
Sol–gel glasses
  optical fibers from, 328–329
    proposed technique (diagram), 330
Sol–gel method of glassmaking, energy savings in, 6
Solid wastes, from glass industry, recycling of, 136
Sols, polymerization steps leading to, 216
Soot formation, in internal thermal oxidation process for optical fibers, 306–307
Soots in optical-fibers technology, open-pore model of, 235
Sphere glass, homogenization by diffusion in melts of, 29
Spheres, from specialized materials, 166–167
Spherical inclusions, in glass melts deformation of, 33
Spinel, synthesis, 276
Sputtering, glass coating deposition by, 257–259

Stokes's law, 18
Stove tops, glass-ceramic anticorrosion coatings for, 280
Strain, definition, 50
Stress
  definition, 50
  in flat glass sheet while drawing, 50–51
Sulfate
  as fining agent, 24
  in gels, 219
  in glass, limits, 7
  solubility of
    in glass melts, 27
    in oxidized glasses, 26
  sources of, for glass raw materials, 111
Sulfide, solubility in reduced glasses, 26
Sulfide coatings, for glass, 274
Sulfite, solubility in glass melts, 26, 27, 28
Sulfur, contamination of tin, in float process, 98
Sulfuric acid, effect on solder glasses, 191, 193
Sunglasses, vacuum-deposited coatings for, 256, 258
Surface tension
  deformation limited by, 32–34
  effects on bubbles in glass melts, 22

T

Table cast process for flat glass, 48
  description and history, 58
Tammann–Jander equation, 10
Television industry, solder glass use in, 170
Temperature cycling test, on cerdip packages, 186
TG191BF (KC402) solder glass
  chemical durability of, 193
  emission spectrographic analysis of, 180
  properties of, 178
Thermal gradients, in furnaces for glass container manufacture, 118
Thermal polymerization, of glasses, 210
Thermometer tubing, manufacture, 143
Thermoplastic elastomers, as coatings for optical fiber waveguides, 320
Thomson's relation, 230

SUBJECT INDEX

Thorium, in solder glasses, 202
Tin
  molten, in float glass process, 85–86, 97–98
  oxygen solubility in, 99
Tin plating, of cerdip packages, 192
Tin speck, in float glass, 98
Tinted glass, for radiation control, 102
Torque testing, of cerdip packages, 186
Transition metals, effect on optical fiber waveguide attenuation, 290
Transverse restraint, in flat glass, 52–53
Tritium, glass spheres filled with, 165
TTT diagrams, in studies of gel glass devitrification kinetics, 240–241
Tubing, interior coatings, 264–265
Tubing and rod manufacture, 137–147
  controlling dimensions, 146–147
  Danner process, 138–140, 145
    disadvantages, 139–140
  drawing operations in, 144–147
  updraw process, 140–141
    disadvantages, 141
  Vello and downdraw processes, 141–144, 145
    advantages, 142–143
Tuckstones, in furnaces for container manufacture, 120
Tyler equipment for glass raw material testing, 127

U

Ultraviolet curing, of optical fiber waveguides, 319–320
Ultraviolet fiber optics, Teflon-coated glass fibers in, 279
Ultraviolet radiation, effects on glass color, 16
Updraw process
  for flat glass, 49
    basic science of process, 54–55
    stress and stretch in, 51
  for tubing and rod manufacture, 140–141
    disadvantages, 141
Uranium, in solder glasses, 202
Urbach tails, 288
Urea, use in glass gel preparation, 235

V

Vacuum coating method for radiation-control glass, 103
  glass properties, 104–105
Vacuum processes for glass coatings, 256–259
Vapor-phase techniques for optical fiber waveguides, 299–323
Vello and downdraw processes for rod and tubing manufacture, 141–144
  advantages of, 142–143
Venice, glass development in, 108
Vertical tube furnace, for drop-generator process, for glass spheres, 157–158
Viscosity, of flat glass, while being drawn, transverse variation, 53–54
Vitreous glasses, devitrifying glasses compared to, 174–180
Vitreous silica rods, behavior in glass melts, 10
Volatilization from glass melts, 36–40
  experimental data, 38–40
  kinetics, 38
  theory, 36–38

W

Water
  as outgassed product from solder glass, 196
  oxygen and carbon dioxide bubbles in, 23
Water pollution, from glass manufacture, 135
Weighing, of raw materials, for container manufacture, 113
Wet chemical processing method, for radiation-control glass, 102–103
  glass properties, 104
Wet reduction deposition, of glass coatings, 254, 265
Willemite, as glass filler, 178
Window glass
  coatings for, 253, 255, 271–272
    spray deposition, 260
  manufacture, 71

Wired rolled glass
  processes for, 67–70
  use of, 47
Wire mesh, for wired glass, 69–70
Working points of various glasses, 172
WRD, *see* Wet reduction deposition

X

Xenon, glass spheres filled with, 165
Xerogels, in gel glass preparation, 221
X-ray fluorescence analysis, of glass composition, 128
XS1175-M1 solder glass
  chemical durability of, 193
  emission spectrographic analysis of, 180
  moisture sorption mechanism in, 197–199
  outgassed products from, 196
  properties of, 181
  scanning electron micrographs of, 177
  thermal analysis of, 188

Y

Young's modulus
  definition, 51
  of glass, composition and, 3

Z

Zircon
  as glass filler, 174, 178
  as paving material for glass furnaces, 121, 122
  as refractory heavy mineral, 112
Zirconia, in solder glasses, α activity in, 202, 203
Zirconia furnace, for optical fiber waveguide processes, 317
ZnO scintillation method, cosmic-ray background measurement by, 204